T0335603

STAR NOISE

Until Karl Jansky's 1933 discovery of radio noise from the Milky Way, astronomy was limited to observation by visible light. Radio astronomy opened a new window on the Universe, leading to the discovery of quasars, pulsars, the cosmic microwave background, electrical storms on Jupiter, the first extrasolar planets, and many other unexpected and unanticipated phenomena. Theory generally played little or no role – or even pointed in the wrong direction. Some discoveries came as a result of military or industrial activities, some from academic research intended for other purposes, some from simply looking with a new technique. Often it was the right person, in the right place, at the right time, doing the right thing – or sometimes the wrong thing. *Star Noise* tells the story of these discoveries, the men and women who made them, the circumstances that enabled them, and the surprising ways in which real-life scientific research works.

KENNETH I. KELLERMANN studies radio galaxies, quasars, cosmology, and the history of radio astronomy. He is the former president of the IAU Commission on Radio Astronomy, former chair of the National Academy of Sciences Astronomy Section, and former chair of the IAU Working Group on Historical Radio Astronomy.

ELLEN N. BOUTON is the NRAO Senior Archivist, overseeing an extensive collection of historical radio astronomy materials. She manages the web page for the IAU Working Group on Historical Radio Astronomy, and is a co-author, with Kellermann, of *Open Skies*, on NRAO and its impact on US radio astronomy.

STAR NOISE

Discovering the Radio Universe

KENNETH I. KELLERMANN
National Radio Astronomy Observatory

ELLEN N. BOUTON
National Radio Astronomy Observatory

Shaftesbury Road, Cambridge CB2 8EA, United Kingdom

One Liberty Plaza, 20th Floor, New York, NY 10006, USA

477 Williamstown Road, Port Melbourne, VIC 3207, Australia

314–321, 3rd Floor, Plot 3, Splendor Forum, Jasola District Centre, New Delhi – 110025, India

103 Penang Road, #05–06/07, Visioncrest Commercial, Singapore 238467

Cambridge University Press is part of Cambridge University Press & Assessment, a department of the University of Cambridge.

We share the University's mission to contribute to society through the pursuit of education, learning and research at the highest international levels of excellence.

www.cambridge.org
Information on this title: www.cambridge.org/9781316519356

DOI: 10.1017/9781009023443

First published 2023

A catalogue record for this publication is available from the British Library.

Library of Congress Cataloging-in-Publication Data
Names: Kellermann, Kenneth I., 1937- author. | Bouton, E., author.
Title: Star noise : discovering the radio universe / Kenneth I. Kellermann, National Radio Astronomy Observatory, Charlottesville, Virginia, Ellen N. Bouton, National Radio Astronomy Observatory, Charlottesville, Virginia.
Description: Cambridge, United Kingdom ; New York, NY : Cambridge University Press, 2023. | Includes bibliographical references and index.
Identifiers: LCCN 2022049614 (print) | LCCN 2022049615 (ebook) | ISBN 9781316519356 (hardback) | ISBN 9781009023443 (epub)
Subjects: LCSH: Radio astronomy–History.
Classification: LCC QB475.A25 K45 2023 (print) | LCC QB475.A25 (ebook) | DDC 522/.682–dc23/eng20230117
LC record available at https://lccn.loc.gov/2022049614
LC ebook record available at https://lccn.loc.gov/2022049615

ISBN 978-1-316-51935-6 Hardback

When one group had almost completed making a radio telescope carefully designed for a specific purpose, according to legend one member woke up in the middle of the night in utter horror because it had suddenly come to him that in calculating what signal they might expect they had multiplied by light-speed c where they ought to have divided by c. When he hastened to point this out to his colleagues they shared his horror, but decided that having gone so far they might as well finish the instrument. One of them is supposed to have said that in the whole universe there might be something that would give a signal c^2 times stronger than the one that they had thought they were going to detect – and there was!

W. McCrea, 1984, The Influence of Radio Astronomy on Cosmology. In The Early Years of Radio Astronomy, *ed. W. T. Sullivan III (Cambridge: Cambridge University Press), p. 272*

Contents

Foreword

Ken Kellermann and I obtained PhDs in radio astronomy in the 1960s. We were both supervised by John Bolton, one of the leading scientists making early discoveries in radio astronomy. Ellen Bouton was the NRAO librarian during my period working at NRAO in the 1980s and she has built up an extensive archive of radio astronomy documents that complements Ken's personal knowledge of many of the discoveries discussed in this book.

During the 3C 273 occultations that led to the discovery of quasars in 1962/63 I was a summer student at the CSIRO Division of Radiophysics in Australia and attended radio astronomy lectures given by the scientists who were making the discoveries. In the early 1980s I attended a talk by Derek de Solla Price at Yale and was impressed with his analysis of the exponential growth of scientific discoveries when a new field of research opens up. This sparked my interest in the role of the research environment in the discovery process. In 1983 Ken Kellermann and I organized a meeting on *Serendipitous Discoveries in Radio Astronomy* for the 50th anniversary of Jansky's discovery (Kellermann and Sheets 1984). Later, I became aware of the book *The Travels and Adventures of Serendipity* by the renowned sociologist of science, Robert Merton, and Elinor Barber, that was written in the 1950s, but not published until 2004. This book, and additional research papers by Merton, included many examples of serendipitous discoveries in science. Merton identified a "serendipity pattern" that was an excellent match to the discoveries in radio astronomy, although none of his examples were taken from this field. Martin Harwit had reached similar conclusions in his analysis of the discoveries across all astronomy in his series of books, starting with *Cosmic Discovery* (1981). The authors have adopted Harwit's discovery classification criteria for their analysis in the concluding chapter.

Robert Merton emphasized that chance discoveries depend on an impressive list of estimable qualities in a scientist: enterprise, courage, curiosity, imagination, determination, assiduity, and alertness. The background stories about how these discoveries were made provide many examples that reinforce this view. There is "nothing fortuitous" in so-called serendipitous discoveries. As Pasteur famously quoted "In the field of observation, chance favours only the prepared mind."

As the authors point out in their concluding summary, astronomy is a technique-oriented observational science. Astronomers can't do experiments; they can only observe. In 1933, less than 100 years ago, radio technology opened a new window on the universe that revealed an incredibly rich plethora of previously unknown phenomena. The authors have written the stories behind these discoveries, taking 10 chapters to cover the entire universe. The book focuses on the nature of the discoveries and Ken Kellermann uses his personal knowledge of the experiments and the scientists involved to tell the inside stories. This is not a straight history of the astronomical developments but an exploration of the circumstances leading up to each discovery, including many stories not generally known, but which provide the background and context. These details are often excised from the standard scientific narrative.

When we say "discovery" it often does not mean a single event. Contra the conventions in science that award prizes, professional respect, and recognition to individuals, discovery can be a lengthy process involving many actors, and many kinds of contributions over a period of time. Accounts of science are seriously distorted by emphasizing individuals and presenting discoveries as single eureka moments. Examples of the difficulties in identifying and classifying discoveries include aperture synthesis, that the authors attribute to Martin Ryle in 1962 and for which he was awarded a Nobel Prize in 1974. But, as pointed out in Chapter 3, the concept was first published by Pawsey's team in 1947 and was demonstrated by Christiansen and Warburton in 1955. As also noted in Chapter 3 and in Table 12.1 it was the use of the EDSAC electronic computer in 1962 that made aperture synthesis imaging in radio astronomy practical. So, who or what date do we record? The case for the existence of dark matter, discussed in Chapter 10, is another example of a discovery for which evidence accumulated over decades and involved many players. While radio astronomy played a decisive role by extending the HI rotation curves to the outer reaches of galaxies, the first indications had been found by Zwicky from the dynamics of galaxy clusters and by Babcock and later Rubin using optical rotation curves in the inner parts of galaxies.

The discussion in Chapter 8 on interstellar atoms and molecules detected using radio spectroscopy provides a rather different story, and one that I have not seen discussed in any detail before. While some of these spectral lines, such as the famous 21 cm hydrogen line, were predicted, the actual discovery processes were quite chaotic and rarely a direct result of the predictions. Unusually, for radio astronomy, there was intense competition and secrecy between some of the early observers of molecular lines, that even resulted in unprofessional behavior as scientists strived for the initial discovery credit. However, the field has long since matured and astrochemistry and the ALMA telescope are outcomes from this unruly beginning.

Chapter 11 discusses why telescopes are built, a very good adjunct to discussion of the nature of discoveries. The authors are in an excellent position to do this based on their 2020 publication *Open Skies* describing the radio telescopes constructed by the NRAO in the United States, although this does lead to a greater emphasis on the role of US radio telescopes. While discoveries are serendipitous, they do depend on the development of new technology. So, is it the telescopes, the instruments connected to the telescopes, or the data analysis that leads to a new discovery? Table 12.1 includes a column specifying the key instrument or technology involved in each discovery. Although it is conventional to assign credit to the scientists involved in the discovery, in some cases the new technology would have enabled the discovery for whoever happened to make the first observations. One extraordinary consequence, that is strongly emphasized throughout this book, is that the scientific discoveries for which facilities become famous are rarely those they were built for. Given the nature of many of the discoveries described here we can see that this outcome is not unexpected. But what is surprising is that this obvious fact has had so little influence on the discussions of future facilities and concepts like "exploring the unknown" have had little emphasis.

Chapter 11 also includes the unusual case of the Sugar Grove radio telescope, that was started by the US Defense Department but never completed. The authors provide intriguing details and an alternate interpretation of "highly classified."

The final chapter on future discoveries is based on lessons to be learned from past discoveries and speculates on the unexpected future discoveries. The authors have a warning "beware theoreticians," noting that they "played no role and sometimes delayed discoveries." But obtaining a theoretical understanding of discoveries that have been made is another matter and is hugely important. Without explanation the observed effects have no context and the contribution to knowledge can't be assessed. The visionary theoretical physicist Freeman Dyson explained "why heretics who question the dogmas are needed." He pointed out that "scientists often make confident predictions ... Their predictions become dogmas which they do not question ... [but] it may sometimes happen that they are wrong."[1] A view very well corroborated in many of the stories behind the discoveries described in this book.

The distribution of the ages of the scientists making discoveries (Figure 12.4) is quite dramatic, with a sharp decline at age about 40, and the authors note that it is well known that scientists generally do their best work as young men and women. But this is also the age when researchers tend to leave basic research as they become more involved in research management, writing grant applications, and serving on committees.

The cumulative discovery plot (Figure 12.5) is particularly intriguing and begs explanation. This plot covers the entire period from the beginning of the new science "radio astronomy," that did not exist before the first entry in 1933. The initial burst of discoveries heralded the beginning of a new technology enabled science, as was discussed by Harwit. But in this early period the small number of players with the Second World War technology would have limited the growth rate. Then as the radio wavelength technology and digital signal processing developed in many more countries and organizations, especially in the United States, more scientists were involved and there was rapid growth. This burst of discoveries in radio astronomy was certainly one of the most dramatic periods of discovery in all the physical sciences (note the number of Nobel Prizes in radio astronomy). But why has the curve flattened so dramatically since the 1980s? This fascinating puzzle is addressed in the last chapter. Maybe it's the Harwit effect as we reach a finite number of things to discover. Technology has continued to develop exponentially during this period (see Figure 12.1) so the discovery rate is not constrained by lack of new technology. Is it the consequence of the movement of astronomers out of radio astronomy as multiwavelength astronomy opened up in the space age? However, the authors make a rather compelling case that it may be a consequence of the change in culture, with the introduction of proposal reviewing and greater risk adversity in the current era. If so, this may be the most important message in this book for the future of basic research.

Ron Ekers, FRS, NAS, FAA
Fellow, CSIRO Space and Astronomy, former director of the VLA and of the ATNF radio observatories, and past president of the International Astronomical Union
June 2022, Sydney

1 Excerpt from *Many Colored Glass: Reflections on the Place of Life in the Universe* (Page Barbour Lectures) by Freeman Dyson, University of Virginia Press, 2007, p. 44.

Preface

Following an earlier remark by Bernie Burke, sometime in 1982 Ron Ekers and one of us (KIK) remembered that the following year would be the 50th anniversary of Karl Jansky's first public report that he had detected radio emission coming from our Milky Way Galaxy, and that we should do something to celebrate what is now recognized as the beginning of radio astronomy. We were joined by Bob Wilson and Tony Tyson from Bell Labs to organize a conference on Serendipitous Discoveries in Radio Astronomy held in Green Bank, West Virginia, in May 1983. We were fortunate in being able to bring many still living radio astronomy pioneers to Green Bank to relate their stories. All of the presentations and discussions were recorded and later transcribed with only limited editing by the speakers. *Serendipitous Discoveries in Radio Astronomy* (Kellermann and Sheets 1984) was privately published by the National Radio Astronomy Observatory.

Star Noise: Discovering the Radio Universe builds on this earlier publication by providing background to the historic discoveries, showing the connections among the discoveries, and bringing the stories up to date with the new discoveries made over the last four decades. We hope that *Star Noise* will be of interest not only to practicing scientists and students, but also to historians and others interested in how scientific research really works.

Over the near century of time covered in this book, many technical definitions and terminologies have changed. Until the middle of the twentieth century, most workers used "kilocycles per second (kc) and megacycles per second (Mc/s) to describe radio frequencies, although some, especially Grote Reber, used the more casual Mc. Except when in a direct quotation, we have used the modern terminology of Hz, kHz, MHz, and GHz, independent of what was originally used by the investigators at the time.

Radio astronomers carelessly switch between the use of frequency (f) and wavelength (λ), which is equal to the speed of light (c) divided by frequency ($\lambda = c/f$). We generally use whatever form was used by the original investigators, with the alternate representation given in parentheses the first time. Similarly, we use imperial or metric representations as appropriate to the circumstances, with the alternative representation shown in parentheses the first time a quantity is mentioned. In general, when the original number is an approximate value, we show the alternative value to equal precision, rather than using the precise numerical ratio that would give an inappropriate impression of accuracy.

In researching the relatively brief, but rich history of radio astronomy we have greatly profited from the definitive work by Woody Sullivan, *Cosmic Noise* (2009), his earlier books on *Classics in Radio Astronomy* (1982) and *The Early Years of Radio Astronomy* (1984a), as well as his collection of research notes and oral interviews with many of the pioneers of radio astronomy that Sullivan generously donated to the Historical Archives of the National Radio Astronomy Observatory.[1] We have also been aided by the extensive bibliography of radio astronomy publications covering the period 1898 to 1983 that was compiled by Sarah Stevens-Rayburn.[2]

Karl Jansky's sister, Helen, and Karl's son, David, gave us valuable insight into Karl's early life. Mary Harris kindly shared with us memories of her mother, Elizabeth Alexander, who, in 1945, was one of the first to document the detection of radio noise resulting from solar activity.

Some parts of this book, including Chapters 1, 2, and 8, make use of material previously published in *Open Skies* (Kellermann et al. 2020). For more extensive information on the life and career of Grote Reber discussed in Chapter 1, see Kellermann (2004). Chapter 4 is based in part on an earlier publication on the discovery of quasars that appeared in the *Journal of Astronomical History and Heritage* (Kellermann 2014) used here with the permission of the *Journal*. John Bolton, Ron Ekers, Jesse Greenstein, Cyril Hazard, Tom Matthews, Allan Sandage, and especially Maarten Schmidt all kindly shared with us their recollections of the complex events leading to the discovery of quasars. Maarten Schmidt generously made available his written communications with Bolton and Hazard from 1962 and 1963. We also acknowledge many discussions with Miller Goss and Dave Jauncey that have helped clarify the sequence of events surrounding the 1962 occultations of 3C 273 that led to Schmidt's identification of the redshift.

The book by Jim Peebles, Lyman Page, and Bruce Partridge (Peebles et al. 2009) provided valuable background to the complex history surrounding the development of the early theoretical speculations and experiments concerning the cosmic microwave background (CMB), and we are grateful to Jim Peebles and Bob Wilson for helping to clarify the history of the discovery of the CMB.

Frank Drake, Dale Frail, Duncan Lorimer, Richard Porcas, Lew Snyder, Joe Taylor, Bob Wilson, Alex Wolczcan, and Ben Zuckerman shared their reminiscences of their discoveries, which we describe in succeeding chapters. Imke De Pater, Christian Ho, Paul Horowitz, Duncan Lorimer, and Jim Moran gave us images for use in this book. Ron Ekers also gave us valuable advice on how to obtain permission to reproduce third-party material. Lance Utley and Kristy Davis were of enormous help in tracking down obscure references. We are grateful to all of them. We also thank Tim Bastian, Alan Bridle, Don Campbell, Marshall Cohen, Imke de Pater, Ed Fomalont, Miller Goss, David Jansky, Dave Jauncey, Jim Moran, Jim Peebles, Richard Porcas, Scott Ransom, Mort Roberts, Richard Schilizzi, Irwin Shapiro, Joe Taylor, Tony Tyson, Jasper Wall, and Ben Zuckerman for reviewing individual draft chapters, and especially Ron Ekers and Peter Robertson who read the whole manuscript and made valuable suggestions that helped improve our presentation.

Continued discussions and debates with Ron Ekers have contributed enormously to clarifying our interpretation of many of the topics that we discuss. Jim Condon kindly provided our front cover image. Heather Cole and Sarah Stevens-Rayburn also carefully read our manuscript and corrected an embarrassing number of remaining errors. Sheila Marks and Nicole Thisdell helped us solve thorny issues with Microsoft Word. Vince Higgs, Sarah Armstrong, Liz Steel, and India Priyadarshini Siddharthan at Cambridge University Press provided valuable advice and support during the long process of preparing the manuscript, and we acknowledge the continued support of the National Radio Astronomy Observatory throughout this project.[3] We thank them all, but note that any remaining errors of fact or in our presentation are due to us alone.

Finally, we acknowledge the support of Michele and Ron who have persevered from the beginning of this long project and encouraged us with their love and patient counsel; they look forward to a celebratory dinner upon publication.

1 www.nrao.edu/archives/sullivan-finding-aid (last accessed 21 November 2022).
2 https://rahist.nrao.edu/History_of_Radio_Astronomy_Bibliography.pdf (last accessed 21 November 2022).
3 The National Radio Astronomy Observatory is a facility of the National Science Foundation operated under cooperative agreement by Associated Universities, Inc.

Introduction

Doesn't expecting the unexpected make the unexpected expected?[1]

Astronomy is an observational science. With the possible exception of a few nearby Solar System objects, astronomers are unable to do experiments. All they can do is observe. When Galileo turned his primitive telescope toward the sky he did not plan to discover mountains on the Moon, sunspots, the rings of Saturn, the moons of Jupiter, or the phases of Venus. His discoveries during the years 1609 to 1611 were all accidental, the result of looking at the sky with a powerful new tool, the telescope.

For the next 400 years, astronomers built ever larger telescopes to explore the stars and galaxies and recognized the vastness of the expanding Universe. But the view of the Universe was limited to the narrow visible band of light between roughly 350 and 700 nanometers or 3,500 to 7,000 Angstroms. (Figure I.1).[2] Until 1933 we remained unaware of the plethora of objects that radiate primarily outside the familiar visual band corresponding to the sensitivity of the human eye.

The discovery, by Karl Jansky in 1933, of radio emission from the Milky Way Galaxy, while looking for the source of interference to transatlantic telephone communication, opened a new window on the Universe. Later, the advent of space-based astronomy gave astronomers access to the infrared, X-ray, and gamma-ray parts of the electromagnetic spectrum. But as the first of the new astronomies, it was radio observations that, within a few decades, resulted in the remarkable series of new discoveries that changed our view and understanding of the Universe and its contents. It is not surprising that the first six recipients of the Nobel Prize awarded for work in observational astronomy received their Prize for their work in radio astronomy.

After Jansky's detection of galactic radio emission, the following years saw the discovery of a wide range of previously unrecognized cosmic phenomena, including powerful bursts of radio emission from the Sun, electrical storms on

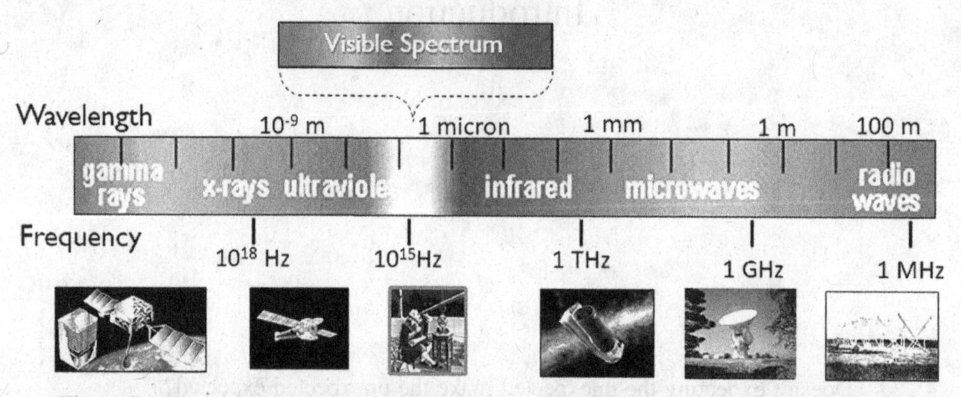

Figure I.1 The electromagnetic spectrum from radio to gamma rays. The insets left to right along the bottom show the Compton Gamma-Ray Observatory, the Chandra X-ray Observatory, an astronomer at an early optical telescope, the Spitzer IR Observatory, the Green Bank 140 Foot radio telescope, and Karl Jansky's 21 MHz (15 m) Bruce Array. Credit: Jeff Hellerman NRAO/AUI/NSF.

Jupiter, radio galaxies, quasars, pulsars, the cosmic microwave background, interstellar atoms and molecules, cosmic masers, apparent faster-than-light-motion, along with the first evidence for the big-bang and a Universe of finite age. The scientists responsible for these discoveries had little or no background in astronomy. They were instead skilled radio and electronic engineers and physicists, many of whom had received their training during the Second World War either in laboratories developing advanced radar systems or, especially in Great Britain and Australia, in the field deploying radar defense systems. Very few had earned doctorates in any field. This led to a new breed of astronomers, the radio astronomers, who for many years struggled for recognition from their conventional optical counterparts.

At first, radio astronomy developed nearly independently from optical astronomy. One reason was that the techniques were so different from those of traditional astronomy. At least within the United States, with the growing pressure of the Cold War, military patronage played an important role in defining research topics that were very different from those being pursued by conventional astronomers using large optical telescopes. Instead of the traditional studies of thermal radiation from the Sun, stars, and galaxies, radio astronomers studied a new type of nonthermal radiation from the Sun, the Galaxy, and from other galaxies. Also, many of the discoveries by radio astronomers came from outside the customary academic environment and, due to the many collaborations initially created during wartime radar research, radio astronomers formed a small but close-knit clique seeking recognition from the more traditional astronomical community. There was often intense competition and disagreements, sometimes heated, over observational results or recognition of credit in this rapidly developing new science of radio astronomy. However, for the most part the radio astronomy pioneers developed lifelong friendships leading to the free exchange of ideas along with their unprecedented postwar mobility.

Some of their discoveries came as a result of astronomical investigations, others from military or industrial studies, some from looking for something else. Most came from just looking with new and more powerful instruments. Generally, theory played no role; in other cases, the theory was wrong and delayed the discovery. The discoveries were almost all accidental or serendipitous.

The probably now overused word *serendipity* did not appear in the common English lexicon until the middle of the twentieth century, although it was first introduced by Horace Walpole, in a letter to his friend Horace Mann, back in 1754. As Walpole, the Earl of Orford and son of the first British Prime Minister Sir Robert Walpole, wrote to Mann, then the British Minister to Florence, he had coined the word from the title of a "silly" Persian fairy tale called *The Travels and Adventures of Three Princes of Serendip*, and remarked "as their Highnesses traveled they were always making discoveries, by accident and sagacity, of things which they were not in quest of." The three Princes were the sons of King Giaffer, the King of Serendip, the ancient name for Ceylon, now called Sri Lanka.[3]

The Oxford English Dictionary (OED) defines *serendipity* as "the faculty of making happy and unexpected discoveries by accident." Many of the classical scientific discoveries and inventions, starting with the apocryphal story of Newton's observation of a falling apple, may be said to have been lucky or serendipitous. Later examples included Wilhelm Röntgen's discovery of X-rays; the invention of the photographic plate by Louis Daguerre; and the development of penicillin by Alexander Fleming. These well-known accidental discoveries stand in contrast to the "scientific method" that we all learned in high school: hypothesis, followed by experiment, then results, discussion, and finally conclusions. As William Shakespeare wrote, "Fortune brings in some boats that are not steered."[4] However, Louis Pasteur, who developed the first vaccination for rabies and the pasteurization of milk and beer, recognized the fine line between luck and scientific skill, and famously pointed out, "In matters of observation, chance favors only the prepared mind."[5]

The word *serendipity* was first introduced into the English vernacular in 1958 by Robert Merton.[6] According to Merton, prior to that time the word *serendipity* had been used in print only 135 times, but by the year 2000 it had become the tenth most popular name for privately owned pleasure yachts. Merton claimed that he stumbled across the word accidently while looking up another word in the Oxford English Dictionary (OED) starting with *se*. He then described his *serendipitous* acquisition of the OED following Franklin Roosevelt's closure of US banks on 6 March 1933, two days after he took office as the 32nd President of the United States. Without any access to his limited savings, Merton, then an impoverished Harvard University student, took advantage of the generosity of the Harvard Square Phillips Book Store owner who accepted Merton's credit, allowing him to purchase the 13 volume set of the 1933 edition of the OED. Merton claimed he acquired the OED with an IOU for approximately $90. He anticipated that he would never get the funds to pay the bookseller.

Our story of *Star Noise*: *Discovering the Radio Universe* begins a month later in April 1933.

1

A New Window on the Universe[1]

> . . . a very steady hiss type of static, the origin of which is not yet known.[2]

Karl Guthe Jansky was the first person to look at the Universe from outside the traditional visible light window. His presentation, with the innocuous title "A Note on Hiss Type Atmospheric Noise," stunned the small group of radio scientists who heard his talk at the 27 April 1933 Washington meeting of the US National Committee of the International Union of Radio Science (URSI). Although cautioned by his employer, Bell Laboratories, to avoid any sensationalism, following several years of meticulous research aimed at improving the reliability of the AT&T long distance telephone circuits, Jansky announced that he had detected radio noise from the Milky Way Galaxy. Subsequently, the effect of the US economic depression followed by Jansky's increasing defense-related responsibilities limited any further work at Bell Labs on what Jansky called his "Star Noise." However, starting in 1937, Grote Reber, a young radio amateur and recent engineering graduate, working alone with a homebuilt antenna, followed up Jansky's pioneering work and made the remarkable discovery that Jansky's "Star Noise," unlike the light from the Milky Way, did not come from any normal thermal processes that dominate the light from the planets, stars, galaxies, and other celestial objects. Reber made the first maps of this new nonthermal radio emission from the Milky Way that was later understood to be synchrotron radiation from ultra-relativistic electrons moving in weak cosmic magnetic fields.

The serendipitous detection of nonthermal radio emission from the Milky Way by Karl Jansky, followed by the theoretical interpretation by Soviet astrophysicists, set the stage for the plethora of later – also serendipitous – discoveries of the powerful nonthermal radio emission from radio galaxies, quasars, pulsars, cosmic masers, and other cosmic radio sources that would be discovered by radio astronomers over the next half century and which are described in later chapters.

Star Noise

In 1933, the world was in the midst of the Great Depression. One quarter of Americans were out of work. Two months before Franklin Roosevelt became the 32nd president of the United States on 4 March 1933, Adolph Hitler rose to power in Germany, and two days later he dissolved the German Parliament, beginning the path to global turmoil and destruction. But the turmoil and destruction of the Second World War also led to unprecedented advances in radio technology that would make possible the new science of radio astronomy.

In 1933, the United States passed the 21st amendment to the Constitution repealing prohibition; Wiley Post became the first person to fly solo around the world; Albert Einstein emigrated to the United States; based on James Chadwick's discovery of the neutron a year earlier, Leó Szilárd conceived the idea of a nuclear chain reaction; the movie *King Kong* premiered in New York City; the United States went off the gold standard; the legendary gangster John Dillinger robbed his first bank; Paul Dirac and Erwin Schrödinger shared the Nobel Physics Prize; Babe Ruth hit a home run at the first major league All-Star baseball game; and Edwin Armstrong obtained a patent for the invention of FM radio. A gallon of gas cost 10 cents, a loaf of bread 7 cents, but a 3 minute transatlantic phone call using shortwave radio cost $75 and was plagued by noise and fading.

By 1915, using cables and regularly spaced vacuum tube repeaters, AT&T had extended telephone service to reach across the North American continent, but transatlantic service presented more of a challenge. Following experimental systems built in 1923, AT&T started the first commercial transatlantic telephone communication between New York and London in 1927, using very long wavelength 5 km (60 kc) radio transmissions from a 200 kilowatt water-cooled transmitter located in Maine.[3] Two operators were required at each end, one "to make contact with the telephone network in her country" while the "other operator directs her attention to the transatlantic link." But the system was unreliable due to the effects of varying propagation and static, especially during the summer months. Also, at 60 kc, the available bandwidth was limited and there was persistent interference from powerful telegraph transmitters.

In 1925, AT&T created the Bell Telephone Laboratory as its research arm to complement Western Electric, which manufactured the telephone equipment used by AT&T. Three years later, AT&T introduced a new shortwave telephone service between the United States and England and later South America. But the shortwaves brought a new set of difficulties related to the uncertain propagation connected with the time of day, the seasons, and solar activity. Static, especially over the tropical path to South America, and locally generated noise from thunderstorms, electrical equipment, and automobile ignition were all problems.[4]

Bell Labs engineers were aware that even in the absence of locally generated noise, the noise level was greater when an antenna was connected to a shortwave receiver than when the antenna was replaced by a resistor.[5] When Karl Jansky arrived at the Bell Telephone Laboratories on 20 July 1928 at the age of 22, he was assigned the task of understanding the propagation of shortwave radio signals and studying the noise that was limiting the effectiveness of long distance telephone calls, both for the existing long wave operations and the newly inaugurated short wavelength service.

Karl Guthe Jansky (Figure 1.1) was born in the Oklahoma Territory on 22 October 1905, where his father, Cyril, was Dean of the College of Engineering at the University of Oklahoma, and he grew up in Madison, Wisconsin, after his father became a Professor of Electrical Engineering at the University of Wisconsin. Karl was an outstanding student at the University of Wisconsin where he studied physics, and he excelled at sports. During his student years, he starred at right wing for the Wisconsin Badgers ice hockey team (Figure 1.2) and graduated Phi Beta Kappa in 1927. Starting in his college years, Karl first began to suffer from high blood pressure and the chronic kidney disease that would ultimately take his life. After a year in graduate school, Jansky applied for a job at the AT&T Bell Telephone Laboratories but, due to his questionable health, he was turned down. Fortunately, Karl's older brother, Moreau, who had previously worked at Bell Labs, pulled some strings and convinced AT&T to take a chance and hire Karl. Due to concerns about the possible effect of city pollution on his health, AT&T management sent Jansky to live and work at the Bell Labs' Cliffwood Beach site in rural New Jersey, instead of at the AT&T Wall Street office in New York City. As it fortuitously turned out, not only was the air quality in New Jersey better than in New York City, but the reduced radio noise from automobiles and industry would facilitate Jansky's research that resulted in his remarkable serendipitous discovery of radio noise from the Milky Way.[6]

While working at Bell Labs, Karl played tennis and golf and had the highest batting average on the Laboratory softball team. He was the table tennis champion of Monmouth County, New Jersey, and enjoyed bowling, skiing, tennis, and playing chess. Jansky was also a passionate bridge player and avid Brooklyn Dodger baseball fan and was described by his colleagues as very competitive in sports and bridge but friendly, modest, and easy to get along with.[7] In 1929, after a year at Bell Labs, Karl married Alice Larue Knapp. Karl and Alice had two children, Anne Moreau and David, and, in spite of the economic uncertainties of the depression, Karl and Alice enjoyed an active social life with fellow Laboratory staff (Figure 1.3).

During his entire career at Bell Labs, Karl's boss was Harald Friis, a well-known Danish engineer who had immigrated to the United States (Figure 1.4). Although Friis and his wife Inge became close personal friends of Karl and Alice and were the

Figure 1.1 Karl Jansky. Courtesy of David Jansky.

godparents to their daughter Anne Moreau, Karl developed a strained relationship with Friis over his work assignments. Fortunately, Jansky's historic accomplishment is chronicled through the entries in his laboratory notebooks and monthly work reports, both of which have been partially preserved at the Bell Labs Archives, as well as a series of detailed letters that he and Alice regularly wrote to his parents and younger siblings back home in Wisconsin. In particular, in writing to his father, who understood and appreciated the significance of Karl's work, Karl was able to go into significant technical detail and at the same time discuss the personal challenges of living and working under the prevailing economic constraints.[8]

For his long wavelength (near 4,000 m) investigations that partly occupied his first few years at Bell Labs, Jansky adopted the instrumentation and techniques previously developed by Friis.[9] However, the short wavelength study was new territory, and he had to design both his antenna and receiver. Since he had studied physics at Wisconsin, Jansky was unfamiliar with radio and electronic terminology, and reflected to his father, "that is what I get for not taking engineering."[10] In order to locate the source of radio noise affecting the shortwave telephone communications, Karl needed a directional antenna, which he based on a design developed by his Bell

Figure 1.2 1928 photograph of Karl Jansky as a University of Wisconsin ice hockey player. Courtesy of David Jansky.

Labs colleague, Edmund Bruce. The laboratory shops constructed the radiating element from standard 12 foot sections of ⅞ inch diameter brass plumbing pipe as the ¼ wavelength vertical elements of a Bruce Array. A second set of elements acted as a reflector giving Karl a directional antenna having more than a factor of 300 front-to-back ratio (25 db). As shown in Figure 1.5, the whole arrangement was mounted on a turntable using tires from a Ford Model T automobile, and was driven by a motor to make a complete rotation every 20 minutes. Karl's children, David and Ann Moreau, enjoyed playing on their father's "merry-go-round."

Jansky initially decided to work at a frequency of 20,689.7 kc (14.5 m) which he found was relatively free of radio signals. While continuing his analysis of long

Figure 1.3 1938 photograph of the Jansky family. Left to right: Karl's wife Alice, family friend, son David, Karl, daughter Anne Moreau. Courtesy of David Jansky.

Figure 1.4 Harald Friis, Karl Jansky's long time supervisor at Bell Laboratories. Credit: AIP Emilio Segrè Visual Archives, Physics Today Collection.

Figure 1.5 Karl Jansky and his Bruce Array at Holmdel, NJ, used for the first detection of cosmic radio emission. Credit: Nokia Corporation and AT&T Archives.

wavelength noise, as well as developing a system designed to study ultra-short waves (3 to 6 m), he designed and built the instrumentation needed to study the source of 14.5 m noise, putting particular emphasis on minimizing the noise contribution from his receiver and reducing instabilities and spurious oscillations. He used a bandwidth of 26 kHz, much greater than the 3 to 6 kHz normally used for voice communication, and he averaged the output of his receiver over a period of about 10 seconds to minimize the noise fluctuations. Although his work suffered a setback with the move of the Bell Labs site to a new location near Holmdel, New Jersey, that offered space to accommodate the growing Bell Laboratories staff, there was less local radio noise at the new site, which turned out to significantly benefit his research.

By mid-1930, Karl's rotating array had been relocated to a new circular track that was constructed at the Holmdel site, and he was able to begin work. He quickly located radio transmissions originating in England and South America. By November, he had located a source of noise toward the southwest. As he wrote to his parents on 5 December, although he was plagued by static or noise from local thunderstorms, he was aware of static that appeared to come from a direction unrelated to any known weather disturbance.[11]

During the first half of 1931, Jansky upgraded his receiving system and obtained more systematic records of his data by attaching a moving chart recorder to the output of his receiver (Figure 1.6). By this time, he realized that he was detecting static unrelated to weather activity over a range of directions from the southeast to the southwest. Following the summer thunderstorm activity, during the 1931/1932 winter, Jansky realized that the noise came from a direction that appeared to change with the time of the day. Although distracted by other Laboratory responsibilities, Jansky recognized that in the morning the noise appeared to come from the east, moved toward the south around noon, and then came from the west in the late afternoon, suggesting an origin associated with the Sun. Around this time, Jansky recognized that his noise, that he called a "hiss," was not characteristic of the kind of static normally associated with electrical noise or thunderstorm activity. After the

Figure 1.6 Karl Jansky in 1933 at the chart recorder output of his 21 MHz radiometer. Credit: Nokia Corporation and AT&T Archives.

birth of his daughter and proudly telling his parents that Anne Moreau "is the best natural baby I have seen in a long time," Karl excitedly wrote,[12]

> I have been receiving a very weak and very steady static lately that is so very steady that it could be easily mistaken for a signal but which I can definitely show is not a signal. The peculiar thing about the static is that the direction from which it comes changes gradually and what is most interesting always comes from a direction that is the same or very nearly the same as the direction the sun is from the antenna. I have not had much time to put on this problem but as soon as I get two more reports written am going to concentrate on it alone and see what I can find. Sounds interesting doesn't it.

And in his January 1932 Work Report Karl wrote,

> Thunderstorm static was almost completely absent, but there was, and still is, present a very steady continuous interference; the term static doesn't quite fit it, that changes direction continuously throughout the day going completely around the compass in 24 hours. During the month of December this varying direction of arrival followed the sun almost exactly making it appear that perhaps the sun causes this interference or at least has something to do with it.

Having no background in astronomy, Jansky did not yet appreciate that, during his winter observations, the Sun was coincidently in the same direction as the Milky Way. Although he continued to spend time with his ultra-short wave system, in preparing for a demonstration to AT&T executives, and in preparing an introductory speech for Friis to give at an URSI Symposium on Static, during early 1932 Jansky recognized that the hiss came earlier each day and was no longer associated with the direction of the Sun. In his April 1932 presentation to the US National Committee for the International Union of Radio Science (URSI)[13] and in the first of his three famous papers published in the *Proceedings of the Institute of Radio Engineers* (*Proc. IRE*), Jansky described the following three distinct types of static:[14]

> The first group is composed of the static received from local thunderstorms and storm centers. Static in this group is almost always of the crash type. It is very intermittent . . . The second group is composed of very steady weak static coming probably from [ionospheric] refractions from thunderstorms some distance away. The third group is composed of a very steady hiss type static the origin of which is not yet known.

Apparently, he had not recognized the third group of static until January 1932, but, after going back to examine his earlier data, he reported that:

> The static of the third group is also very weak. It is, however, very steady, causing a hiss in the phones that can hardly be distinguished from the hiss caused by [receiver] noise, [and] the direction of arrival of this static coincided with . . . the direction of the sun. However, during January and February, the direction has gradually shifted so that now [March 1] it precedes in time the direction of the sun by as much as an hour.

Jansky was puzzled as he tried to reconcile the apparent shift of the direction of the static with the lengthening of the day and he eagerly awaited the summer solstice when he expected the direction of the hiss to reverse direction. Throughout this period, Jansky's monthly work reports showed that he was still spending more time on monitoring the long wavelength recorder than studying the mysterious shortwave hiss. He also spent time each month instructing technical assistants on mathematics, attending committee meetings, and in other assigned Laboratory duties. Due to mechanical breakdowns or overhauls of the rotating array, or to circuit revisions characteristic of his technical background, during the summer and autumn months, the shortwave rotating array was apparently used only over weekends. Only in September did he shut down the long wavelength recording system, "to allow more time to be spent on more pressing problems."[15]

After 21 June, when the Sun reached its most northerly declination, instead of reversing as expected, the direction of the static continued to fall behind the Sun at a rate of two hours per month. Now, even further mystified, Jansky consulted his boss, Harald Friis, and other Bell Labs colleagues, who were equally perplexed.[16] The breakthrough came from Karl's close friend, Melvin [Mel] Skellett, a graduate student at Princeton University working for a PhD in astronomy, who recognized the sidereal nature of Jansky's data. With the help of Skellett, and after consulting several astronomy text books, Jansky finally understood the implications of what he had found and, on 21 December 1932, he wrote to his father,[17]

> I have taken more data which indicates definitely that the stuff, whatever it is, comes from something not only extraterrestrial but from outside the solar system. It comes from a direction that is fixed in space . . . I've got to get busy and write another paper right away before someone else interprets the results in my other paper in the same way and steals my thunder from my own data.

In February, he wrote again (Figure 1.7), "My records show that the 'hiss type static' mentioned in my previous paper comes, not from the sun as I suggested in that paper, but from a direction fixed in space. The evidence I now have is very conclusive and, I think, very startling."[18]

Excited about his finding, Karl approached Friis about presenting his work at the June 1933 meeting of the Institute of Radio Engineers (IRE). The meeting was to be held in Chicago and would give him an opportunity to visit his family in Wisconsin. Instead, Friis told Jansky to give his talk at the April meeting of the US National Committee for URSI in Washington, which was closer to Bell Labs, and he cautioned Jansky not to generate undo attention. But, as Karl later wrote to his father,[19]

> I have not the slightest doubt that the original source of these waves whatever it is or wherever it is, is fixed in space. My data proves that, conclusively as far as I am

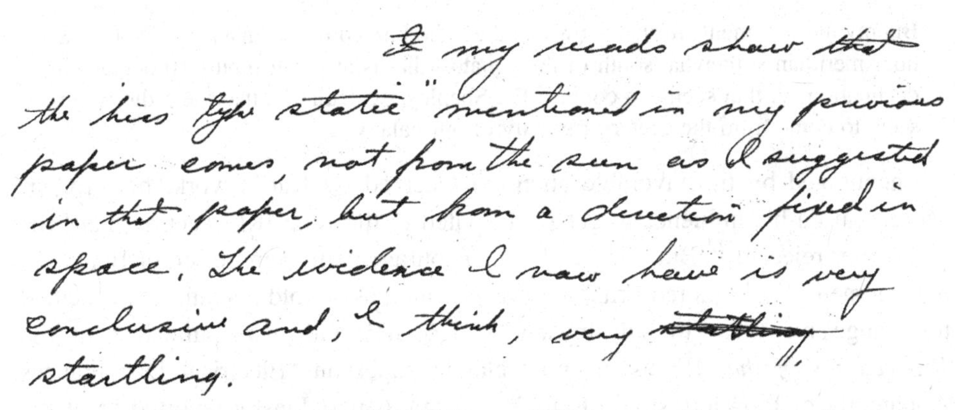
I my records show that the hiss type static" mentioned in my previous paper comes, not from the sun as I suggested in that paper, but from a direction fixed in space. The evidence I now have is very conclusive and, I think, very ~~startling~~ startling."

Figure 1.7 Excerpt from Jansky's February 1933 letter to his parents where he writes "my records show that the 'hiss type static' mentioned in my previous paper comes, not from the sun as I suggested in that paper, but from a direction fixed in space. The evidence I now have is very conclusive and, I think, very startling." Credit: University of Wisconsin Archives, Cyril M. Jansky Papers.

> concerned. Yet Friis will not let me make a definite statement to that effect but says I must use the expression '<u>apparently</u> fixed in space' or '<u>seems</u> to come from a fixed direction' etc. etc. . . . But I suppose it is safer to do what he says. (underlining original)

Following Friis' strict instruction, Karl simply titled his short 12 minute talk on 27 April 1933, "A Note on Hiss Type Atmospheric Noise," that, as he complained to his father, "meant nothing to anybody but a few who were familiar with my work." He was disappointed in the lack of any reaction to his talk and described URSI to his family as "an almost defunct organization . . . attended by a mere handful of old college professors and a few Bureau of Standards engineers."[20]

Karl's older brother, Moreau, who recognized the significance of Karl's discovery, was an influential leader of the US National Committee for URSI and stepped in to get Karl some recognition. Through his former contacts at Bell Labs, Moreau arranged for AT&T to issue a press release that drew widespread attention throughout the national and international news media. The 5 May 1933 edition of *The New York Times* featured a front page "above the fold" article titled, "New Radio Waves Traced to the Centre of the Milky Way." The next day, the *Times* "Week in Science" section and the 15 May edition of *Time Magazine* carried their own stories of Karl's discovery, and on the evening of 6 May, radio station WJZ in New York and the NBC Blue Network carried a live broadcast of Jansky's star noise received at Holmdel and sent over the AT&T Long Lines to the studio in New York City.[21] In describing his star noise during his radio interview, Jansky explained,

> The observations show definitely that the maximum of hiss comes from somewhere on the celestial meridian designated by astronomers as "18 hours right ascension" . . .

> But my measurements further show that the radio hiss comes from a point on that 18-hour meridian somewhat south of the equator, that is at about minus 10 degrees in declination ... that seems to confirm Dr. Shapley's calculation that the radio waves seem to come from the center of gravity of our galaxy.

Encouraged by the favorable attention received by Karl's work, his brother, Moreau, used his influence to get Karl invited to the June IRE meeting that Friis had earlier rejected.[22] But now, with the encouragement of more senior Bell Labs management, Karl ignored Friis' reservations and, as he told his father, he decided to change his title "to suit myself."[23] His IRE talk was published in the *Proceedings of the IRE* as his now classic paper on "Electrical Disturbances Apparently of Extraterrestrial Origin."[24] In this paper, Jansky reported "that the direction of arrival of these waves is fixed in space, i.e., that the waves come from some source outside the solar system," and he gives this direction as the "center of the huge galaxy of stars and nebulae of which the sun is a member."

Aware that his first *Proc. IRE* paper published the previous year had suggested that the hiss type noise originated in the Sun,[25] Jansky also published a short note in *Nature*, with the title, "Radio Waves from Outside the Solar System." In this paper, he explained that "the direction of arrival of this disturbance remains fixed in space, that is to say the source of this noise is located in some region that is stationary with respect to the stars," and gave the direction of the radio noise as "right ascension of 18 hours and declination of –10 degrees," although, as he pointed out, the declination "might be in error by as much as ±30°."[26] Later, in an effort to reach a broader audience, he published an article in *Popular Astronomy* with a similar title, "Electrical Phenomena that Apparently Are of Interstellar Origin."[27] In these two later papers, we note an increased use of astronomical terminology which Jansky had now mastered. Then, in October, Karl was invited to lecture at the American Museum of Natural History in New York where he boldly used the provocative title, "Hearing Radio from the Stars" and proudly played his "hiss" noise sent over the AT&T lines from Holmdel to New York.

Although Jansky spent most of the next few years working on other problems, including the direction of arrival of ultra-short wave radio transmissions, he did find time to further analyze his star noise data. In July 1935, he traveled to Detroit, Michigan, where he presented his new analysis at the annual IRE meeting. His talk was one of the last in the late afternoon and only a few dozen people stayed at the end of a hot summer day to hear him explain that the radio noise came from throughout the Galactic Plane and that there is an important contrast between the visual and radio sky.[28] In the published version of his Detroit talk, "A Note on the Source of Interstellar Interference," Jansky referred to his 16 September 1932 data as a "typical day's record" that showed "Beside varying gradually in height throughout the day, the peaks obtained for each revolution of the antenna also change decidedly in shape" (Figure 1.8). In this, the third of his *Proc. IRE* publications, Jansky concisely summarized the known characteristics of his star noise:[29]

(a) The radio noise is distributed throughout the galactic plane with the strongest emission coming from the center of the galaxy.
(b) The characteristics of star noise are very similar to the noise generated by vacuum tube circuits suggesting that "it is produced by the thermal agitation of electrical particles."
(c) Visually the Sun appears brighter than the radiation from all the stars combined, but the reverse is true at radio wavelengths.

Later, radio astronomy historian Woodruff Sullivan III reanalyzed Jansky's data taken on 16 September 1932 and, as shown in Figure 1.9, displayed his analysis in the form of a radio intensity image.[30]

Around this time, according to Grote Reber, Jansky proposed building a 100 foot diameter transit dish to continue his investigation of star noise at 60 MHz (5 m). However, as Jansky explained to Reber, he was told, "that the proposal was outside the realm of company business."[31]

Bell Labs was proud to share in the publicity surrounding Karl Jansky's unexpected discovery of cosmic radio noise, but it didn't really contribute to making telephones work better. For a period, Jansky's "merry-go-round" was used by others for testing antennas, but it ultimately fell into disrepair and the remnants were destroyed. In 1964, a replica of Karl's antenna was erected at the entrance of the NRAO Green Bank Observatory. Years later, former Bell Labs scientists J. Anthony (Tony) Tyson and Robert (Bob) Wilson located the site of the antenna at Holmdel and, on 8 June 1998, Bell Labs dedicated a memorial model antenna on the site of Jansky's merry-go-round (Figure 1.10).

Jansky did return briefly to a measurement of his star noise, using a variety of different antennas to measure its frequency dependence between 5 MHz (60 m) and 23 MHz (13 m). Although he commented on a marginally greater noise at 16.7 m wavelength compared to 14 m, due to interference from diathermy machines and the effects of ionospheric absorption associated with increasing sunspot activity, Jansky's measurements of the frequency dependence of galactic radio noise were inconclusive. About the same time, Friis and C. B. Feldman, using Friis's Multiple Unit Steerable Antenna (MUSA), reported the detection of what they referred to as "star static," with up to a factor of six above receiver noise over a wide range of frequencies down to 10 MHz.[32] However, since they were apparently interested in testing MUSA, not in investigating the properties of star static, they did not report any quantitative dependence of how the star noise varied with frequency.

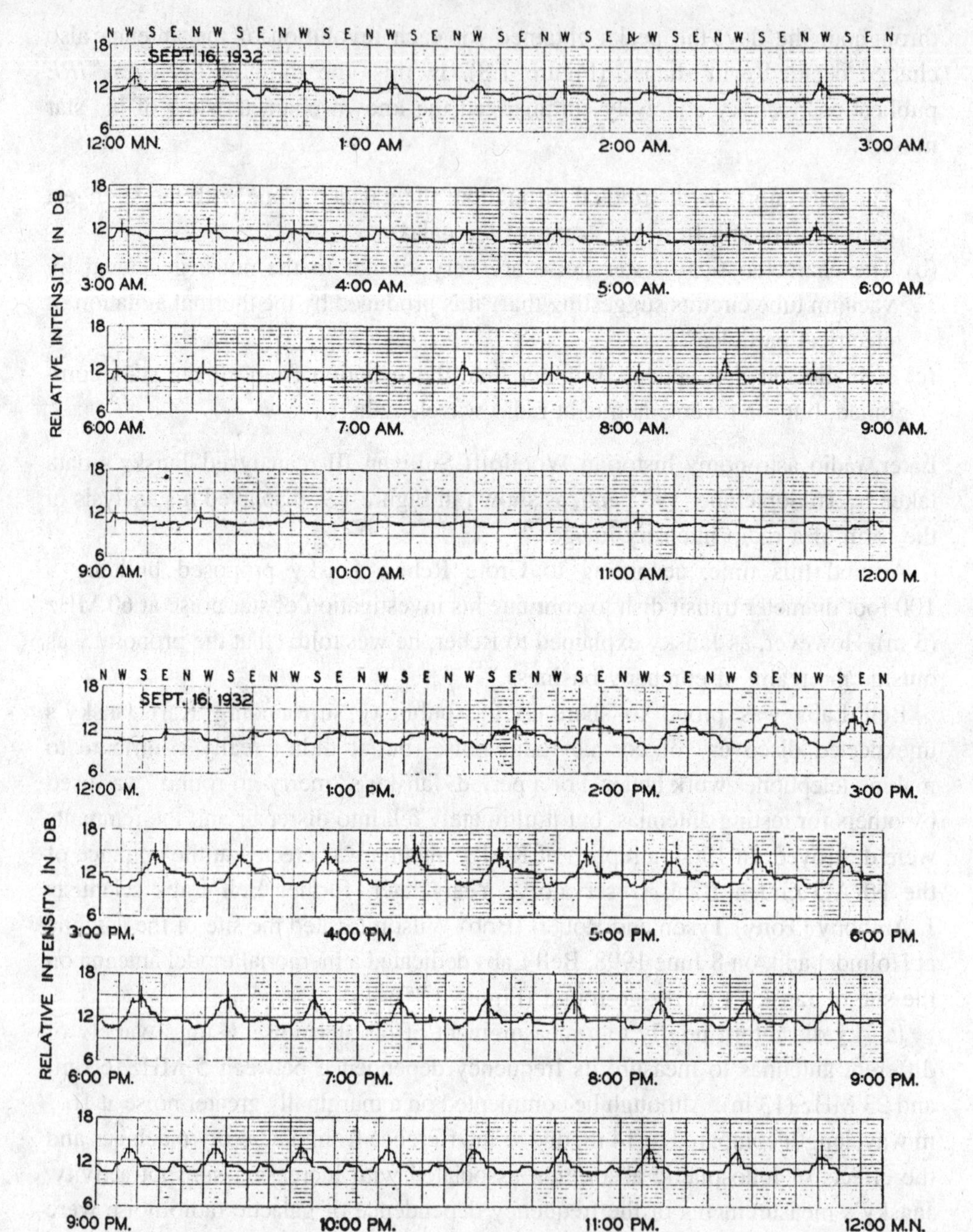

Figure 1.8 Reproduction of Jansky's 16 September 1932 chart recorder output showing the radio emission from the Milky Way in a 12 hour period as his antenna beam swept across the Galaxy three times an hour. The different response profiles reflect the varying angle of Jansky's fan beam alignment with the plane of the Milky Way. The lettering at the top of each plot indicates the direction of the antenna beam at that time. Credit: NRAO/AUI Archives, Papers of W.T. Sullivan III.

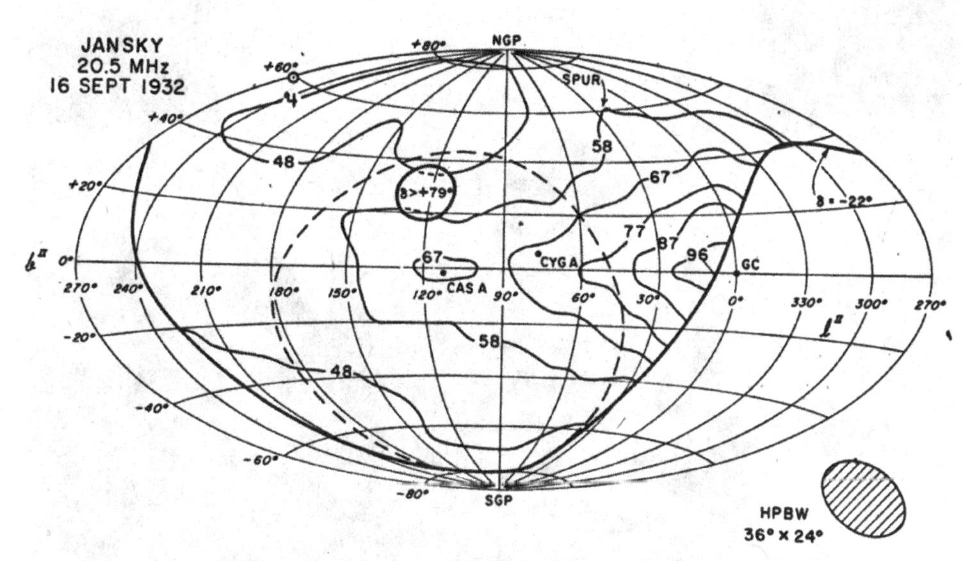

Figure 1.9 Contour map plotted in galactic coordinates showing the distribution of radio intensity over the sky, constructed by Woodruff T. Sullivan III from Jansky's 16 September 1932 data. Courtesy of W. T. Sullivan III.

Karl Jansky spent most of the remainder of his career at Bell Labs continuing his research on shortwave radio propagation, as well as investigating the sources of noise that limited radio communication. As was later recognized by Second World War radar operators, Jansky showed that, in the absence of man-made interference, the sensitivity of shortwave radio systems was often limited by cosmic noise, and not by receiver circuit noise.[33] In the 1940s, he helped to develop the first microwave repeater systems used by AT&T to carry television signals from New York to Boston.

Increasingly toward the end of the decade, Jansky was called on to work on classified defense-related programs. He was recognized by an Army–Navy citation for his development techniques for the electronic detection of submarines. After the Bell Labs invention of the transistor in 1947, Jansky was one of the first to use transistors to build low noise preamplifiers and received several patents on a radio direction finder or sextant based on the radio emission from the Sun, that was later developed by the Collins Radio Co. for the Naval Research Laboratory. However, his deteriorating health limited his activities and, after a series of strokes, on 14 February 1950, Karl Jansky succumbed to his long illness and died at the young age of 44, before the impact of his discovery of cosmic radio noise was fully appreciated by the astronomical community.

Figure 1.10 Grote Reber (left) and Jesse Greenstein meet again at the 1998 Bell Labs dedication of a memorial monument marking the original location of Jansky's antenna. Credit: NRAO/AUI Archives, Papers of K. I. Kellermann.

The early 1930s coincided with the 11 year minimum in the level of sunspot activity and a correspondingly low level of ultraviolet radiation that ionizes the upper regions of the earth's atmosphere, known as the ionosphere. When heavily ionized during periods of enhanced solar activity, the ionosphere reflects radio waves, permitting global shortwave radio communication when radio waves can bounce off the ionosphere and return to the Earth thousands of miles away, well beyond the direct line of sight distance. Similarly, near sunspot maximum cosmic radio waves are reflected back into space, so only near sunspot minima can shortwave cosmic radio emission reach the Earth. The reflectivity of the ionosphere depends not only on solar activity but on the radio frequency (wavelength) as well.[34] Above some critical frequency that depends on solar activity the ionosphere is transparent. Radio waves much below the critical frequency are absorbed by the ionosphere.

Jansky was fortunate that he started his experiments just near the beginning of an extended period of low solar activity when the ionosphere was transparent to the incoming galactic radio emission. Had he worked five years earlier or five years later, near periods of enhanced sunspot activity, he likely would not have been able to detect cosmic radio emission. He was also fortunate in choosing to work at a wavelength near 15 m. Had he chosen to work at a longer wavelength, the ionosphere would have appeared opaque even at sunspot minimum; had he worked at shorter wavelengths, the incoming cosmic radio waves would have been too weak for Jansky to detect with the equipment then available to him.

Jansky's most productive period, in the early 1930s, coincided with the depths of the economic depression with many people out of work. In May 1932, Bell Labs, "discharged" some staff and cut the remaining staff to a four day work week with a corresponding cut in pay, although Jansky and others continued to work five and even six days a week while getting paid for four. He borrowed money from his father, which he paid back in installments, to help buy their new house, and was concerned throughout his career about his medical expenses. Looking for opportunities to continue his star noise research, Jansky wondered about a university position but, aware of his chronic illness and growing family responsibilities, he could not relinquish his relatively secure position at Bell Labs. However, by 1944, to meet the demands of the defense effort, Jansky was working overtime and was thankful for thc extra pay he needed to meet the increasing cost of medical care.

Historians, along with Karl's colleagues and family members, have asked whether or not he was restricted by Friis from continuing his star noise research. In his 1956 book, *The Changing Universe: The Story of the New Astronomy*, author John Pfeiffer provocatively wrote that, "Rarely in the history of science has a pioneer stopped his work completely, at the very point where it was beginning to get exciting. Yet Jansky did just that . . . He did all he could to convince his associates and superiors that the work was worth pursuing for practical reasons. But his arguments failed to produce results."[35] In a review of Pfeiffer's book in *Science*, Frank Edmondson, from the National Science Foundation, further stirred the pot and suggested that "Jansky's failure to secure support for continued pure research at Bell Laboratories" may have led the United States to be "lagging far behind other countries in the development of radio astronomy."[36] A few years later, when writing an introductory article in a special issue of *Proc. IRE* devoted to radio astronomy, Karl's brother, Moreau, declared that, "his superiors transferred his activities to other fields. He would have preferred to work in radio astronomy."[37]

Harald Friis vigorously refuted these assertions, claiming that Karl "was free to continue work on star noise if he had wanted to," and that Karl had never indicated to him "a desire to continue his star noise work."[38] Moreover, noted Friis, "during this period there had been no interest or encouragement from astronomers and it was not clear in what directions such research should go and what kind of equipment was needed." Essentially, he argued that Karl felt that having found the source of noise, he had completed that particular project.[39]

When she learned of Friis' reaction, Karl's widow, Alice, animatedly wrote to Karl's brother,

> Harald says that Karl never expressed to him a desire to continue work on his star noise. How incredible, how preposterous, how positively unbelievable. Periodically, over the years that Karl worked under Friis, he would come home and say, "Well, Friis and I had a conference today to discuss what my next project should be, and, as usual, Friis asked what I'd like to do, and as usual, I said, "You know I'd like to work on my star noise," and as usual, Friis said, "Yes, I know, and we must do that someday, but right now I think — and — is more important, don't you agree?"[40]

Although Bell Labs management later took a very defensive position regarding Jansky's activities, other Bell Labs colleagues called the management response a "cover-up."[41] George Southworth explained that, "Most of us, without knowing why, thought intuitively that he had not been dealt with fairly,"[42] and another colleague added that Friis was a "dictator [who] wanted things done exactly the way he said."[43]

It is difficult after many years to fully understand the accuracy of these later, clearly not unbiased, recollections of who told what to whom, the impact of the economic depression along with the increasing threats of war, the natural financial pressures faced by Bell Labs management, and why certain research topics were pursued or not pursued. However, it is clear from his contemporaneous letters with his family that Karl Jansky clearly understood the scientific importance of his discovery, that he wanted to continue his star noise research, and that he was discouraged by Friis and most other Bell Labs management from presenting his discovery in a way that could best be understood and appreciated. On the other hand, Harold Arnold, a senior AT&T executive, encouraged Jansky to publish and was instrumental in arranging *The New York Times* and *NBC* publicity.[44] Unfortunately, Arnold died later in 1933 and Jansky lost perhaps his only strong supporter among AT&T management.

Karl Jansky was nominated for the 1948 Nobel Prize in Physics for his discovery of cosmic radio emission, but he died before the explosive growth of radio astronomy in the 1950s and before the importance of his work became widely appreciated.[45] Karl Jansky's name was, however, honored at the 1973 General Assembly of the International Astronomical Union, that passed a resolution declaring "that the name 'Jansky,' abbreviated 'Jy' be adopted as the unit of flux density in radio astronomy and that this unit, equal to 10^{-26} $Wm^{-2}Hz^{-1}$, be incorporated into the international system of physical units."[46] Also, each year, the National Radio Astronomy Observatory, a federally-funded radio astronomy research center, appoints several young scientists as Jansky Fellows to further develop their careers in radio astronomy along with naming a distinguished scientist to give the annual "Jansky Lecture." In 2012, the NRAO Very Large Array was rededicated as the "Karl G. Jansky Very Large Array" and it continues to operate as the most powerful radio telescope in the world. In 1998, the Bell Telephone Laboratories dedicated a monument to Karl's memory located at the site of his "merry-go-round." In a message read at the ceremony, Sir Bernard Lovell wrote, "There can be few occasions when such observations have not only had such profound consequences but also belong totally undisputed to one man – Karl Jansky."

Cosmic Static[47]

In their early experiments with radio waves, both Heinrich Hertz and Guglielmo Marconi used parabolic structures to focus their radio waves. But it would take a young engineer, Grote Reber, working by himself in Wheaton, Illinois, to make the important step of designing, building, and then using a large parabolic dish to study Jansky's star noise.

Grote Reber was born in Chicago, Illinois, on 22 December 1911 and grew up in the Chicago suburb of Wheaton. Apparently, his parents were slow to decide on a name so his birth certificate merely listed his name as "Baby Reber." He was called Grote by his family, his mother's maiden name, and it was not until he was 20 years old that he officially had his name verified on a revised birth certificate by the authority of the Cook County Clerk, Richard E. Daley, who later became the infamous major of Chicago.

Grote's father, Schuyler Colfax Reber, who was named after Schuyler Colfax, the US Vice President during the Grant administration, was a lawyer and part owner of a canning factory. Before her marriage, his mother, Harriet Grote, was an elementary school teacher. Among her seventh and eighth grade students at Longfellow School in Wheaton was Edwin Hubble, who, she later informed Grote, was "a bright boy." As a youth, Red Grange, the legendary football star for the University of Illinois and then the Chicago Bears, delivered ice to the Reber home before the family acquired an electric refrigerator.

While still in high school, Reber obtained his amateur radio license, W9GFZ,[48] signed by the then Secretary of the Interior, Herbert Hoover. Reber later recalled that, in the late 1920s and 1930s, he and other radio amateurs knew that if they connected an antenna to their receiver, when the various stages were tuned to the same frequency the noise level would increase; but no increase was noted when this same procedure was repeated with the antenna disconnected.[49] Probably they had detected galactic radio noise, but did not realize this until many years later. After contacting more than 60 countries with his amateur radio station, Reber was looking for new challenges. He was intrigued by Jansky's papers, which he read in *Proc. IRE*, and had listened intently when Jansky's "star noise" was rebroadcast by the NBC Blue Network in May 1933.

After graduating in 1933 from the Armour Institute of Technology (now the Illinois Institute of Technology) with a degree in electrical engineering, Reber held a series of jobs with various Chicago companies developing consumer broadcast radio receivers and later military electronics. While working in Chicago he wrote to Jansky, asking if he could come to Bell Labs to work on star noise,[50] but he was surprised and disappointed to learn that Bell Labs did not plan any further work on cosmic radio emission. Reber then contacted various observatories and university departments but he found little interest among the astronomers of the time, who were busy with their own projects. He tried to interest Otto Struve and other astronomers at the nearby Yerkes Observatory in Williams Bay, Wisconsin, but Struve was dismissive of Reber's inquiry. As Reber later described it, "The astronomers were afraid, because they didn't know anything about radio, and the radio people were not interested, because it was so faint it didn't even constitute an interference – and so nobody was going to do anything. So I thought, well if nobody is going to do anything, maybe I should do something"[51] (Figure 1.11).

Reber was aware that attempts to interpret and understand Jansky's observations failed by a large factor. So he did not want to merely confirm Jansky's work, but wished to understand the nature of the galactic radio emission by investigating "how does the intensity at any wavelength change with position in the sky," and "how does the intensity at any position change with wavelength?"[52] He recognized that instruments used by Jansky and others were effectively monochromatic, that is they worked only at one frequency or over a very small range of frequency. As he later explained, Reber decided to build on techniques used in optical astronomy, asserting that, "I consulted with myself and decided to build a dish."[53] To supplement his

Figure 1.11 Grote Reber at age 27. Credit: NRAO/AUI Archives, Papers of G. Reber.

background in engineering and to enhance his understanding of optics and astronomy, Reber took classes at the University of Chicago. As part of the requirements for Philip Keenan's class in astrophysics, Reber wrote a report on "Long Wave Radiation of Extraterrestrial Origin," in which he tried to interpret the surprisingly strong Milky Way radio emission reported by Jansky.[54]

So that he could better estimate the sensitivity he would need to detect interstellar radio emission, in April 1937, Reber wrote again to Jansky inquiring about the absolute intensity of the galactic radio emission.[55] After receiving Jansky's response, Reber took leave from his job at Stewart, Warner, and Belmont Radio in Chicago and, that summer, at the age of 25, using his own funds, he designed and built a 31.4 foot (9.6 m) parabolic dish in a vacant lot next to his mother's house.[56] Except for some help from his next door neighbor with the heavy lifting of the metal surface plates, Reber worked entirely by himself with no outside contractors. To keep the construction simple, Reber's dish moved only in elevation and was oriented to observe along the meridian. In contrast to Jansky's antenna, that moved only in azimuth, Reber's antenna could look at any elevation and depended on the rotation of the Earth to scan across the sky. Except for the galvanized iron surface plates and fasteners, Reber constructed his antenna entirely out of wood.[57] Like Jansky, Reber made use of scrapped parts from an old Model T truck as part of the elevation drive system. Curious neighbors could only speculate about the purpose of the unfamiliar structure rising in the small town of Wheaton, but Reber's mother found it a convenient place to hang her washing (Figure 1.12).

Jansky's observations were made at 20.5 MHz (14.5 m). Reber understood that a heated body would radiate over a wide range of the electromagnetic spectrum, and that for any given temperature the level of radiation should increase as the square of frequency.[58] Also, he wanted to go to as high a frequency (short wavelength) as possible, to obtain the best possible angular resolution to study the distribution of galactic radio noise.[59] He initially decided to observe at 9 cm (3,300 MHz) which was the shortest feasible wavelength for the then existing technology. At that frequency, he anticipated that galactic radio noise would be about 25,000 times stronger than Jansky had observed at 20.5 MHz and where he would have a resolution of about 0.5 degrees compared with Jansky's resolution of about 35 degrees.

Using his experience and skills as an electrical engineer and radio amateur, Reber meticulously designed and built a radio receiver using a homemade crystal detector followed by an amplifier. He placed his receiver and half wave dipole feed at the focal point of his antenna, with the connecting wires running through a coal chute to his observing room in the basement of his mother's house. By the spring of 1938 he had completed the construction of his antenna and receiver and eagerly anticipated being able to detect radio emission from the Galaxy. But to his surprise and disappointment,

Figure 1.12 Reber's 32 foot Wheaton antenna, with the Reber family home in the background. Credit: NRAO/AUI Archives, Papers of G. Reber.

Reber was unable to detect any cosmic radio noise. After pointing his antenna toward the Milky Way, he also tried to detect radio noise from a few bright stars, the Sun, the Moon, and several nearby planets, all with negative results.

Although disappointed by his inability to confirm Jansky's detection of galactic radio noise, as well as the failure to find radio emission from the other likely targets,

Reber drew the important conclusion that "Perhaps the actual relation between the intensity of celestial radiation and frequency was opposite from Planck's law,"[60] and so was not ordinary thermal radiation. This was the first recognition of what is now called nonthermal radio emission.

Undeterred, Reber rebuilt his receiver to operate at the longer wavelength of 33 cm (910 MHz) where more sensitive and more stable instrumentation was available. But still, he had no success. Finally, he reluctantly decided to accept the reduced resolution of going to an even longer wavelength and built yet another receiver that operated at 1.9 m (162 MHz) closer to the 20 MHz system used by Jansky. In order to improve the sensitivity of the 160 MHz receiver, Reber used a regenerative vacuum tube detector preceded by a newly-developed RCA tube as a radio frequency amplifier. With this more sensitive equipment, in late 1938 Reber finally succeeded in detecting Jansky's galactic radio noise at 1.9 m wavelength, which he called "cosmic static." But he was only able to study this cosmic static at night when interference from passing automobile ignition was at a minimum and when the stability of his receiver was not degraded by solar heating. Having no recording equipment, Reber laboriously wrote down the output of his receiver every minute. In the daytime Reber returned to his job designing broadcast radios in Chicago, where he commuted one hour each way by train.[61] Upon arriving home, Reber would catch a few hours' sleep each evening before returning to his night's observing. On weekends, he analyzed his data. By the summer of 1939, just before the start of the Second World War, he was able to confirm Jansky's observations that the maximum radiation came from a direction close to the center of the Galaxy. Again, he attempted to detect radio emission from a few bright stars such as Vega, Sirius, and Antares, as well as Mars and the Sun, and he made the important conclusion that there is little correspondence between the brightness of the sky at radio and optical wavelengths.

Grote's younger brother, Schuyler, who was a student at the Harvard Business School, put Grote in contact with Fred Whipple and Harlow Shapley at the Harvard College Observatory.[62] Although Whipple expressed interest in Reber's accomplishments,[63] Shapley, who had previously been in contact with Jansky, remained reluctant to get involved with something that no one at Harvard knew anything about and claimed that they could not start any new activities as they were already over-committed to other programs.[64]

Wanting to reach out to astronomers, Reber submitted his paper, with the title "Cosmic Static," to the *Astrophysical Journal* (*ApJ*), where it was received with skepticism by the editor, Otto Struve. According to Jesse Greenstein, then a young astronomer at the Yerkes Observatory, since Reber had no academic connection and unclear credentials, his paper produced a flurry of excitement and uncertainty at the *ApJ* editorial offices at the Yerkes Observatory located north of Chicago.[65]

Reber later complained that, since the astronomers didn't understand how the radio waves could be generated, they felt that "the whole affair was at best a mistake and at worse a hoax."[66] At various times, Bart Bok, Otto Struve, Subrahmanyan Chandrasekhar, Philip Keenan, Jesse Greenstein, and Gerard Kuiper traveled to Wheaton to evaluate Reber's radio observations and equipment and also to evaluate Reber. According to Greenstein, Kuiper apparently only reported that Reber's apparatus "looked modern" and that his work "looked genuine."

Fortunately, Harvard professor Bart Bok wisely cautioned Struve that he could not afford to turn down the paper because it might "be a great success."[67] Reber's former professor at the University of Chicago, Philip Keenan, put in a good word for Reber and, after extensive exchanges of letters between Reber and Keenan, Struve finally published "Cosmic Static" as a short note in the *ApJ*.[68] But the *ApJ* paper did not include Reber's theoretical speculations that the galactic radio emission was due to electron–ion interactions in the interstellar medium.[69] Meanwhile, fed up with the delays at the *ApJ*, Reber submitted a very similar paper to the *Proc. IRE* where Jansky had published most of his pioneering papers. *Proc. IRE* also questioned Reber's interpretation, but nevertheless published his manuscript in full, four months before the *ApJ* paper appeared, ironically including his theoretical analysis that had been rejected by the *ApJ*.[70] As Reber later claimed,[71]

> Otto Struve didn't reject my 160 MHz paper. He merely sat on it until it got moldy. I got tired of waiting, so I sent some other material to the Proceedings of the IRE. It was published promptly in the February, 1940 issue. From a much slower start, this beat the ApJ by four months. During the early days of radio astronomy, the astronomy community had a poor track record. The engineering fraternity did much better!

Struve finally became convinced that Reber's work had promise and encouraged him to seek external funding. With the encouragement of Struve and Jesse Greenstein, Reber unsuccessfully tried to negotiate with the University of Chicago and the Office of Naval Research (ONR) to move his antenna to a quieter site at the McDonald Observatory in Texas. But they could not agree on how to recover the cost of moving the antenna and operating a radio observatory in Texas. Greenstein had also been fascinated by Jansky's discovery of cosmic radio noise and, following his visit to inspect Reber's equipment, Greenstein and Reber became "moderately good friends"[72] and together wrote what was the first review of the new field of radio astronomy.[73] Greenstein and Struve suggested that Grote receive an appointment at the University of Chicago, so the university could administer the program and collect overhead costs from ONR. But Reber insisted on preserving his independence and was not interested in working for the university. He explored the possibility of continuing his astronomy research while remaining an employee of his company, that he proposed would administer ONR funding, but this never came to fruition. Later, following an exchange of letters with George Southworth, Reber explored the possibility of funding from Bell Laboratories, but that too turned out to be unrealistic.

Due to a chronic hearing impairment, Reber was exempt from Second World War military service. From about the age of 40, he wore a hearing aid, that he found convenient to turn off if he did not want to bother hearing what others were talking about. During the War, he worked for a limited time at the Naval Ordnance Laboratory in Washington on the protection of naval vessels from underwater explosive devices. After returning to Wheaton, Reber made several improvements to his system. He built a new, more sensitive receiver with an RF amplifier, modified the feed, increased the bandwidth, and importantly purchased a chart recorder so he did not need to be present to write down the meter readings. He reported these technical improvements in a new paper published in *Proc. IRE*,[74] along with a discussion of the impact of automobile ignition noise. As in his two previous papers, as well as those to follow, Reber again used the title, "Cosmic Static" (Figure 1.13).

Using his new instrumentation, Reber went on to systematically map the entire sky visible from Wheaton by changing the elevation of the antenna each day and letting the rotation of the Earth scan the sky. He converted his fixed elevation scans of the sky to a two-dimensional 160 MHz map that he published in the *ApJ*.[75] Reber's maps (Figure. 1.14) clearly showed the pronounced maxima at the Galactic Center and what were later recognized as the Cygnus A/Cygnus X complex of sources, the Cassiopeia A (Cas A) radio source, and evidence for galactic structure. Reber was also finally able to detect radio emission from the Sun. Although he noted that the Sun was a surprisingly strong radio source, like Jansky, he realized that even if all the stars in the Galaxy radiated with the same radio intensity as the Sun, this would fail to account for the radio emission from the Milky Way by a factor of 10 billion.

As a result of his Chicago job as a radio engineer, Reber had access to state-of-the-art test equipment and the latest microwave vacuum tubes, so he set out to build a new receiver and feed to work at a shorter wavelength of 62 cm (480 MHz) with the goal of improving his angular resolution. But he was surprised that the 62 cm radiation was weaker than at 1.9 m, although he later appreciated that he made the important discovery that the celestial radiation was not due to thermal radiation described by the Raleigh-Jeans radiation formula, but was a new type of nonthermal radiation.[76] Over a 200 day period in 1946, Reber repeated his observations at 480 MHz and produced a more detailed map of the radio sky. His new higher resolution observations now separately revealed what was later recognized as the strong radio galaxy Cygnus A, the supernova remnants

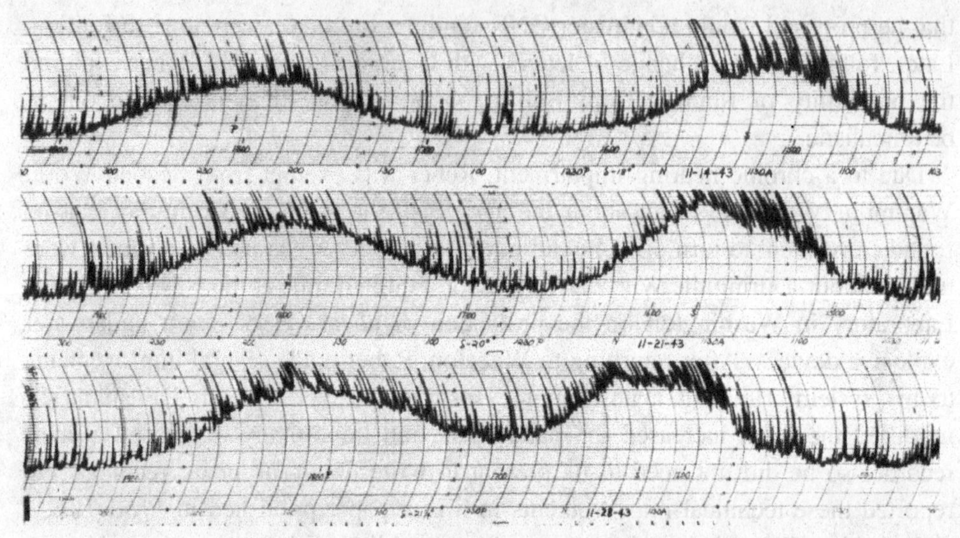

Figure 1.13 Reproduction of Reber's 160 MHz chart recordings from three separate days in late 1943. Each scan is at a different declination. The two bumps indicate the plane of the Milky Way on the left and the Sun on the right. The multiple sharp spikes show ignition noise from passing automobiles.
Credit: NRAO/AUI Archives, Papers of G. Reber.

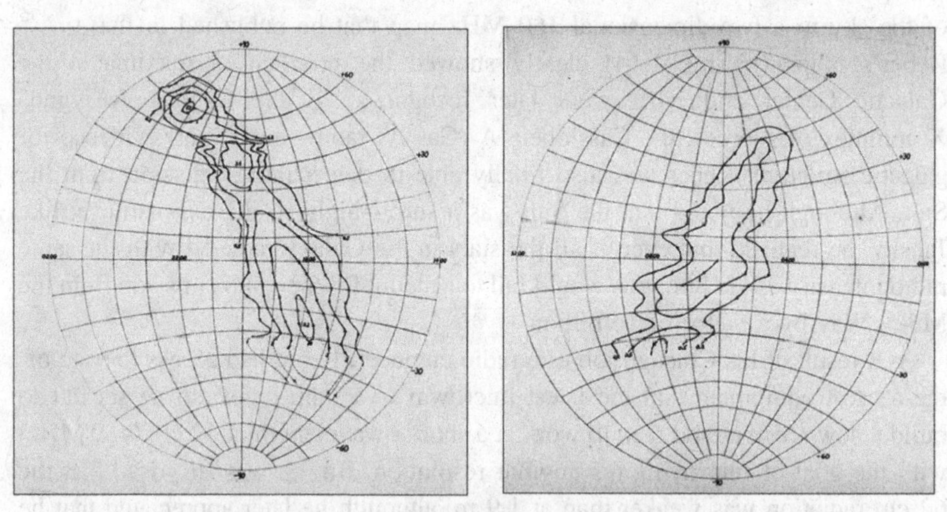

Figure 1.14 160 MHz maps of galactic radio emission. Left image, looking toward the central region of the Galaxy. Right image, looking toward the outer parts of the Galaxy away from the Galactic Center. Credit: NRAO/AUI Archives, Papers of G. Reber.

Cassiopeia A and the Crab Nebula, a structure in the Milky Way, and showed the presence of the strong radio source at the center of the Galaxy.[77] The identification of the strong Cas A radio source and the Crab Nebula with the remnants of supernovae that occurred near the end of the seventeenth century and in 1054 CE, respectively, is discussed in Chapter 3.

Encouraged by his successes and increasing interest from the astronomy community, Reber developed an ambitious plan to build a 220 foot (61 m) fully steerable version of his Wheaton antenna. But he was not able to raise the money for his dish, which he estimated would cost $100,000. Nevertheless, his design became the basis for the next generation of large radio telescopes that were built much later at Jodrell Bank, Effelsberg, and in Green Bank.

By 1947, Reber realized he could no longer effectively compete with the increasing government funded radio astronomy activities in the United Kingdom and Australia as well as in several US laboratories. In 1948, he went back to Washington to work at the National Bureau of Standards (NBS) and sold his antenna and receivers to the Bureau. He had his antenna disassembled and moved to Sterling, Virginia, near the present location of the Dulles International Airport, where it was re-erected on a turntable, so that it could be steered in azimuth as well as elevation. But there is no record of any successful use of Reber's dish at the Bureau of Standards. In 1959, the dish was re-erected, with Reber's supervision and help, near the entrance to the National Radio Astronomy Observatory in Green Bank, West Virginia.

While at the Bureau of Standards, Reber began to build a set of matched dipoles covering the frequency range from 25 to 110 MHz to measure the absolute intensity of the cosmic static, with the important goal of determining how the intensity of galactic radio emission varies with frequency, and the equally important (to the Bureau) goal of establishing the best frequencies to be used for radio communication. At the same time Reber planned to obtain better resolution using his dish at frequencies up to 1,420 MHz and to construct a larger 100 foot diameter dish.[78] However, he never got approval to build the larger dish and there is no record that he ever completed the frequency-dependent measurements to study the variation of galactic radiation with frequency, or attempted to observe the predicted transition of interstellar atomic hydrogen at 1,420 MHz discussed in Chapter 7.

During this time, Reber tried to reach out to a broader audience with accounts of his "Cosmic Radio Noise" and other recent postwar discoveries in the semi-popular magazine *Radio-Electronic Engineering*, followed by articles in *Sky and Telescope*, *Scientific American*, and in the popular *Leaflets of the Astronomical Society of the Pacific*.[79] In these articles, Reber not only described his own work, but speculated, with some imagination, on the origins of cosmic static, that he emphasized could not be understood in terms of free–free interactions of thermally agitated electrons.

But Reber did not fit in well with the Washington government laboratory culture. He apparently left NBS one day, for a vacation in Hawaii, and never bothered to return. In Hawaii, Reber erected a rotating antenna near the 10,000 foot summit of Haleakala on the island of Maui. This was the first of what would subsequently be many mountain-top telescopes in Hawaii. But he detected more interference than cosmic noise, and his antenna fell down in a storm before he could get any useful observations. Realizing that mountain tops were not the best place for radio telescopes, Reber then turned his attention to very long wavelengths, at the same time that the growing mainstream radio astronomy effort was moving toward shorter wavelengths.

When Reber arrived in Tasmania in 1955, it was near the minimum of the 11 year solar cycle. Although he was encouraged by his initial observations made between 0.5 and 2 MHz (600 and 150 m) at a time of unusually low level of ionospheric absorption,[80] subsequent sunspot minima were less favorable for low frequency radio astronomy, and he faced increasing levels of broadcast interference. Nevertheless, Reber built what was then, and probably still is, the largest radio telescope ever built, with a square kilometer of collecting area. But, despite initial optimism, it proved difficult to get any useful astronomical results. He tried, unsuccessfully, to obtain a demobilized ICBM to carry a load of liquid hydrogen into the ionosphere where it would combine with free electrons to make the ionosphere transparent to long wavelength cosmic radiation. For over two decades he relentlessly wrote letters to colleagues, friends, observatory directors, NASA officials, and congressmen to solicit their help, and refused to accept no for an answer. Finally, in 1985, he was able to arrange for the ill-fated NASA Challenger space shuttle to ignite its engines when passing over Tasmania, hoping to create a temporary hole in the ionosphere that would allow the 144 m cosmic radio waves to pass through. The results were encouraging, but it was unrealistic to follow up with more extensive studies.[81]

Unlike Karl Jansky, whose work was supported by arguably the world's leading industrial laboratory, Grote Reber worked alone as an amateur, relying on his deep curiosity along with his imagination and skills as a professional engineer, combined with a persistent forceful personality and stubborn disregard for conventional views. While Jansky was uniformly described by his colleagues as quiet and modest, Reber, by contrast, proudly believed in himself and paid no attention to establishment science, except to express his disdain. "There were no self-appointed pontiffs," he explained, "looking over my shoulder giving bad advice. The kinds of things I want to do are the kind establishment men will not have any part of."[82] As his nephew Jeff Reber (Schuyler's son) later noted, "He obviously learned to ignore the opinions of those he didn't agree with."[83] Reber had no patience for negotiation or compromise, and he was always forcefully direct in choosing his words. Grote Reber always made it clear what he wanted and why he wanted it.

Working by himself, Reber relentlessly pursued his own study of Jansky's important discovery, and for a decade he was the only person in the world devoting significant effort to this new and unexplored field. After designing and building his own antennas, receivers, and test equipment, he worked in a previously unexplored region of the electromagnetic spectrum. His 31.4 foot home-built radio telescope was the largest parabolic dish ever built at that time, and remained so until 1951, when the US Naval Research Laboratory built a 50 foot dish in Washington, DC. Although Karl Jansky was the first to detect cosmic radio noise, it was Grote Reber who finally convinced astronomers that the radio spectrum was a new window to understanding the Universe. Through his meticulous investigations at 160 MHz and then 480 MHz, Reber

(a) built the first facility intended to investigate cosmic radio emission;
(b) demonstrated the concentration of radio noise along the Galactic Plane with evidence for structure possibly associated with the location of the spiral arms in the Milky Way;
(c) provided the first clear demonstration that the galactic radio emission was nonthermal in origin;
(d) showed the first evidence for the discrete radio sources, including what later were identified with two supernova remnants, a radio galaxy, and the center of the Milky Way;
(e) independently discovered the nonthermal emission from the quiet Sun and later the intense solar radio bursts;
(f) understood that if all the stars in the Galaxy were like the Sun, it could not explain the observed level of galactic radio emission;
(g) was the first to recognize and advocate the importance of long wavelength radio astronomy, although it would be more than half a century before the technology was sufficiently mature to make possible instruments such as LOFAR in the Netherlands and the Murchison Widefield Array (MWA) in Australia; and
(h) brought radio astronomy to the attention of future leaders of the field such as Jesse Greenstein, who later started the radio astronomy program at Caltech and was instrumental in the creation of the National Radio Astronomy Observatory (NRAO), and Otto Struve, who became the first director of NRAO.

Like Jansky, Reber never won a Nobel Prize, although he was nominated in 1950, two years after Jansky.[84] His first maps of the nonthermal galactic radio emission provided much of the incentive for the discoveries discussed in later chapters of this book. He remained active in research nearly until he died in 2002, two days before his 91st birthday, but his most productive years were those in Wheaton before age 35. Except for the three years that he spent in Washington at the

National Bureau of Standards, Reber never received a salary for his radio astronomy work, or support from any of the conventional funding sources such as the National Science Foundation, NASA, or the Department of Defense. Except for a total of about $200,000 in grants he received from the privately operated Research Corporation from 1951 to 1981, his research was supported entirely from his personal funds. As a result of a small inheritance from his father, wise investments, and a frugal lifestyle, Reber was financially comfortable and was able to fund his own research.

To save the cost of postage, Reber reused stamps by writing "return to sender" in replying to correspondence, and rather than hire someone to cut his grass, he allowed his neighbor's sheep to graze in his yard. In order not to waste paper, he drew circuit diagrams and wrote letters and reports on the back of old bank statements or received letters. Being independent of conservative referees and committees that constrain modern research (Chapter 12), and driven by his own curiosity, engineering skills, and uncanny insight, Reber was able to initiate his unconventional research that has changed our understanding of the Universe in a fundamental way.

From the time he left Wheaton in 1947, Reber became increasingly outspoken against the growing trend to build large expensive facilities for radio astronomy, which he felt was a waste of money.[85] Although, unlike Jansky, Reber was recognized with several major prizes for his pioneering discoveries, his determined rejection of mainstream astronomy, and his reluctance to accept conventional wisdom without detailed scrutiny led to his increasing isolation from the astronomical community. But his broad curiosity about everything he came in contact with steered him to additional research in a variety of fields ranging from radio circuitry and ionospheric physics, to cosmic rays, meteorology, botany, geophysics, and archeology. He held several patents, including one on a radio sextant that could be used to determine terrestrial positions on cloudy days. His controversial experiments growing beans and observing the difference between vines that wound clockwise and those that wound counter-clockwise brought considerable notoriety.[86]

Other Early Investigations of Galactic Radio Noise

While Karl Jansky's discovery of galactic radio noise generated a lot of interest within the astronomical community, professional astronomers, who had little understanding of radio or electronics, considered his star noise as more of an interesting curiosity rather than an opportunity for further research. There were, however, a few important exceptions. One astronomer who took notice of Jansky's discovery was Joel Stebbins, the Director of the University of Wisconsin's Washburn

Observatory. Karl's father proudly informed him that in a lecture that he attended in Madison, Stebbins compared Jansky's discovery to Charles Lindbergh's solo flight across the Atlantic, although greater.[87] But Stebbins, like other astronomers of the time, did not appear to entertain any thoughts of extending Jansky's observations to other frequencies or to other possible cosmic sources of radio emission.

After his 1933 Chicago talk, Jansky sent copies of his paper to Princeton astronomer Henry Norris Russell, and to Harlow Shapley, Director of the Harvard College Observatory. Later he met with both Russell and Shapley to discuss the implications of his discovery and how his work might be extended. Shapley considered repeating Jansky's experiments at Harvard, but, discouraged by the cost and Harvard's unfamiliarity with anything to do with radio or electronics, he apparently lost interest and did not pursue it. However, Fred Whipple, who was a young Harvard professor, had a clever idea on how to confirm Jansky's discovery. Whipple proposed building a directional rhombic antenna on the dome of the Harvard 60 inch telescope so he could use the rotation of the dome to scan his antenna around the sky but was unable to get permission from Shapley and never got to try his scheme.[88]

Caltech Professors Fritz Zwicky and Gennady Potapenko were also impressed by Jansky's work and similarly thought about building a steerable rhombic antenna to detect galactic radio emission. But they were apparently unable to convince Caltech president Robert Millikan to support the project.[89] Potapenko gave a talk at the Caltech Astronomy and Physics Club on "The Work of the Bell Laboratories on the Reception of Shortwave Signals from Interstellar Space,"[90] and in the spring of 1936, Potapenko and his student Donald Folland tried to reproduce Jansky's work. First, they built a small loop antenna on the roof of the Caltech Physics Building to detect Jansky's star noise at 20.55 MHz (14.6 m), close to the frequency used by Jansky. But automobile ignition noise from their site overlooking busy California Boulevard frustrated their experiments. To get away from the noise of Pasadena, they moved their experiment out to the nearby desert, where they fastened one end of a 35 foot wire to a 25 foot mast. One person walked the slanted wire around the pole to exploit the directivity of the arrangement while the other took data. They were able to detect a maximum radio signal toward the direction of the Galactic Center and later a second maximum in the Cygnus constellation, but Millikan discouraged them from publishing their results and was unwilling to support their proposed joint research with Zwicky (Figure 1.15).

Other early confirmations of galactic radio noise were by John H. DeWitt in the United States, Kurt Fränz in Germany, and K. F. Sander in the United Kingdom.[91] DeWitt, who had previously been an associate of Jansky at Bell Labs, tried unsuccessfully to confirm Jansky's results in 1935 using a simple 300 MHz (1 m)

Figure 1.15 Potapenko and Folland's late 1930s experimental setup confirming Jansky's reported galactic radio noise. Credit: NRAO/AUI Archives, Papers of W.T. Sullivan III.

antenna. But in 1940, he succeeded in detecting galactic radio emission at 111 MHz using a large rhombic antenna and confirmed that the maximum signal came from a direction close to that of the Galactic Center. DeWitt later went on to head Project Diana, that successfully bounced the first radio signals off the Moon in January 1946 (Chapter 9).

During the early years of the Second World War, the German Luftwaffe used a 30 MHz navigational aid known as Knickebein to support aircraft raids on Britain. Kurt Fränz was a German radio engineer who, like Jansky, was an expert on radio noise and had read Jansky's *Proc. IRE* papers. Fränz had studied physics, chemistry, and mathematics at Berlin University under people like Einstein, Heisenberg, Schrödinger, and Planck, so he knew his physics. While working at the large German Telefunkenen Laboratories, Fränz received reports from military authorities of excess noise in their 30 MHz radar receivers, which he recognized as probably related to Jansky noise. Fränz followed up by building a simple dipole array, using it to confirm that the direction of maximum sky noise was in the direction of the Galactic Center. His wartime paper reported his measurements of the apparent radio temperature of the Milky Way as between about 10,000 K and 100,000 K and significantly noted that it was considerably stronger than Reber had observed at 160 MHz. In this surprising wartime German publication of potentially sensitive radar-related technology, Fränz was the first to describe cosmic noise in terms of an astrophysical meaningful temperature.[92]

Shortly after the end of the Second World War, K. F. Sander, working at the British Radar and Research Development Establishment, observed galactic radio noise at 60 MHz using a simple 4-element dipole array, and demonstrated the rapid decrease in intensity with increasing frequency. Fränz and Sander were primarily trying to understand under what conditions external (cosmic) noise limited radio communication and when it was limited by internal noise generated in the receiver circuits. Grote Reber, however, was primarily motivated by trying to understand the implications of galactic radio emission to astrophysics.

Understanding Star Noise and Cosmic Static

With little serious interest shown by the astronomical community, and trained as a physicist, Jansky was probably the first person to seriously try to understand the source of his star noise. He made the important connection between properties of his comic "hiss" and "the sound produced in the receiver headset." As he wrote to his father,[93]

> I now have what I think is definite proof that the waves come from the Milky Way. However, I am not working on the interstellar waves any more. Friis has seen fit to make me work on the problems and methods of measuring noise in general. A fundamental and necessary work, but not very interesting as the interstellar waves, nor will it bring me near as much publicity. I'm going to do a little bit of theoretical research of my own at home on the interstellar waves however ...

At the July IRE meeting in Detroit and in his 1935 *Proc. IRE* paper Jansky made the first crude steps toward understanding the nature of the radio emission from the Milky Way writing,[94]

> One is immediately struck by the similarity between the sounds they produce in the receiver headset and that produced by the thermal agitation of electrical charge. In fact the similarity is so exact, that it leads one to speculate as to whether or not the radiations might be caused by the thermal agitation of charged particles. Such particles are found not only in the stars, but also in the very considerable amount of interstellar matter that is dispersed throughout the Milky Way.

Also, with great perception, Jansky realized that if all the stars in the Milky Way are like our Sun, it could not explain his observed noise from the Milky Way, and so there may be "other classes of heavenly bodies found in the Milky Way," whose ratio of radio emission to that of heat and light must be much greater than from the Sun. As we now understand, galactic radio noise is indeed generated by the motion of charged electrons, although not by their thermal agitation in the interstellar medium as suggested by Jansky but by synchrotron radiation from electrons accelerated to nearly the speed of light moving in weak magnetic fields. However, Jansky's bold prediction that there are "classes of heavenly bodies" that radiate more radio emission than they do in the familiar "form of heat and light" has been dramatically confirmed with the subsequent discoveries of such entities as radio galaxies, quasars, pulsars, and cosmic masers that are discussed in Chapters 3, 4, 7, and 8.

As already mentioned, Reber's interpretation of his 160 MHz observations of cosmic noise as free–free transitions in ionized interstellar hydrogen was deleted from his published paper by the *ApJ* editor, Otto Struve. However, after an exchange of letters between Reber and his former University of Chicago teacher, Philip Keenan, Struve included a note by Luis Henyey and Keenan, following Reber's paper, that refined Reber's calculations. Henyey and Keenan confirmed Reber's analysis that his cosmic static could be understood in terms of emission from electron–ion interactions in ionized interstellar hydrogen at a temperature of about 10,000 degrees, but they also stressed that "in the case of Jansky's data the discrepancy is serious."[95]

Even earlier, Caltech Professor R. M. Langer was one of the first scientists to try to understand possible mechanisms to explain Jansky's radio emission from the Milky Way. Langer wrote to Jansky to get a better understanding of the radio observations, and in 1936, he gave a talk to the American Physical Society suggesting that Jansky's star noise was the result of free electrons combining with ionized dust particles.[96]

Two astronomers who thought hard about Jansky's star noise were Harvard graduate student Jesse Greenstein, and Fred Whipple, then a young Harvard faculty member. Whipple and Greenstein correctly noted that, since Jansky had detected galactic radio emission over a range of wavelengths, it could not be the result of any atomic or molecular transition but had to be "a continuum of radiation." They then went on to consider blackbody radiation, and following up on Jansky's suggestion they tried to interpret the radio noise as thermal radiation from interstellar cold dust.[97] However, as in the case of Henyey and Keenan, their model failed by a factor of 10,000 to account for the intensity of galactic radiation reported by Jansky. Charles Townes, who later won a Nobel prize for his invention of the maser, was a contemporary of Jansky at Bell Labs. Following discussions with Feldman, Jansky and Southworth, Townes was puzzled by extraterrestrial radio noise and, after detailed theoretical calculations, noted that, while Reber's results might be explained as thermal radiation from interstellar ionized gas at 10,000 degrees, Jansky's measurements and those of Friis and Feldman indicated temperatures more than an order of magnitude greater.[98]

None of the calculations based on variations of free–free radiation involving thermally agitated motion of electrons or thermal emission from interstellar dust could explain the large observed intensity of galactic radio emission. Grote Reber was probably the first to recognize that the systematic decrease in intensity with increasing frequency was inconsistent with any thermal radiation process and later commented, "If the data doesn't fit the theory, change the theory not the data."[99] Although, in 1945, Henk van de Hulst, like any good theoretician, rather than modify the theory, suggested instead that Jansky's measurement was "too high by a factor of 10 and more."[100]

The observation of intense radio emission from the Milky Way by Reber at 160 and 480 MHz, and especially by Jansky at 20 MHz clearly presented a problem for astrophysicists. Curiously, the first clues to breaking the puzzle of the origin of cosmic radio emission came not from any observatory or university but from the General Electric (GE) Company Research and Development Center Laboratory in Schenectady, New York, where GE operated a synchrotron that accelerated electrons to velocities very close to the speed of light. The GE scientists noticed that the so-called ultra-relativistic electrons emitted a bright beam of white light within a narrow cone directed along their direction of motion.[101] The complex theory of what later came to be called synchrotron or magneto-bremsstrahlung radiation from ultra-relativistic electrons was independently derived by Julian Schwinger at Harvard and Vasiliĭ Vladimirsky in the USSR based on a long forgotten paper by the British mathematician George Schott.[102]

The frequency, ν_{max}, of maximum synchrotron radiation depends on the strength of the magnetic field, B, and the square of the electron energy, E, and is given by

$$\nu_{max} = 1.6 \times 10^4 BE^2 \text{ GHz},$$

where E is expressed in GeV, and B in Gauss. For typical values of E and B found in particle accelerators, ν_{max} is in the visible spectrum as observed by the GE scientists. However, with the much weaker cosmic magnetic fields, synchrotron emission occurs in the radio spectrum.[103]

Scandinavian scientists, Hannes Alfvén and Nicolai Herlofson, made the connection and suggested that cosmic radio emission might be due to synchrotron radiation from ultra-relativistic electrons moving in weak stellar magnetic fields, an idea that was extended to interstellar magnetic fields by Karl-Otto Kiepenheuer.[104] But it was not until the more quantitative papers by the Russian astrophysicists,

Vitaly Ginzburg and Iosef Shklovsky, that the astronomical community accepted that synchrotron radiation could explain the observed radio emission from the Milky Way as well as from the newly-discovered radio galaxies, and later quasars, as discussed in Chapters 3 and 4, respectively.[105] A crucial step came from Ginzburg's student, German Getmansev, who showed that if the cosmic ray electron energy distribution is a power law with an exponent, γ, the spectral energy distribution of the radiation is also a power law, with a spectral index $\alpha = (1 - \gamma)/2$.[106] For cosmic ray electrons with $\gamma = 2.5$, the corresponding radio spectral energy distribution is then a rapidly decreasing power law with an exponent –0.75, consistent with the observed spectral index of the Milky Way. Since the middle of the twentieth century, synchrotron radiation has been the basis for understanding galactic radio emission, as well as that from extragalactic radio sources. In an extreme example of the competition that sometimes drives scientists, until their deaths in 1985 and 2009, respectively, Shklovsky and Ginzburg bitterly argued over who was first and who deserved the credit for recognizing the importance of synchrotron radiation in understanding cosmic radio emission.

2

Radio Emission from the Sun and Stars

> It was clear to me that the Sun must be radiating electromagnetic waves directly.[1]

Although the Sun is the brightest radio source in the sky, neither Karl Jansky nor, until 1943, Grote Reber were able to detect any solar radio emission due to the wide beam of their simple antenna systems. Radio emission from the Sun was finally detected during the Second World War by a number of independent Allied and Axis radar operators when they reported unexplained interference to their radar systems. But these accidental discoveries of solar radio noise, as well as a planned experiment to detect solar radio emission by one of Jansky's Bell Labs colleagues, remained classified military secrets until the end of the War. Meanwhile, in 1943, Grote Reber, who was unrestricted by wartime security constraints, became the first to recognize and report radio emission from the Sun in the scientific literature. After the War, with rapidly improving instrumentation, the Sun became a major target in the emerging field of radio astronomy. Observations made over a wide range of frequencies measured the thermal emission corresponding to the 6,000 degree K temperature of the "quiet" Sun at shorter wavelengths, the slowly varying component due to the rotation of active regions in the corona, as well as the intense bursts lasting from seconds to hours observed at longer wavelengths (Figure 2.1). Later observations with instruments of increasing sophistication traced the complex time, frequency, and spatial dependence of the solar radio emission which corresponded to a wide variety of emission mechanisms. But it would be another two decades before radio noise was detected from any star other than the Sun.

Early Attempts to Detect Solar Radio Noise[2]

Soon after Heinrich Hertz first demonstrated in 1888 the existence of the electromagnetic waves predicted by James Maxwell, Thomas Alva Edison proposed

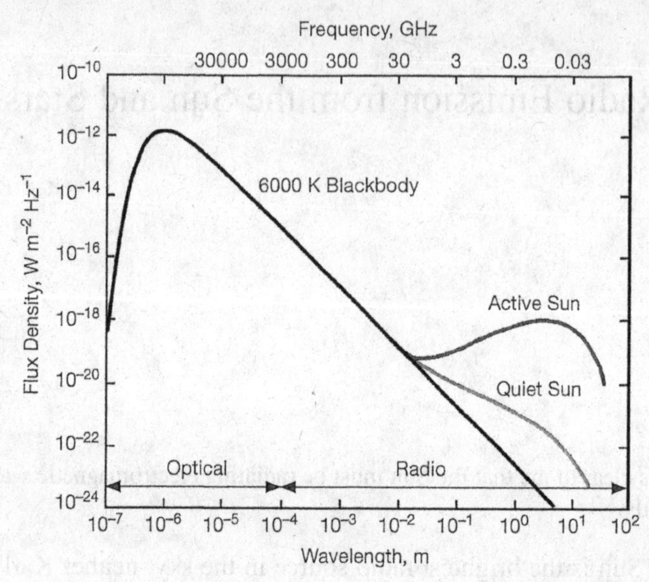

Figure 2.1 Schematic plot showing the typical solar flux density observed for a 6,000 degree K black body, the quiet Sun, and the active Sun. Courtesy of Christian Ho.

a bold but ill-conceived experiment to detect radio emission from the Sun. Assuming that electromagnetic radiation from the Sun would excite large currents in the huge mass of iron ore found in a nearby New Jersey mine, Edison proposed to wrap large loops of wire around the mine that were connected to a telephone where Edison hoped to hear sounds from the Sun associated with solar disturbance. But there is no record that he ever carried out this unrealistic experiment.

Sir Oliver Lodge, who had detected electromagnetic radiation independently of Hertz, also speculated that the Sun might be a source of radio emission. Between 1897 and 1900, Lodge unsuccessfully tried to detect radio emission from the Sun from a location in Liverpool, England. But, as would plague future generations of radio astronomers, Lodge's experiments were limited by the high level of man-made electrical noise. Although Lodge recognized the advantage of working from a more remote location, away from the big city, there is no evidence that he ever attempted any such experiments.

About the same time, in Potsdam, Germany, two German astronomers, Johannas Wilsing and Julius Scheiner, used a somewhat more sophisticated apparatus to try to detect "electric" waves from the Sun. Their observations, made over an 18 day period in the summer of 1896, were unsuccessful, which they suggested was due to absorption in the upper atmosphere.[3] However, Charles Nordmann, a young French PhD student, followed up on the possibility that the solar radio waves might be absorbed in the Earth's atmosphere and took his apparatus to Mont Blanc, where he erected a 175 m (574 feet) long antenna at an elevation of 3,100 m (10,170 feet) using the underlying ice as a nonconducting medium to support his antenna 25 m (82 feet) above the true ground. In order to avoid interference, he placed his antenna so that "it was sheltered from any disturbances originating in the high-voltage electrical line from Chamonix."[4] Nordmann was also unable to detect any solar radio emission and concluded, "If the sun does emit such radiation, it is completely absorbed by the solar atmosphere and the upper sections of the terrestrial atmosphere." Possibly, if Nordmann had not been working at the time of solar minimum, and if he had not been working at such a long wavelength (300 m) where the ionosphere is opaque to external radiation, he might have been more successful.

Based on negative results reported by Nordmann, as well as by Wilsing and Scheiner, and supported by the observation that wireless telegraphic systems do not detect any continuous radiation from the Sun, the French astronomers Henri Deslandres and Louis Décombre concluded that "the Earth does not continuously receive detectable Hertzian waves of length similar to those used in wireless telegraphy (between 10 m and 1000 m)," but speculated that there might be detectable short-term Hertzian waves associated with solar prominences.[5] Realizing that it would be difficult to distinguish radio waves generated by the Sun from those ascribed to terrestrial storms, Deslandres and Décombre proposed that, "the most reliable method will be to distribute receivers or automatic wave recorders over several points on the earth, and then to investigate the signals which are simultaneous and thereby attributable to a single source," and that, "a long series of observations will be necessary to decide in the end if the surface of the earth receives waves from the Sun similar to the Hertzian waves."

In spite of the rapid developments in radio technology that occurred during the early twentieth century, there appear to have been no other efforts to detect radio emission from the Sun until three decades later after Jansky and Reber had revived interest and a new flurry of activity was generated by the unexpected remarkably intense solar radio bursts accidently discovered by wartime radar facilities. Apparently, the failure of these early attempts to detect radio emission from the Sun, and the realization that the expected thermal radiation from the 6,000 K solar photosphere was below the sensitivity limits of even the best equipment, discouraged any new experiments, until Karl Jansky stimulated interest in the radio sky.

Probably the first person to try to extend Jansky's observations of cosmic radio noise to the Sun was John Kraus (Figure 2.2). When still a graduate student at the University of Michigan, Kraus had read about Jansky's discovery of cosmic noise. After receiving his PhD in June 1933, Kraus and fellow former student Arthur Adel attempted to detect the Sun at 1.5 cm (20 GHz) wavelength using a 1 m diameter search-light mirror and a receiver left over from a physics lab experiment to measure the absorption spectrum of ammonia. Although they were unable to

Figure 2.2 John Kraus searching for radio signals. Credit: NRAO/AUI Archives, Papers of J. Kraus.

detect any radio emission from the Sun, this was probably the first time that a parabolic dish had been used as an antenna for radio astronomy.[6] In Australia in 1939, Jack Piddington and David Martyn looked for excess noise from the Sun with their 60 MHz ionospheric array. Although the expected radio emission from the 6,000 K Sun was far below their sensitivity, they might have easily detected a strong solar radio burst, but unfortunately the Sun was not active during the few days that they tried. The following year, John DeWitt, who was later the first person to receive radio echoes off the Moon, was similarly unfortunate when he tried to detect 111 MHz (2.7 m) radio emission on one day, 25 September 1940, when the Sun was apparently quiet.[7]

However, during the 1920s and 1930s, US and British amateur radio operators, mostly operating in the 10 m (28 MHz) amateur band, reported noise or hissing that was likely due to radio emission from the Sun during periods of intense solar activity.[8] Although the sensitivity of their instruments was inadequate, by many orders of magnitude, to detect the expected thermal emission from the 6,000 K Sun at 28 MHz, apparently no professional astronomers or radio scientists took any notice of these anomalous shortwave observations. Not until after the War did Sir Edward Appleton and J. Stanley Hey point out that, in order for radio amateurs to have detected radio noise from the Sun, it would have had to exceed the expected thermal noise by a factor of more than 10,000.[9]

In 1938, two Japanese scientists came close to success. Minoru Nakagami and Kenichi Miya were working at the International Telecommunications Company in Tokyo investigating the fading of short wavelength radio signals and the associated increase in received noise at the time of solar flares, a phenomenon known as the "Dellinger effect." Although their 15 MHz (20 m) antenna system was intended to measure incoming radio signals close to the horizon, fortuitously, their antenna had a large sidelobe response at higher elevations, and Miya concluded that the increase in noise that they observed on 1 August 1938 came from the general direction of the Sun. However, his senior and more cautious colleague, believing that the noise was generated in the ionosphere, apparently rejected Miya's interpretation, and their paper did not mention the possibility that they had received radio noise from the Sun.[10]

Wartime Discoveries of Solar Radio Emission

The Second World War brought a breakthrough in solar radio astronomy and a major impetus to the new science of radio astronomy. The adversaries on both sides of the conflict deployed a series of radar systems of rapidly increasing sophistication that were used for both defense against incoming aircraft and as navigational aids for fighter and bomber pilots. As a result of the widespread use of radar systems, on multiple occasions radar operators in the United Kingdom and in Germany, as well as Allies in the South Pacific, reported strong intermittent interference to their radar systems that appeared to be associated with the Sun.[11] Curiously, there is no known documentation of similar reports from American radar facilities.[12]

During the early years of the War, when German fighter pilots expressed concern about the unexplained noise on their radar systems, it was mostly dismissed as inconvenient interference to their military operations. Perhaps the earliest documented reports of intense solar radio bursts came from the German 175 MHz (1.7 m) radar on the Baltic Sea island of Fehmarn in 1939, and from the German radar station in Skagen, on the northern tip of occupied Denmark, in the spring of 1940. Although the original records were destroyed during the War, E. Schott, a German radar operator, later recalled that, in May and June 1940, the solar interference, which was variable from day to day, was so strong that it disabled his 125 MHz (2.4 m) radar set, even when the Sun was in the antenna sidelobes.[13]

Probably the most dramatic discovery of solar radio interference came in early 1942. By this time, Germany had conquered and occupied nearly all of Europe, and Britain was on high alert for an expected German invasion by air and by sea. The south coast of England was heavily fortified, and protected by the "Chain

Figure 2.3 British Chain Home radar station during the War. Credit: UK Royal Air Force.

Home" 4.2 m (71.4 MHz) radar stations that were constantly on the alert for enemy activity on the English Channel and from the opposite French coast (Figure 2.3). Two of Germany's powerful battle-cruisers, the *Scharnhorst* and the *Gneisenau*, along with the cruiser, *Prinz Eugen*, were pinned in the German occupied port of Brest on the far northwestern Atlantic coast of France[14] (Figure 2.4).

The *Scharnhorst* and the *Gneisenau* had wreaked havoc on Allied shipping in the North Atlantic and had also supported the German invasion of Norway in June 1940. For nearly a year, while in the French port, they were under continuous attack from British bombers. Although they had escaped significant damage, Adolph Hitler wanted both ships to return to Germany, where they would be safe from British attack[15] and where they could be sent on to the North Sea to counter the expected British invasion of Norway, as well as attack Allied shipping headed for the northern Russian port of Murmansk. On 12 February 1942, under orders from Hitler, the *Scharnhorst*, the *Gneisenau*, and the *Prinz Eugen* left Brest, accompanied by a fleet of German destroyers, minesweepers, and torpedo boats. Hidden by fog and smoke screens, the German squadron passed apparently unnoticed through the English Channel, undetected by the British coastal radar that had been made useless by newly

Figure 2.4 The German battleship *Scharnhorst*. Credit: Bundesarchiv, DVM 10 Bild 23-63-07, CC BY-SA 3.0.

Figure 2.5 J. Stanley Hey. Credit: NRAO/AUI Archives, Papers of W.T. Sullivan III.

developed German jamming apparatus. They were finally intercepted by the British aircraft and ships at Dover Straight between Dover and Calais. But by then it was too late. In a front page story, *The New York Times* reported that the British lost 60 aircraft along with 41 vessels, and by the following morning, the *Scharnhorst*, the *Gneisenau*, and the *Prinz Eugen* had all reached the German port of Kiel.[16] The three German warships had suffered significant damage, but this was unknown to the public and, perhaps, to branches of the British military, as the UK government did not want to disclose that they had intercepted the coded German communications. As a result, there was a great public outcry over the failure to contain the German warships that passed undetected by the patrolling British aircraft or coastal radar stations. According to Winston Churchill, "The news astonished the British public, who could not understand what appeared to them, not unnaturally, to be proof of the German mastery of the English Channel."[17]

According to J. Stanley Hey, following an official inquiry, the British War Office decided to increase its efforts to mitigate radar jamming and seek the assistance of the Army Operational Research Group (AORG) to deal with this tough issue.[18] At the time, Hey was a civilian scientist with a background in X-ray crystallography, working with AORG (Figure 2.5). Until he received his six-week intensive training under J. A. Ratcliffe, Hey claimed that his "knowledge of radio was negligible." Nevertheless, after his brief radio course, he was considered a "radio expert" and was responsible for advising the army on the performance of radar installations throughout Britain. Following the 12 February embarrassment, Hey was instructed to investigate the cause of jamming of the British radars, to advise on anti-jamming measures, and to organize an "immediate reporting system," to facilitate the collection of jamming data.

Just two weeks after the events of 12 February, on 27 and 28 February, the British defense radars again became inoperable, which was again thought to be due to apparent German anti-radar activity. According to Hey, "Reports from sites in many parts of the country described the daytime occurrence of severe noise jamming experienced by anti-aircraft radar working at wavelengths between 4 and 8 m, and of such intensity as to render radar operation impossible, [and] the alarm was widespread." Was this a new form of enemy jamming? However, there were no German air raids in progress. The noise never appeared at nighttime, and at all sites the noise apparently came from the direction of the Sun. As Hey later reported, he telephoned the Royal Observatory and learned that "there was an exceptionally active sunspot" and that "It was clear to me that the Sun must be radiating electromagnetic waves directly." In his secret report that became known to scientists in other Allied countries, including Joseph Pawsey in Australia, Hey wrote that, "the interference was associated with the sun and the recent occurrence of sun spots."[19] Wartime security precluded any publication of his discovery in the open scientific literature until more than a year after the close of hostilities, and even then with very few technical details. In his very brief 1946 letter to *Nature*, Hey noted that on 27 and 28 February 1942, the Sun was a source of powerful 4–6 m (37–50 MHz) radio emission about 100,000 times greater than expected from a 6,000 K blackbody and that the emission "appears to have been associated with the occurrence of a big solar flare."[20]

Although it seems that Sir Edward Appleton was aware of Hey's detection of solar radio bursts, he made no mention of Hey's discovery in his earlier *Nature* paper where he reached similar conclusions about the intensity of solar nonthermal radiation based on the reports he received from the radio amateurs.[21] Appleton and Hey later joined forces to further analyze the February 1942 event and to report on a new series of radio outbursts that occurred in February 1946.[22] The following year, Appleton received the Nobel Prize in Physics for his discovery of the ionosphere two decades earlier, although by this time Hey believed that by 1946 Appleton was "essentially an administrator," so he was annoyed at Appleton's "incursion into our research findings," and felt "that the liaison could not survive."[23] The historian Woodruff Sullivan has further suggested that Appleton's paper was prompted by Hey's 1945 classified report that had been sent to Appleton,[24] and that Appleton, who had a controversial reputation for taking credit for the work of others, had deliberately ignored Hey's paper and delayed its publication.[25] Indeed, the radio amateur D. W. Heightman responded to Appleton's paper that he, together with several colleagues, not only investigated the "hissing" noise from the Sun during the period 1936–1939 but also showed that "the appearance of the [solar] radiation coincided with the times of chromospheric eruptions on the sun."[26]

Subsequent *Nature* letters pointed to the "magnificent limb eruption or flare," "radio fadeouts," "magnetic storms," and world-wide changes in cosmic ray density that were observed on 21 February and during the following week.[27] The connection between bursts of intense radio emission and other solar-terrestrial phenomena was firmly established, although the physical process involved was unknown and would be a major area of investigation by postwar radio astronomers.

The intense radio bursts from the Sun were independently discovered on at least two occasions during the War by radar operators in the South Pacific. Toward the end of the War, Les Hepburn, the officer-in-charge of a Royal New Zealand Air Force radar station on Norfolk Island, observed a large increase in his 200 MHz (1.5 m) receiver noise when his radar antenna was pointed at the Sun on 28 March 1945, even when the radar transmitter was turned off. Ian Stevenson, the director of the Radio Development Laboratory (RDL), a euphemism for radar, of the New Zealand Department of Scientific and Industrial Research assigned Dr. Elizabeth Alexander to investigate what was called the "Norfolk Island Effect" (Figure 2.6). Alexander had been trained as a geologist, but had worked on radar while serving as a captain in the British Royal Navy Intelligence Service in Singapore before it fell to Japan in 1942.[28] After returning to New Zealand with her three children, and

Figure 2.6 Elizabeth Alexander. Courtesy of Mary Harris.

Figure 2.7 South Pacific Chain Overseas Laying 210 MHz radar site. Courtesy of Wayne Orchiston.

unable to join her husband in Singapore, she became head of the Operations Research Section of the RDL, where she was studying the effect of meteorological conditions on the propagation of radar signals. Taking up the challenge, in April, Alexander organized a two-week series of solar radio observations with up to four New Zealand radar stations, as well as the one on Norfolk Island (Figure 2.7). Although her study was compromised by the operational requirements of the radar stations, she reported that the "radiation in question is much more intense than would be expected from blackbody theory."[29] As it later turned out, during the period of intense solar activity in March, radar stations in the Philippines as well as in northern Australia reported "jamming" when their radars were pointed at the Sun. Due to staff reductions associated with the end of hostilities, Alexander was unable to implement her more extensive study of solar radio emission as planned. She spent the rest of her career in England, Singapore, and finally in Nigeria, working in various fields of geology. She never returned to radio astronomy, but the investigation and reporting of the Norfolk Island Effect established Elizabeth Alexander as the world's first woman radio astronomer.

Meanwhile, near Darwin, on the north central coast of Australia, 21-year-old Bruce Slee was the officer-in-charge of Radar Station 59 (Figure 2.8). In October 1945, he noticed an increase in noise when his 200 MHz radar antenna was pointed at the Sun, and correctly realized that he was seeing radio emission coming from the Sun.[30] Slee also noted that, when observed near the horizon, the amplitude varied periodically and understood that, as the Earth rotated, he was seeing the alternating constructive and destructive interference between the directly received signal and its reflection from the sea, a phenomena that was apparently widely observed by shipboard radar operators and would later form the basis of exciting postwar Australian radio astronomy discoveries. Slee later sent a report of his observations to the Australian Council of Scientific and Industrial Research (CSIR) Radiophysics Laboratory, where Joe Pawsey was so impressed that he hired Slee to work in their new radio astronomy program. Starting in 1946 and continuing until he died in 2016, Slee's 70-year career covered almost every area of radio astronomy.[31]

Just a few months after Hey's dramatic wartime discovery of intense meter wavelength solar radio bursts, on 29 June 1942, in a planned experiment at Bell Labs, George Southworth (Figure 2.9) succeeded in detecting thermal radio emission from the quiet Sun. Southworth, who had pioneered the development of waveguides for microwave radio signals, was an associate of Jansky and had followed with interest Jansky's discovery of galactic radio noise. Knowing the temperature of the Sun to be close to 6,000 K, Southworth and others understood that the expected radio emission can be calculated from the basic laws of radiation from hot bodies, although it had never been demonstrated that Planck's blackbody

Figure 2.8 Bruce Slee examining one of the chart recorders, with other racks of Mills Cross instrumentation in the foreground, November 1955. Credit: CRAIA.

radiation law necessarily applies at radio wavelengths. But until the wartime development of sensitive microwave receivers, the expected radio emission from the Sun was below existing sensitivity levels. When Southworth pointed his 5 foot diameter parabolic antenna at the Sun, he detected radio emission at 9,400 MHz (3.2 cm) and the following week at 3,060 MHz (9.8 cm), which he claimed to be close to the level expected from the 6,000 K Sun. A year later he successfully extended the measurements to 24,000 MHz (1.25 cm). Cognizant of Jansky's work, Southworth also tried, unsuccessfully, to detect the centimeter wavelength radio emission from the direction of the Galactic Center, confirming the earlier discovery by Reber that cosmic radiation varies inversely as the frequency.

Southworth prepared a detailed report of his investigation on 1 February 1944. However, as with Hey's discovery, due to wartime secrecy, Southworth was not allowed to publish his detection of the thermal radio emission from the Sun until nearly the end of the War, and then with only very limited technical details.[32] Unable to disclose the size of the antenna or actual frequencies, Southworth only referred to "low, intermediate, and high" microwave frequencies. Only in 1956 did he disclose the frequencies he had used and the details of his antenna and receiver.[33] However, copies of his classified report were circulated to colleagues at

Figure 2.9 George Southworth. Credit: AIP Emilio Segrè Visual Archives, Physics Today Collection.

Harvard and to the British War Office for further distribution within the British Commonwealth. During this period, many Bell Labs visitors became aware of Southworth's detection of solar radio emission. Among the visitors to Bell Labs were E. G. (Taffy) Bowen and Joe Pawsey from Australia who learned of the successful Bell Labs solar observations during their visits to the Holmdel, New Jersey laboratories. Bowen and Pawsey would later lead the radio astronomy program at the Australian CSIRO Radiophysics Laboratory.

In an erratum note to his 1945 paper, Southworth reported that Charles Townes, a Bell Labs contemporary, had pointed out a factor of about three error in Southworth's calculation of the thermal emission from the 6,000 K solar disk. As it turned out this fortuitously nearly canceled a different major conceptual error in Southworth's calculation so that his measurement accidently agreed with the predicted value. As Sullivan remarked, for several years Southworth's offsetting errors remained unnoticed, presumably because the measurement seemed to agree so well with theory, no one thought to check.[34] Moreover, the effective temperature of the Sun at 3.2 cm is close to 12,000 K, not 6,000 K as assumed by Southworth. During the summer of 1945, at the Massachusetts Institute of Technology's Lincoln Laboratory, as part of an experiment to determine how the atmospheric attenuation would impact short centimeter wavelength radar, Robert Dicke and Robert Beringer measured a solar temperature of about 10,000 K at 1.25 cm.[35] From a 9 July 1945 partial solar eclipse, Dicke and Beringer also confirmed that the size of the 1.25 cm solar radio emission "is not greatly different from that at optical wavelengths."

Meanwhile, back in Wheaton, Illinois, Grote Reber independently detected radio emission from the quiet Sun and discovered the strong radio bursts associated with solar flares. Unconstrained by wartime restrictions, in 1944 Reber became the first person to publish his radio observations of the Sun in the scientific literature. In his 160 MHz (1.9 m) survey paper, Reber commented only in passing that, in spite of daytime interference, he was able to detect radio emission from the Sun, which he only noted "had the rather surprising center intensity" that he associated with the solar corona.[36] Nevertheless, with great perception, he pointed out that even if all the stars in the Galaxy radiated with the same intensity as the Sun, that would fail to account for the observed radio emission from the Milky Way. "Since this is not the case, [he explained] some other cause must be found to make up the difference of 20 or 30 mag. [factors of 10^8 to 10^{12}]."[37] Reber's later 480 MHz observations showed that the solar radio emission was several hundred times more intense than expected from a 6,000 K blackbody with the apparent size of the Sun, and corresponded to an apparent solar temperature of about one million degrees.[38] He later claimed that he had understood much earlier that his observed emission from the Sun was much greater than could be explained by the 6,000 K surface temperature but was worried that if he expressed his findings in units that astronomers could understand, they might not believe his results.[39]

Rediscovering Solar Radio Bursts

By the mid-1940s, Reber began to realize that he could no longer effectively compete with the rapidly growing and relatively well financed radio astronomy programs in Australia and Britain, as well as the burgeoning interests of the US government, and he decided to sell all of his Wheaton telescope and instrumentation to the National Bureau of Standards (NBS). While demonstrating his equipment to NBS visitors on 21 November 1946, Reber was surprised to observe that intense radio bursts from the Sun received through the side-lobes of his antenna drove his chart recorder off scale and varied on time scales as short as one second. When he reviewed his old unattended chart record, Reber noticed that he had recorded a similar, but much weaker phenomenon on 17 October. In surprising contrast to his earlier slow meticulous work and his 1944 apparent indifference to solar radio emission, but apparently fearing competition from Australia, Reber dashed off a hastily written letter to *Nature* reporting on his new discovery.[40] In this *Nature* letter, that was published in the 28 December 1946 issue and carries a submission date of 24 November, Reber reports on observations made as late as 23 and 24 November. In the paper Reber also noted that "the apparent solar temperature [of the quiet Sun] was about a million degrees," in good agreement with the Australian work that had been reported by Pawsey.[41] When listening to the output of his receiver through headphones he described the audible effect as "much like wind whistling through the trees when no leaves are on the limbs," and estimated that the "great swishes," "grinding noises," "hisses," and "bursts" probably rose to several thousand times normal and appeared to be associated with sunspots."

In a second short paper submitted seven months later to the *Proceedings of the Institute of Radio Engineers*, Reber commented on the characteristic timescales of the different phenomena he was hearing.[42] Although he was not the first to report intense radio bursts from the Sun, and he characterized his observations only in terms of their audible sounds, lasting from seconds to an hour or more, Reber was probably the first to recognize the wide range of different solar phenomena later classified by Paul Wild and others.

The work of Hey, Southworth, and Reber became widely known throughout the radiophysics and radar communities, and after the War many investigators set out to follow up on these exciting discoveries. Many of these embryonic radio astronomy projects used abandoned radar facilities or, particularly in the case of European scientists, captured German equipment, especially the infamous 7.4 m (24 ft) diameter German Würzburg-Riese (Giant Würzburg) antennas. The German military deployed 1,500 of these Würzburg-Riese antennas throughout Europe for directing artillery fire. One German scientist, Horst Wille, later recalled hearing during the War about the increase in noise when one of these Würzburg antennas was pointed at the Sun. Shortly after the War, Wille built a receiver and antenna to use to follow up the reports of radio emission from the Sun as part of his doctoral thesis at the University of Friberg in occupied Germany. However, his installation was confiscated by the occupying French military forces and sent to Paris where it was used by French scientists for their own solar radio research program.[43]

Postwar Discoveries

The late 1940s were a time of great solar activity that provided unparalleled opportunities for the new science of radio astronomy to examine what was by far the strongest radio source in the sky, with a complex time and frequency dependence. Radio observations of the Sun began in the United States at Cornell,[44] Stanford,[45] and the Naval Research Laboratory (NRL), as well as in the United Kingdom, Australia, Canada,[46] France,[47] Japan,[48] and other countries, mostly using war surplus radar equipment.

To a large extent, the postwar radio observations of the Sun developed primarily as a result of the serendipitous wartime detections of solar radio bursts. Additionally, most investigators were aware of the work of Jansky and Reber, which probably also influenced their postwar investigations. However, equally important was the availability of surplus radar equipment, both from abandoned Allied radars as well as from captured German installations, along with a cadre of experienced radar scientists who were looking for opportunities to exploit their hard-earned radio skills acquired during the War.

An event that happened as early as November 1940 would turn out to have a particular impact on radio astronomy as well as on the Allied War effort. At the time Britain was dealing with daily German bombing attacks. Edward G. (Taffy) Bowen (Figure 2.10), who was part of the team that built the first British air-warning radars, traveled to the United States and Canada as part of the secret Tizard mission to bring a prototype 10 cm cavity magnetron, needed to generate the high-power transmissions required for Allied centimeter wavelength radar systems. American scientists had developed sophisticated radar systems, but they did not have the power needed to detect distant aircraft. British scientists had developed the powerful cavity magnetron oscillator, but Britain did not have the industrial power to build them in the quantity needed by the military. Following the Tizard mission, the United States and Britain traded radar secrets and collaborated during the War in the further development of radar. After spending three years in the United States to help coordinate US–British radar development, Bowen was recruited to help the Australian radar program. In 1946 Bowen became Chief of the CSIR (later CSIRO) Radiophysics Laboratory, where he created the very important and effective radio astronomy program.[49]

Figure 2.10 E. G. [Taffy] Bowen. Credit: CRAIA.

Although the coincidence of strong radio emission from the Sun at the time of solar radio flares suggested a causal relation, for the most part the simple Second World War radar systems did not have sufficient angular resolution to pinpoint the size or location of the radio bursts or to determine the size of the apparent million degree radio emission that was thought to be associated with the solar corona. As discussed further in Chapter 10, radio astronomers developed several innovative solutions to improving their angular resolution. American and Russian radio astronomers used the timing of the passage of the Moon across the Sun's disk at times of a total eclipse of the Sun, while teams in Australia and the United Kingdom developed ever more sophisticated interferometric systems.

During the War, while working for the Radiophysics Lab, Joseph (Joe) Lade Pawsey (Figure 2.11) was responsible for radar defense systems in northern Australia. In this capacity, he became aware of the still classified detection of the quiet Sun at 9.4 GHz reported by Southworth, and of the serendipitous discoveries of strong solar radio bursts associated with the occurrence of sunspots reported by Hey.[50] Following the end of the War, Pawsey became Assistant Chief of the Radiophysics Laboratory and head of radio astronomy research. Inspired, in part,

Figure 2.11 Joe Pawsey. Credit: CRAIA.

by the 1945 report that Elizabeth Alexander sent to the Radiophysics Lab, and by Bruce Slee's independent report of solar radio emission, Pawsey established a radio astronomy program at two former defense radar sites at Collaroy and then at Dover Heights overlooking the Pacific Ocean. Following the suggestion of David Martyn, the first Chief of the Radiophysics Lab, Pawsey tried to detect radio emission from the solar corona. His 200 MHz observations during late 1945 and early 1946 were dominated by strong bursts from the active Sun, but Pawsey realized that even the lowest levels of solar radio emission were close to that which had been predicted by Martyn from the one million degree solar corona.[51] In a separate paper, Pawsey and colleagues confirmed the association of solar radio bursts with "the passage of large sunspot groups across the meridian." However, they erroneously challenged the important conclusion of both Jansky and Reber that stars like the Sun could not account for the observed galactic radio emission, writing "It seems more reasonable to attribute it to similar bursts of radiation from stars which because of their large number, could yield an approximately constant value for any one area of the sky."[52]

Pawsey and colleagues went on to exploit the sea interferometer effect to enhance their angular resolution. In their classic paper they discussed their radio observations of the Sun in early 1946, at the time of the largest sunspot ever recorded, and reported several important discoveries:[53]

(a) The meter wavelength radio emission from the sun consists of a slowly varying component as well as intense bursts lasting from a fraction of a second to a minute.
(b) Observations with antennas spaced up to 160 miles apart showed the same time dependence, effectively ruling out any interpretation based on terrestrial atmospheric effects.
(c) From the deep nulls in their interference pattern, they could demonstrate that the radiating region was much smaller than the diameter of the Sun.
(d) The location of the radio bursts coincided with the location of a prominent group of sunspots.

In their final remarks, McCready et al. concluded that "the intensity of cosmic noise is vastly greater than it should be if the stars emitted the same ratio of radio-frequency energy to light as does the sun," so it could not account for the intensity of the galactic radio emission reported by Jansky and Reber. However, clinging to the idea of a stellar origin of galactic radio emission, they went on to speculate on "the possibility of vastly greater output from stars differing from the sun."

In a separate Radiophysics investigation, Ruby Payne-Scott, working with two young radio astronomers, Donald Yabsley and John Bolton, discovered that individual solar radio bursts arrived later at longer wavelengths (higher frequencies).[54] Subsequently, fellow Radiophysics colleague Paul Wild took up the challenge and built a series of sophisticated spectrographs to trace the time and frequency dependence of solar bursts. Together with several colleagues, Wild was able to demonstrate that the bursts originate in the corona well above the visible surface of the Sun and are seen first at high frequencies. Based on the time and frequency dependence of the bursts, Wild and others identified five different types of solar bursts that they classified as Type I, II, III, IV, and V bursts.[55]

Ruby Payne-Scott's contributions to radio astronomy and her checkered career as a female scientist in Australia, including the impact of her undisclosed marital status, were highlighted in the 29 August 2018 edition of the *New York Times* as one of the "Overlooked . . . stories of remarkable people whose deaths went unreported in the Times"[56] (Figure 2.12). As described in Chapters 3 and 4, John Bolton abandoned solar radio astronomy, but went on to discover radio galaxies as well as making important contributions to the discovery of quasars. For the next 15 years, Joe Pawsey led the highly productive cosmic radio astronomy program at the CSIRO Radiophysics Laboratory. In 1961, following increasing tensions among the ambitious but competitive Radiophysics staff, Pawsey accepted an offer to be the Director of the US National Radio Astronomy Observatory in Green Bank, West Virginia, but he died of a brain tumor before he could assume the directorship. During the 1950s, Bowen was instrumental in getting Caltech to establish a radio astronomy

Figure 2.12 Ruby Payne-Scott. Credit: CRAIA.

program which, for a decade, became the most prominent radio observatory in the world.[57] He retired as Chief of the Radiophysics Laboratory in 1971 and returned to the United States as the Australian Scientific Liaison Officer in Washington. Paul Wild succeeded Bowen as Chief of the Radiophysics Laboratory. Wild had developed his radio skills as a Second World War radar officer on board the HMS *King George V*. As head of Radiophysics, he concentrated on the development of a sophisticated aircraft instrument landing system that was used in several countries but was soon superseded by the system that used the network of Global Position Satellites (GPS). In 1978, Wild became the Chairman of all of CSIRO and led the organization through a restructuring, bringing about a better balance between serving Australian industries and providing scientific and technical leadership.[58]

Meanwhile, at the Cambridge Cavendish Laboratory in the United Kingdom, Martin Ryle also started a program to study the radio emission from the Sun. Ryle had gained his expertise in radio physics, first as a radio amateur, and then during Second World War, where he helped develop radar countermeasures to neutralize German radar and to jam the V2 rocket control system. After the War, Ryle returned to Cambridge, where he had lived before entering Oxford University, and took up a fellowship at the Cavendish Laboratory, working initially on ionospheric research. Then, using confiscated German radar equipment that he collected from a German radar control center in Denmark as part of an Allied assessment team, Ryle began a program in radio astronomy that would become one of the most productive in the world.[59]

One of Ryle's earliest observations came about as a result of the July 1946 massive sunspot and corresponding great burst of radio emission. Using a radio analogue of the two element interferometer that Albert Michelson used in 1920 to measure the diameter of the star α Orionis (Betelgeuse),[60] Ryle and Derek Vonberg were able to show that the radio emission on 26 July 1946 came from a region less than 10 arcminutes in diameter with a corresponding brightness temperature of more than a billion degrees, clearly demonstrating the nonthermal nature of solar radio bursts and confirming the results obtained a few months earlier in Australia using the sea interferometer. Then, to verify the nonthermal nature of the solar radio bursts, Ryle and Vonberg made an innovative modification of their interferometer to demonstrate that the solar radio emission was 40–100 percent circularly polarized, providing some of the first indications of the role of strong magnetic fields in solar activity.[61]

While the Australian radio astronomy group was composed of young but still relatively experienced men, who were able to set out and develop their own individual programs of research and the development of new instrumentation, Ryle's group included mostly younger and inexperienced students who responded to Ryle's strong centrally managed philosophy. Although the Cavendish group continued their solar radio observations for several more years under Ryle's controlling leadership, as discussed in Chapters 3–5, they increasingly turned their attention to study radio galaxies and their implications for cosmology.

In Russia, Nicolai Papalesky planned an ambitious expedition to Brazil to observe the total solar eclipse of 20 May 1947, intending to carry out a broad program to investigate the effects of the eclipse on solar–terrestrial relations. But, after learning of the existence of a hot solar corona that was predicted by Vitaly Ginzburg and Iosef Shklovsky,[62] independently of Martyn's work, he decided to include a program of direct observations of the radio emission from the Sun during the eclipse. The goal was to check the prediction that the strong meter wavelength radiation came from the larger region of the Sun's corona and would not be completely obscured by the Moon when it covered the visible solar disk.

Both Shklovsky and Ginzburg were included as members of the expedition. Although they were both to later make important theoretical contributions to radio astronomy, their roles in the 1947 expedition to Brazil were to participate in the optical and ionospheric experiments, respectively. As the administrative assistant to the expedition, the crew also included a well-known polar explorer, G. A. Ushakov. Just as in Western countries, left-over Soviet radar equipment was used for the radio astronomy antenna and other instrumentation. Unfortunately, the expedition leader, Papalesky, suddenly died in February, only shortly before the expedition was scheduled to depart for Brazil. Alexander Mikhailov was appointed as head of the expedition, and his assistant Semen E. Khaiken assumed supervision of the radio astronomy and ionospheric groups. The 1946/1947 winter was a particularly severe one in northern Europe, and heavy ice covered the Latvian port of Liepaya where the expedition ship, the *Griboyedov*, was to depart for Brazil. An icebreaker had to be summoned from Riga to lead the *Griboyedov* to clear waters. But another obstacle remained. The Baltic Sea was still littered with live mines left over from the War, and the *Griboyedov* had to spend two days in a Swedish port to be demagnetized before it could traverse the dangerous waters. The expedition finally left Sweden on 15 April and arrived at the Brazilian port of Salvador, about a thousand miles north of Rio, on 10 May, only 10 days before the eclipse.

Shklovsky, Ginzburg, and other members of the optical and ionosphere teams left by train for the thousand mile journey to the inland town of Araxa where the planned optical eclipse observations were subsequently rained out. The radio group faced their own challenges. There was no provision for moving the 200 MHz (1.5 m) large 96 dipole radar array in elevation, that would vary between 35 and 57 degrees during the eclipse. Khaiken solved the problem by laying the antenna on the ship's deck and hoisting one side by winches to the appropriate elevation. To provide azimuth motion, the ship's captain strung ropes to the shore to rotate the ship by adjusting the tension on the lines, and the antenna was kept pointing toward the Sun with an accuracy of about 2 degrees, well within the antenna beam width. During totality, the

Figure 2.13 John Hagen. Credit: US Naval Research Laboratory.

observed flux density decreased by less than 60 percent, demonstrating that nearly half of the radio emission came from a region larger than the visible Sun. Shklovsky, Ginzburg, and Martyn's prediction that the meter wavelength radio emission comes from the corona about 150,000 miles above the Sun's photosphere was confirmed.[63]

At the same time a team from the NRL, led by John Hagen (Figure 2.13), chose to observe the same eclipse from a ship located in the South Atlantic Ocean, although this meant observing from an unstable platform. Radio astronomy work at NRL evolved from their wartime radar research that focused on the smaller short centimeter and millimeter wavelength instruments so that they could be mounted in airplanes and submarine periscopes. After the end of the War, as in the United Kingdom and Australia, US radar scientists turned their expertise toward radio astronomy, but at shorter wavelengths than used in Australia and the United Kingdom. Since the intensity of most cosmic radio sources decreases with increasing frequency, the thermal radio emission from the Sun, that increases with frequency (short wavelengths), was one of the few extraterrestrial radio sources that could be observed with the NRL short centimeter wavelength equipment. Thus, the NRL team planned their eclipse observations at a much higher frequency of 10 GHz (3 cm) than the Russian group, and so were sensitive primarily to the thermal radiation from the 6,000 K solar photosphere rather than the million degree chromosphere. In contrast to the Russian 1.5 m observations, the NRL observations showed that, at 3 cm, the solar radio emission was nearly completely obscured during totality, confirming that it was coming from a region coincident with the optically visible disk.[64] Hagan, who had studied astronomy at Boston University before the War, subsequently submitted the results of these observations for his 1949 PhD thesis at Georgetown University. Hagen thus probably became the first person in the world, and certainly the first in the United Sates, to receive a PhD based on radio astronomy observations.

NRL continued to make future eclipse expeditions. These often involved huge logistical efforts to transport people and tons of equipment, mostly to far off exotic places like Khartoum, Sudan. One expedition even went to Attu, at the end of the Alaskan Aleutian Island chain, with Grote Reber as part of the 11-man expedition. After selling his equipment to the National Bureau of Standards (NBS), Reber left Wheaton and went to work for the NBS. Unfortunately, as he later recalled, the tail end of a typhoon that had earlier struck Japan arrived at Attu just at the time of totality. But in what was probably the first successful astronomical eclipse observations made during a rainstorm, the Attu eclipse observations showed that only 26 percent of the 65 cm (460 MHz) solar radio emission was obscured by the Moon (Figure 2.14). Curiously, Reber found that the minimum occurred two minutes after totality, which he concluded was "probably due to an asymmetrical excitation of the corona caused by a large group of spots near the east limb of the sun. Another large group of spots was near the center of the solar disk. A marked fall and rise of the solar radio intensity was observed when the moon covered and uncovered this group."[65] Subsequently, French radio astronomers continued to use lunar eclipse observations to understand the connection between solar radio emission and optical features observed in the photosphere and chromosphere.[66] Later, they built a series of innovative multi-element arrays to better isolate the positions of radio bursts and showed that they occurred at altitudes between 20,000 and 30,000 km above the photosphere.[67]

The Canadian solar radio astronomy program begun by Arthur E. Covington at the National Research Council in Ottawa deserves special mention. Covington had a life-long interest in radio and held both amateur (VE5CC) and commercial radio operator's licenses. While a graduate student in physics, during summers he worked as a wireless operator aboard ships plying the west coast of British Columbia. One day, while browsing library journals, he stumbled across reports of the work of Reber and Jansky and wondered if the noise that he heard while listening for distress signals might be "static from the stars."[68] After receiving his degree, Covington joined the Canadian National Research Council in 1942, where he worked on microwave radar systems. After the close of the War, he proposed a program in radio astronomy using a surplus 4 foot (1.2 m) parabola from a gun-laying radar and 10 cm receiver that he had previously developed to test crystal mixers. Observing the Sun at the time of the massive sunspot of 26 July 1946, like Reber and Pawsey, Covington found that the 10 cm wavelength intensity was about 10 times greater than expected or that had been reported by Southworth. The discrepancy was only partially resolved when they learned of Southworth's factor of three error in his reported temperature. Subsequent observations showed that the intense 10 cm solar flux was slowly varying.

Figure 2.14 Grote Reber on Attu island, September 1950. Credit: NRAO/AUI Archives, Papers of G. Reber.

In November, while reading the daily newspaper, Covington's wife mentioned the upcoming partial eclipse of the Sun on 23 November 1946. Although Covington's radio telescope was in the shop undergoing modification, he managed to obtain the necessary priority to complete the modifications in time to test the system on the Sun on the day before the eclipse. As he reported in a letter to *Nature*, the eclipse observations showed that the solar flux dramatically changed when the Moon covered a large sunspot and that the equivalent temperature of the sunspot was about 1.5 million degrees.[69] Inspired by this exciting result, on 14 February 1947, Covington began daily observations of the 10.7 cm solar flux that he showed was closely correlated with sunspot activity.[70] He presented his results at the May 1947 meeting of the International Union of Radio Science in Washington, DC. At the end of the session, Karl Jansky and Grote Reber came up to Covington to introduce themselves. This was the first (and only) time that Jansky and Reber had met in person. At the time, Reber was working in Washington and Jansky had traveled from New Jersey to attend the meeting and to catch up on what was happening in the field that he had begun more than a decade earlier. Reber later recalled that he and Jansky had snuck away from the conference to have lunch together and that Jansky, because of his health concerns, was allowed to eat only rice, that he had preordered from the restaurant.

Covington continued his daily measures of the 10.7 cm solar flux at Ottawa until 1990, after which the measurements were relocated to the Dominion Radio Astronomy Observatory in Penticton, BC, where they continue to this day. As it has fortuitously turned out, the best estimate of solar activity and the ionizing radiation that creates the Earth's ionosphere is the 10.7 cm radio flux density, that is not subject to the subjective effects and interpretation of counting or measuring sunspots and is unaffected by the weather. The daily monitoring of the 2,800 MHz (10.7 cm) solar flux is probably the longest continually operating scientific experiment ever. It is used throughout the world to determine the optimum frequency for shortwave radio propagation that depends on the conditions in Earth's ionosphere and how it is affected by solar activity which varies throughout the 11 year solar cycle, as well as the dependence on the daily fluctuations in solar activity.

Why did Covington choose 10 cm to begin his solar observations? Not because of any theoretical prediction or understanding of the relation between 10 cm flux and terrestrial radio propagation, but because that was where his experimental radar equipment worked. Why was the Canadian radar designed to work at 10 cm? Because the prototype British magnetron that was brought to North America by Taffy Bowen worked at 10 cm. Why did the British scientists choose 10 cm for their magnetron that became the basis for wartime Allied centimeter radar? Here the answer is less clear. One speculation is that the magnetron was fabricated in the same factory in Birmingham, England that built the similarly looking Colt revolvers.[71]

The Further Development of Solar Radio Astronomy

The Sun is a remarkably complex body of hot plasma containing strong varying magnetic fields powered by the nuclear burning of hydrogen and powerful convection flows driven by the Sun's rotation. During the early postwar years, radio astronomers had identified three general sources of solar radio emission:

(a) The thermal radiation ranging from about 6,000 K just above the photosphere to a million degrees high up in the corona.
(b) The intense "bursts" of radio radiation seen on time scales of seconds to hours that are now observed from sub-millimeter to decameter wavelengths but that appear most intense at meter wavelength. Radio bursts occur most often near the peak of the solar sunspot cycle and are observed first at shorter wavelengths and then later at longer wavelengths. Five different types of solar radio bursts were identified, based on their detailed time and frequency dependence.
(c) A "slowly varying component" related to sunspot activity and ionizing UV radiation that varies with the rotation of Sun and phase of the 11 year sunspot cycle.

The observed radio emission from the Sun is caused by a variety of processes including plasma oscillations, synchrotron radiation, cyclotron radiation, gyro-synchrotron radiation, and electron–cyclotron maser emission.[72] Unraveling these emission mechanisms and what it can tell us about the role of strong magnetic fields and physics of the Sun has been a complex adventure that has stimulated the design and construction of a series of sophisticated instruments.

In Australia, Wilbur (Chris) Christiansen made increasing use of what later came to be called "earth-rotation synthesis" by developing a series of multi-element radio interferometers to map out the two-dimensional distribution of solar radio emission. Unlike most of his contemporaries, Christiansen did not serve in the military, but spent the War years as an engineer working at Amalgamated Wireless Ltd in Sydney, where he helped to develop antennas that were used to maintain communications with other Allied countries. After joining the Radiophysics Laboratory, Christiansen used the 1948 and 1949 solar eclipses to confirm that most of the solar emission came from multiple small regions coincident with sunspots. He then built a 32 element east–west array of 6 foot diameter parabolic dishes that formed a one-dimensional beam whose position angle on the Sun changed with the rotation of the Earth. Using a laborious mathematical analysis that took months, Christiansen and his colleague Joe Warburton produced a two-dimensional image of the Sun that for the first time showed the limb brightening that had been predicted by Martyn. A later two-dimensional array, known as the Chris-Cross produced a 3 arc minute resolution image of the Sun every day.[73]

Paul Wild took the next step in solar radio imaging with the construction of the Culgoora Radioheliograph that consisted of ninety-six 13.7 m diameter parabolic dishes equally spaced around the circumference of a 3 km diameter circle that was located in northern New South Wales near the town of Narrabri. The Radioheliograph made movies of the Sun's radio emission at four frequencies (43, 80, 160, and 327 MHz) to study the detailed time, frequency, and spatial evolution of solar radio bursts.[74]

In the United States, Ronald Bracewell, a former member of the Sydney Radiophysics group, built a microwave spectroheliograph at Stanford University, similar to the Chris-Cross subsequently built in Australia. Bracewell's heliograph continued for many years to provide daily images of the Sun.[75] Also, at the Harvard Fort Davis site in Texas, ex-New Zealand radio astronomer Alan Maxwell began a long program to monitor solar activity over a wide range of frequencies, first using a 28 foot diameter antenna and then an 85 foot antenna.[76]

More recently, an international team of radio astronomers used the Atacama Large Millimeter/submillimeter Array (ALMA) in Chile to make high resolution images of the Sun at millimeter wavelengths. In Figure 2.15, we show a 1.25 mm (239 GHz) image of the solar disk showing regions of enhanced emission associated with sunspots. This image shows the temperature differences in the Sun's chromosphere that lies just above the visible photosphere.

US solar astronomers have developed an ambitious plan to build a Frequency Agile Solar Radiotelescope (FASR) to study coronal magnetic fields, solar flares, the drivers of space weather, and the physics of the quiet Sun. To unravel the complex radio solar radio emission reflecting the multiple physical processes in the Sun, the FASR proposal included several hundred antenna elements to obtain instantaneous images of the Sun from 30 MHz (10 m) to 30 GHz (1 cm), with sub-second time resolution and better than 1 percent frequency resolution. FASR would have been a valuable complement to the new generation of ground- and spaced-based optical, UV, and X-ray solar instruments being built in the twenty-first century, but unfortunately has not yet been built, in part due to the competition for funds from the other wavelength bands as well as from other new NSF-funded radio telescopes intended to study cosmic radio sources.[77]

True Star Noise

Although Karl Jansky christened his discovery "star noise," it would be another three decades before the first clear radio emission would be reported from stars, and only after a series of likely false reports, and then, in part, as a result of a comedy of errors.

Even with the most sensitive radio telescopes currently in operation, stellar radio emission at levels comparable to what is observed from the Sun can only be observed from the nearest stars. Probably the first person to consider a serious attempt to detect stellar radio emission was Sir Bernard Lovell. During the 1960s, Lovell extensively used his 250 foot radio telescopes at Jodrell Bank and reported the detection of radio emission from a series of red dwarf flare stars.[78] Most likely, however, these early results were due to interference, and it was not until 1978 that

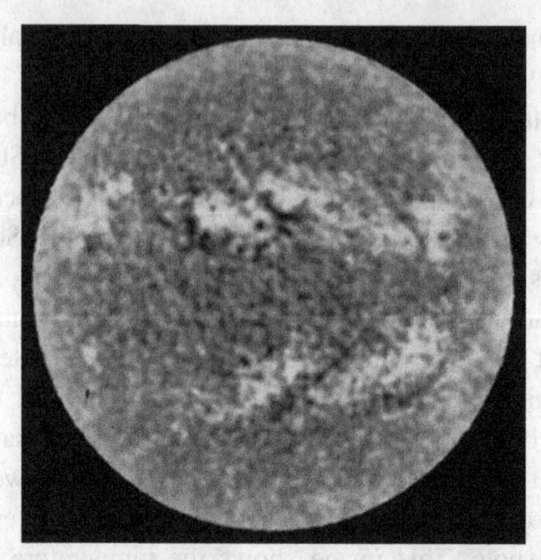

Figure 2.15 1.2 mm image of the Sun's chromosphere mapped with a single ALMA antenna. Credit: NRAO/ESO/NAOJ/ALMA.

Jodrell Bank radio astronomers reported the reliable detection of radio flare emission using interferometric techniques that discriminate against terrestrial interference and acknowledged that "the previous were not trustworthy."[79]

Two other early reports of stellar radio emission are worthy of mention. In 1966, independent observations made with the Green Bank 140 Foot radio telescope and the Canadian Algonquin Park 150 Foot telescope, reported the marginal detection on separate nights of radio emission from the red supergiant star α Orionis (Betelgeuse).[80] Although α Orionis is now known to be a weak radio source, it has never been observed at even close to the level reported in these 1966 observations. As NRAO radio astronomer Robert Hjellming later remarked,[81]

> The problem with evaluating these reports is simple, but insoluble: as with most scientific experiments, results that are not reproducible are assumed (sometimes by the experimenters and nearly always by everyone else) to be the results of unknown problems with the equipment. Because most radio stars are variable on some time scale, or range of time scales, it can be unclear when different results mean some measurements are "bad" and when a difference is "proof" of time variability.

In March 1970, after two weeks of searching for radio emission from red giant stars using the newly upgraded Green Bank three-element interferometer, Hjellming and Campbell Wade found possible weak radio emission from only one star – Antares. Frustrated by their poor success, they decided to try something different, and looked at Nova Herculis, which exploded in 1934 and was the brightest nova of the twentieth century. They were surprised, but delighted, by the strong radio emission that was immediately seen on their chart recorder without the need for any computer averaging. Encouraged by their apparent success, they turned their interferometer toward Nova Aquilae 1918, the second brightest nova of the century, and again saw a strong interferometer signal. Further inspired, over the next week they searched a number of other novae and reported weaker, but convincing radio emission only from Nova Delphini 1967 and Nova Serpentis 1970.[82]

From their interferometer data, Wade and Hjellming were able to determine the position of the observed radio emission with an accuracy of about 1 arcsecond. However, at the time, the optical positions were not well known. So, in order to confirm that their radio sources were indeed associated with the novae, Wade sent their radio positions to Professor Larry Frederick at the University of Virginia, a respected expert in astrometry and head of the University astronomy program. As Wade later described it, a short time afterward, "I got a somewhat agitated call from Larry." Their radio declination for Nova Herculis was wrong by exactly three degrees. They had made an error in punching in the declination of the nova into the computer that pointed the interferometer. Moreover, the source that they had thought was due to Nova Herculis was in fact a previously cataloged radio source that had nothing to do with Nova Herculis. Furthermore, due to a similar error in entering the source coordinates, their position for Nova Aquila had the right minutes and seconds but was wrong by one hour of right ascension. Their apparent detection of strong radio emission from Nova Herculis and Nova Aquila were fortuitous and were unrelated to the novae. But, as a result of these false detections, they were encouraged to go on to detect the first real radio emission from novae, Nova Delphini and Nova Serpentis, that were probably the first reliable detections of radio emission from any star.[83] Later observations in June and November 1970 with better sensitivity also confirmed the marginal 1 March detection of the red giant star Antares.[84]

With the greatly improved sensitivity of radio telescopes during the 1970s and 1980s radio astronomers went on to observe radio emission from a wide variety of stellar types including red and brown dwarf flare stars, T Tauri stars, dwarf novae, magnetic cataclysmic variables, and binary systems such as β Persei (Algol) and β Lyrae. However, not all radio stars are faint radio sources. A few sometime rise to be among the brightest radio sources in the sky. Perhaps the most extensively studied, and probably the most interesting radio star, is the object known as SS 433. Although SS 433 was first observed as a radio source and cataloged as 4C 04.66[85] in 1967, it would be more than 10 years before David Clark and Paul Murdin, as well as others, recognized the association of the radio source with the peculiar star SS 433 and a known X-ray source at the same position.[86] SS 433, which lies in the center of the supernova remnant W50, was one of 455 stars cataloged by Case Western Reserve University astronomers Charles Stephenson and Nicholas Sanduleak to have strong

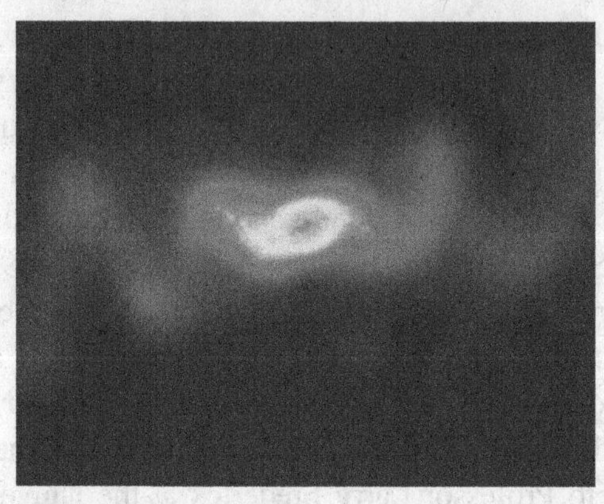

Figure 2.16 5 GHz VLA image of the microquasar SS 433 showing the corkscrew motion of the ejecta due to the precession of the host star. Credit: B. Saxton, R.M. Hjellming, K.J. Johnston NRAO/AUI/NSF.

optical hydrogen emission lines.[87] However, in SS 433 the hydrogen emission is split into two Doppler shifted lines whose corresponding velocities systematically vary with a period of 164 days. The variation suggested a rotating processing jet that was dramatically confirmed when high resolution radio observations revealed a twin beam outward flow from the star with a velocity of 0.26c (Figure 2.16).[88]

SS 433 is one of a small group of what are called microquasars, which, like their extragalactic quasar counterparts (see Chapter 4) are powered by a relativistic outflow that produces intense synchrotron radiation at radio frequencies.[89] More generally, the observed radio emission from stars is the result of a wide range of both thermal and nonthermal processes including synchrotron and gyromagnetic emission, as well as coherent mechanisms such as plasma and cyclotron maser emission. The radio emission from some stars reaches levels as much as 10,000 times that of the most powerful radio bursts observed from the Sun. Even though the suggestion of McCready et al. on "the possibility of vastly greater output from stars differing from the sun" to account for the observed galactic radio emission was not correct, their speculation on the existence of powerful stellar radio emission has turned out to be prophetic. Although radio emission from stars was not even mentioned in the 1965 Very Large Array (VLA) planning report, by the mid-1980s, nearly one-third of the VLA observing time was devoted to the observations of radio emission from stars.[90]

3

Radio Galaxies

. . . the possibility of an unusual object in our own galaxy seems greater than a large accumulation of such objects at great distance.[1]

Even after the discovery of discrete radio sources and the identification of two strong sources with nearby galaxies, the discrete radio sources were for many years still referred to as "radio stars," especially by Martin Ryle and his group at the Cambridge Cavendish Laboratory. Ultimately, the increased precision of radio source positions led to thousands of radio sources identified with distant galaxies. The improvement in radio galaxy imaging showed that most of the radio emission comes from regions hundreds of thousands of light-years from the galaxy itself. Often a very compact radio source was found coincident with the active nucleus of the parent galaxy, and a thin jet-like feature could be seen extending from the active region to the distant radio lobes. Generally, the jets appeared one-sided, although the extended lobes of radio emission are typically symmetrically oriented around the optical counterpart. Observations of the spectral distribution and polarization of radio galaxies provided convincing evidence that the radio emission was due to synchrotron radiation from ultra-relativistic electrons moving in weak magnetic fields, but the source of the enormous energy needed to power the radio galaxies remained a mystery.

The Discovery of Discrete Radio Sources

J. Stanley Hey's interest in radio astronomy was ignited when he filed his 1942 report about solar radio emission and sunspots and was told by his supervisor at the British Army Operational Research Group (AORG) about Karl Jansky's discovery of galactic "star noise."[2] Inspired after reading Jansky and Grote Reber's papers in the *Proceedings of the IRE*, Hey used a 64 MHz radar antenna to make his own observations of galactic radio emission. Together with his colleagues, S. J. Parsons

Figure 3.1 John Bolton checking chart recorder output. Courtesy of Letty Bolton.

and J. W. Phillips, Hey succeeded in confirming the peaks in radio emission near the Galactic Center and several other regions including the one in the constellation of Cygnus first reported by Reber.[3] Continued observations with improved instrumentation led to a surprising discovery: the radio emission from the Cygnus region appeared to fluctuate by about 15 percent on timescales of a minute or less. Hey and his colleagues realized that such fluctuations must come from one or more discrete sources of radio emission and not from a broad background.[4] But their antenna beam was 6 by 15 degrees, too large to isolate the discrete source or to determine its position with sufficient accuracy to see if it was aligned with any known optical counterpart.

Halfway around the world, at the Dover Heights Field Station in Sydney, Australia, John Bolton (Figure 3.1) had been assigned to continue the observations of solar radio activity begun by his boss, Joe Pawsey, who had pioneered the use of the sea interferometer in radio astronomy. Bolton was familiar with the sea interferometer effect from his experience as a radio officer on the Royal Navy HMS *Unicorn* during the Second World War when he could see the changing interference pattern between the direct and reflected radar returns from incoming aircraft. Following the end of the War, Bolton had joined the Australian Council for Scientific and Industrial Research (CSIR) Radiophysics Laboratory where he worked in Joe Pawsey's radio astronomy group.[5] He was assisted by Gordon Stanley, who started work at Radiophysics developing classified radar instrumentation during the War and then became part of the radio astronomy group (Figure 3.2). Later, Stanley would become known as one of the world's experts in radio astronomy

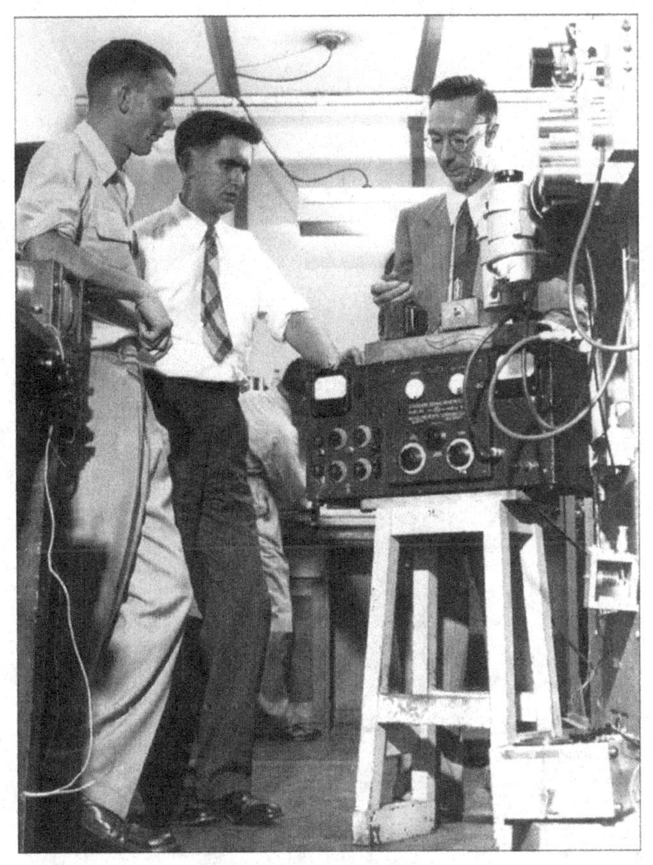

Figure 3.2 John Bolton, Gordon Stanley, and Joe Pawsey at the CSIRO Radiophysics Laboratory Dover Heights Field Station. Credit: CRAIA.

instrumentation.[6] Bolton's interest in cosmic radio noise dated back to his wartime observation that "Jansky Noise" would appear to increase when his radar antenna was pointed toward the Milky Way. Discouraged by the lack of solar activity during their observations of the Sun, to the chagrin of Pawsey, he pointed their antenna toward the Milky Way instead of the Sun. Bolton later claimed that Pawsey was not happy with the change and ordered him back to the Radiophysics Lab to help Stanley prepare instrumentation for a planned expedition to study solar radio emission during an expected eclipse in Brazil.[7]

When the eclipse expedition was canceled, Bolton and Stanley were able to take the instrumentation intended for the eclipse expedition back to the Dover Heights site overlooking the Pacific Ocean. Bolton had heard about Hey's detection of an apparent discrete radio source in Cygnus, and wanted to see if they could better determine the size and location of the fluctuating radio source and determine if the

Figure 3.3 Dover Heights antenna used by Bolton and Stanley as a sea interferometer to isolate the Cygnus A radio source. Credit: CRAIA.

intensity fluctuations were intrinsic or were due to the intervening medium, as well as search for other discrete radio sources.[8] Together with Stanley, he built sea interferometers at several sites on the coast near Sydney (Figure 3.3).

On their first night of observing, Bolton and Stanley saw the interference pattern from Cygnus as it rose in the north-east over the Pacific Ocean. Following continued observations over the next few months, they were able to show that the Cygnus source was less than 8 arcminutes in extent, and they claimed to have located its position to within 7 arcminutes.[9] They also placed limits on the distance of the radio source to between about 10 AU and 1 kpc and commented that the minimum brightness temperature of 4 million degrees "makes a thermal origin of the radiation improbable." But, as it turned out, due to uncertain corrections for refraction, their position was in error by more than one degree, so was insufficient to identify any optical counterpart. Nevertheless, from inspection of the Henry Draper Catalogue, they found no evidence of any bright star, nebula, or variable star that they could associate with the radio source, and they made the important conclusion that "the radio noise from this region is out of all proportion to the optical radiation."[10]

Bolton then went on to discover six additional discrete radio sources,[11] which he named after their constellations, with the suffix A. The letters B, C, D, etc. were reserved for potential new discoveries of other discrete sources in the same constellation, but the nomenclature was never used past A and B. For the galactic sources, the plan was to use the letters, X, Y, Z but, similarly, they never went past source X. Unfortunately, the position accuracy of Bolton's new sources was poor. In fact, what later turned out to be one of the most interesting radio sources, Virgo A, was so poorly located that Bolton wrongly had it in the Coma Berenices constellation.

The detection of multiple sources of discrete radio emission was evidence that at least some of the Jansky noise did not come from interstellar space and they were called "radio stars." But what were these radio stars? Were they really stars or were they associated with some other kind of celestial body? Cygnus A was located in a crowded region of the Milky Way Galaxy with no obvious optical counterpart. Its origin remained a mystery, and both Cambridge and Sydney radio astronomers set out to better localize the Cygnus A radio star, as well as other radio stars that had been discovered at Cambridge and Sydney.

Radio Galaxies or Radio Stars

Although Cygnus A is the second strongest radio source in the sky, the position discrepancies measured among observers were as large as one degree. Bolton realized that, in order to more accurately locate Cygnus A and the other newly-discovered discrete sources with a sea interferometer, he needed to observe sources both rising and setting. Together with Stanley, and now assisted by Bruce Slee, who remained in Sydney, Bolton set up sea interferometers in New Zealand. The first was near the small fishing village of Leigh, on the east coast of the North Island. A second site was near Piha on the west coast, where, for the first time, they could observe the setting source to measure the position along a different position angle in the sky.[12] Slee was then a young technician who had recently joined the CSIR Radiophysics group after being inspired by his wartime detection of solar radio bursts. He died in 2016 after an extraordinary career in radio astronomy lasting more than 70 years.[13]

Simultaneous observations of Cygnus A made from the New Zealand and Sydney sites, some 2,000 miles apart, showed that the observed intensity fluctuations were not correlated, demonstrating that they were not intrinsic to the source but must be due to fluctuations in the Earth's ionosphere, much in the same way that stars twinkle in the Earth's atmosphere.[14] Indeed, for the same reason that only stars, and not planets, twinkle, ionospheric scintillation of radio sources is only observed for small sources.[15]

In addition to their work on Cygnus A, Bolton and his colleagues succeeded in determining the positions of three strong radio sources, Centaurus A (Cen A), Taurus A, and Virgo A, with accuracy better than half a degree. Unfamiliar with the astronomical sky, Bolton consulted Norton's Star Atlas to find that the brightest object by far within the Taurus error box was the Crab Nebula, the nebula Messier 87 was the brightest object near the Virgo source, and the nebula NGC 5128 the brightest object near the Centaurus source. Curious about what was special about the Crab Nebula, Bolton telephoned Harley Wood, the director of the Sydney Observatory, to ask, "What can you tell me about the Crab Nebula?"[16] About the same time, he also began a correspondence with Rudolph Minkowski at the Mount Wilson and Palomar Observatories in the United States and Jan Oort in the Netherlands that led to lifelong collaborations and friendships.

Bolton and Stanley reported their Crab Nebula discovery in a lengthy paper in the *Australian Journal of Scientific Research* that gave a detailed description of their experimental procedures and results, characteristic of radio astronomy papers of that era.[17] They also commented that the equivalent brightness temperature was about 2 million degrees, so the radio emission from the Crab Nebula was unlikely to be thermal radiation. In a separate short Letter to *Nature*, with the modest title "Positions of Three Discrete Sources of Galactic Radio Frequency," Bolton, Stanley, and Slee discussed the optical counterparts of Virgo A and Cen A, as well as Taurus A.[18] Most of the *Nature* letter was devoted to a discussion of the Crab Nebula. In a single paragraph near the end of the Letter, they discussed the identification of Virgo A and Cen A with NGC 4486 (Messier 87) and NGC 5128, respectively, noting that both are "generally classed as extra-galactic nebula" (Figure 3.4).

Although NGC 5128, with its conspicuous dark lane, and M87, with its prominent jet, were well known to optical astronomers as peculiar galaxies, Bolton et al. argued that

> neither of these objects have been resolved into stars, so there is little definite evidence that they are true extra-galactic nebulae or diffuse nebulosities within our own galaxy. If the identification of these objects with the discrete sources of radio-frequency energy can be accepted, it would tend to favour the latter alternative, for the possibility of an unusual object in our own galaxy seems greater than a large accumulation of such objects at great distance.

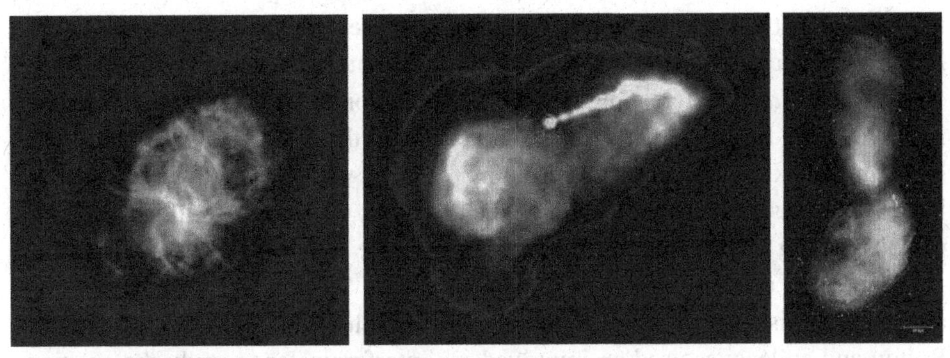

Figure 3.4 From left to right, modern radio images of the galactic supernova remnant Taurus A (Crab Nebula), the radio galaxy Virgo A, and the radio galaxy Centaurus A (NGC 5128). Credits: NRAO/AUI/NSF, F. Owen NRAO/AUI/NSF, Feain et al. (2011).

Given the known distance to the Crab Nebula of about 1,400 parsecs, Bolton would have realized the enormous radio luminosity of the Crab Nebula of about 10^{28} watts needed to produce the observed radio emission. If truly extragalactic, Virgo A and Cen A would be radiating at around 10^{34} watts, so would be a million times more luminous than the Crab Nebula. This huge luminosity apparently led to Bolton's skepticism about the extragalactic nature of Virgo A (M87) and Cen A (NGC 5128). Instead, it seems that he suspected that M87 and NGC 5128 were galactic objects similar to the Crab Nebula, since, if they were extragalactic, their observed radio emission would require that they contain about a million radio sources similar to Taurus A, which he considered unlikely. Apparently, Bolton did not seriously entertain the possibility that a single radio source in these galaxies could be a million times stronger than the Crab Nebula, and the title of their paper, "Positions of Three Discrete Sources of Galactic Radio-Frequency Radiation," reflected their conservative conclusion about the nature of Virgo A and Cen A. To support his point, in a letter to Oort, Bolton drew Oort's attention to a recent paper by D. S. Evans, that "had shown that Centaurus-A (NGC 5128) is probably in our own galaxy."[19] Nevertheless, the 1949 paper by Bolton, Stanley, and Slee is generally credited as the first discovery of radio galaxies and the beginning of extragalactic radio astronomy, although as late as 1954, Bolton and other radio astronomers, particularly at the Cambridge Cavendish Laboratory in England, were still referring to "radio stars" and "galactic radiation."[20] Even a decade later, Bolton noted that the radio contours of Cen A blended into those of the Milky Way, and he had apparently not dismissed the possibility that NGC 5128 was part of the Milky Way.[21]

Many years later, when one of the present authors (KIK) asked Bolton why he didn't stake his claim for discovering the first radio galaxies, Bolton claimed that he knew they were extragalactic, but that he also realized that since the corresponding radio luminosity would be orders of magnitude greater than that of our Galaxy, he was concerned that, in view of their apparent extraordinary luminosity, a conservative *Nature* referee might hold up publication of the paper.[22] Nevertheless, in a 1989 talk, Bolton remarked that he considered the 1949 paper as marking the beginning of extragalactic radio astronomy.[23] As we will see in Chapter 4, a decade later, Bolton, as well as others, missed the discovery of quasars for much the same reasons: wrong preconceptions of what the Universe could and could not contain.

Meanwhile, using a variety of new instruments, Sydney and Cambridge radio astronomers discovered more than 100 new discrete radio sources. In Australia, Bolton and colleagues used an improved sea interferometer to study the position, angular size, and frequency dependence of some of the newly-discovered radio sources.[24] Cambridge, under the leadership of Martin Ryle, used more conventional two-element interferometer systems. Ryle had gained his expertise in radio physics first as a radio amateur[25] and then during the Second World War, where he helped develop radar countermeasures to support the British Bomber Command in neutralizing German radar and later in jamming the V2 rocket control system. Unfortunately, in those areas of the sky where the Cambridge and Sydney surveys overlapped, there was poor agreement on the location and strength of individual sources. At the same time, the debate continued over whether the discrete sources were galactic stars or very much more powerful radio galaxies. A few more discrete radio sources were identified with galaxies and some with supernovae remnants, but many others were misidentified due to inaccurately measured radio positions, or remained unidentified due to their poor position accuracy and the correspondingly large number of galaxies within the error circle. Ryle noted that radio emission had not been detected from any nearby stars and was convinced that the discrete radio sources were a new type of star.[26] Although Ryle and his colleagues expressed difficulty comprehending "some new type of body which produces little visible radiation but very intense radio waves," they wrote that "they are distributed throughout the galaxy with an average population density comparable with that of visual stars," and, "it is therefore concluded that the radio star represents a hitherto unobserved type of stellar body, distributed widely throughout the galaxy."[27] At a meeting of the Royal Astronomical Society, Ryle argued that, since radio emission from even the nearest galaxies was barely detected, the stronger discrete sources could not be extragalactic.[28] But others, especially the astrophysicists Thomas Gold and Fred Hoyle, pointed out the existing radio source identifications with M87 and NGC 5128 and contended that the "radio stars" were extragalactic[29] (Figure 3.5).

Concerned about the increasing competition from Australia and later from the United States, Ryle established a cloud of secrecy or, as John Bolton called it, "an iron curtain around the Cavendish radio astronomy activities."[30] Perhaps as a result of his wartime radar activities where lives depended on secrecy, Ryle was rather paranoid

Figure 3.5 Sir Martin Ryle. Credit: NRAO/AUI Archives, Papers of J. Kraus.

about competition, especially from the United States, which he felt had more money than ideas. His students were routinely instructed to treat visitors courteously and to answer the questions truthfully but not to volunteer any additional information. Unlike at Jodrell Bank, where many graduates went to the United States to help develop programs in interferometry, Cambridge radio astronomy students were discouraged from taking positions elsewhere, especially in the United States.

Meanwhile, in Sydney, using a two-element interferometer that had been built to observe solar radio bursts, Bernard Mills and Adin Thomas were able to observe Cygnus A at nighttime to determine its position to an accuracy of 3 minutes of arc and suggested an identification with a faint optical nebulosity.[31] However, when they learned from Minkowski that their proposed identification was a distant galaxy, they decided that the corresponding large luminosity seemed "unlikely." Instead, they argued, "On the evidence presented it would seem most likely that the source is located in some nearby faint star of abnormal properties." Considering the still relatively large position uncertainty and the handful of faint galaxies in the area,

Minkowski also rejected the suggested association of Cygnus A with such a faint distant galaxy.

Around the same time, Ryle and his student Francis Graham-Smith noted the fluctuations observed in the Cygnus source appeared similar to that observed from the strong solar radio bursts. Since the radio bursts from the Sun are strongly circularly polarized, they suspected that the Cygnus A source might also be circularly polarized. Using a simple two-element 80 MHz interferometer, Ryle and Graham-Smith tried to detect circular polarization but, as discussed next, the Cygnus radio emission is unrelated to solar radio bursts. Although their attempt to detect polarization was unsuccessful, they stumbled across another, even stronger radio source in the constellation Cassiopeia that transited their interferometer pattern after the Cygnus source and was too far north to have been seen from Australia. Similar to the Cygnus source, Cassiopeia A (Cas A) was located in a crowded region of the sky and neither position was sufficiently well determined to identify any optical counterpart.[32]

In order to determine a more precise position of Cygnus A, Graham-Smith then built a two-element interferometer system working at 215 MHz (1.4 m) and measured the position of Cygnus A and three other northern sources with an uncertainty of only about 1 arcminute.[33] Graham-Smith's position for Cygnus A was in good agreement with a newer position determined by Mills.[34] The area of the error box was now about 10 times smaller than the earlier one from Sydney. Graham-Smith also determined the positions of the strong northern source, Cas A, and in August 1951, Graham-Smith communicated these positions to Walter Baade at the Mount Wilson and Palomar Observatories. The following month, Baade photographed these two radio source positions using the Palomar 200 inch telescope, at the time the largest in the world.

With the reduced size of the error box, there was now no uncertainty that the same optical nebulosity identified earlier by Mills was indeed the counterpart to the Cygnus A radio source. Inspecting the 200 inch photograph, Baade noticed that the radio position of Cygnus A was not coincident with a single galaxy but coincided with an apparent double nebulosity, and he reached the startling conclusion that Cygnus A appeared to be two galaxies in collision. Subsequent spectroscopic observations by Minkowski showed the excited broad emission lines that Lyman Spitzer and Baade had earlier predicted would be seen from colliding galaxies, and which they commented were similar to the excited lines seen in the nuclei of Seyfert galaxies (Chapter 4).[35] The redshift was 0.056, which placed Cygnus A at the remarkable distance of 33 Mpc (about 100 light-years) (Figure 3.6). See also Cygnus A in Figure 11.9.

Although the identification of Cygnus A with a pair of colliding galaxies was not published by Baade and Minkowski until nearly three years later,[36] their discovery became well known much earlier. In a footnote to their paper, Baade and

Figure 3.6 From left to right, Bernard Mills (Credit: CRAIA), Walter Baade (Credit: Caltech Archives), and Rudolph Minkowski (Credit: AIP Emilio Segrè Visual Archives, John Irwin Slide Collection).

Minkowski noted that this was the same nebulosity that Mills and Thomas had previously identified with Cygnus A. In the footnote, Minkowski confessed that "it seemed unlikely at that time that such a distant galaxy could be the radio emitter," and he had written to Mills, "that he did not think it was permissible to identify the source with one of the faint extragalactic nebulae in the area, and emphasized that what was wanted was a more accurate radio position." Apparently, Minkowski, like others, did not comprehend that a galaxy could be a source of such powerful radio emission, which he later estimated to be about 10^{36} watts, more than is radiated at visual wavelengths and a million times greater than the Milky Way radio luminosity. We saw earlier that Bolton and colleagues ran into the same mental block with M87 and NGC 5128, although Virgo A and Cen A were less powerful than Cygnus A by a factor of a thousand. In Chapter 4, we will see the problem repeated yet again by the reluctance of Bolton and others to accept the redshift of the radio source 3C 48, that turned out to be the first quasar located at a much greater distance than even Cygnus A.

The idea of a pair of colliding galaxies excited the astronomical community, as well as the general public, and brought a lot of attention to radio astronomy. But it was wrong. Cygnus A is not a pair of colliding galaxies. The double appearance of the magnitude 18 galaxy identified with Cygnus A is due to a dust lane that obscures the central part of a single 18th magnitude galaxy, similar in fact to the NGC 5128 dust lane. At the time, such a faint galaxy was close to the limit of the 200 inch telescope, the largest and most powerful optical telescope in the world. Next to Cas A, Cygnus A was the second strongest radio source in the sky. As it was widely anticipated that many of the fainter radio sources that had been cataloged in Sydney and Cambridge might be at much greater distances and beyond the limits of the

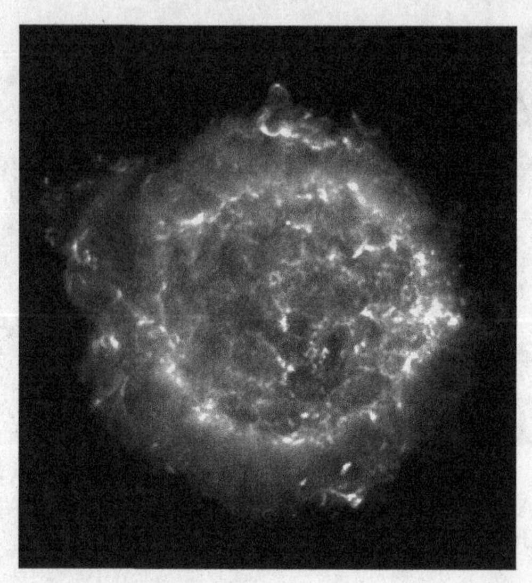

Figure 3.7 Modern VLA radio image of the supernova remnant Cassiopeia A. Credit: NRAO/AUI/NSF.

world's most powerful optical telescopes, radio astronomy suddenly attracted the attention of traditional optical astronomers.

In their same paper, again using Graham-Smith's new radio position, Baade and Minkowski were also able to identify Cas A, the strongest radio source in the sky, with what they called, "a galactic-emission nebulosity of a new type" (Figure 3.7). Earlier, Robert Hanbury Brown and his student Cyril Hazard at Jodrell Bank had detected radio emission from Tycho's supernova of 1572, and Taurus A (the Crab Nebula) was already a well-known supernova remnant.[37] However, due to the lack of any observed systematic expansion motion, Baade and Minkowski argued that "there is every reason to believe that the Cassiopeia source has nothing to do with supernovae." Subsequently, a few radio sources located near the Galactic Plane were identified with historically recorded supernovae events, mostly from ancient observations by Chinese astronomers. Although no record has been found for the supernova event that created Cas A, later observations by Minkowski of the expansion of the filament led him to estimate the age as 256 ± 14 years.[38]

Based on the relatively short estimated age of the Cas A supernova remnant and assuming that the radio emission is due to synchrotron radiation, Shklovsky predicted that the radio flux density should decay by about 2 percent a year.[39] Iosef Shklovsky's prediction was dramatically confirmed by later radio observations in the United Kingdom and the United States.[40] This was one of the few times in radio astronomy history where an important discovery was made as a direct result of observers testing and confirming a theoretical prediction.

Understanding Radio Galaxies

Over the next years, as discussed further in Chapter 5, radio surveys, made primarily in Sydney and especially in Cambridge, cataloged hundreds, then thousands of radio sources. But, except for the strongest sources, agreement among the various catalogs was poor. The most widely used catalog was the Cambridge 159 MHz 3C Catalogue of 471 radio sources,[41] that replaced the earlier but largely erroneous 2C Catalogue (Chapter 5). The 3C Catalogue was later revised to include 328 sources based on additional observations at 178 MHz.[42] But still the radio positions were unreliable and attempts to identify optical counterparts were uncertain except for the strongest radio sources with bright optical counterparts.

In 1955, John Bolton and Gordon Stanley left Australia for Caltech, where they built a two-element radio interferometer (shown in Figure 3.8) specifically designed to accurately measure radio positions and to identify and study their optical counterparts. The Caltech Owens Valley Radio Observatory (OVRO) interferometer became a powerful instrument that made many new discoveries and for a decade Caltech dominated US radio astronomy.[43] Starting in early 1960, Caltech radio astronomers using the OVRO interferometer were able to measure radio source positions accurate to about 10 arcseconds. Tom Matthews, a Canadian by birth and recent Harvard PhD, brought his astronomical background to the Caltech radio astronomers. Assisted by graduate students Edward Fomalont and Richard (Dick) Read, Bolton and Matthews obtained accurate radio positions for hundreds of radio sources and, based on inspection of Palomar 48 inch Schmidt prints and plates, they were able to identify dozens of new radio galaxies. At the Mount Wilson and Palomar Observatories, Baade, Minkowski, and Allan Sandage teamed up with the Caltech radio astronomers to obtain Palomar 200 inch photographs and spectra. At Caltech, Jesse Greenstein, and, after Minkowski's retirement in 1960, Maarten Schmidt provided spectroscopic follow-up to determine the redshifts of the identified radio galaxies.

The radio sources were mostly identified with bright elliptical galaxies, while the radio emission from spiral galaxies like the Milky Way and the Andromeda Nebula (M 31) was found to be much weaker. Generally, the radio galaxy was found to be the brightest galaxy in a cluster of galaxies, and it became clear to astronomers that the best way to find the distant galaxies needed to address the outstanding cosmological problems of the day was to concentrate on galaxies identified with radio sources. Moreover, it was naturally assumed that the smaller radio sources were most likely to be the more distant, so emphasis was given to the smallest radio sources, whose dimensions were determined with the long baseline radio-linked interferometers at Jodrell Bank[44] as well as with the Caltech OVRO interferometer.

The joint Caltech–Palomar program had dramatic success. One of the then smallest known radio sources was 3C 295. Using the 200 inch telescope during his

Figure 3.8 Owens Valley Radio Observatory two-element interferometer. Courtesy of Marshall Cohen.

last observing session before he retired at age 65, Minkowśki determined that the 20.5 magnitude galaxy identified by Matthews and Bolton with 3C 295 had a redshift of 0.46, which was, at the time, by far, the largest known redshift.[45] Although prior to Minkowski's observation the largest measured spectroscopic redshift was less than 0.2, curiously an unrelated foreground galaxy located only a few arcminutes from 3C 295 was observed by Minkowski to have a redshift of 0.24, making it the second largest known redshift at the time. Interestingly, 3C 295 was targeted not because of any special properties, but only because it was at a high declination, where an accurate declination could be measured with the OVRO interferometer, which, until late 1960, had only an east–west baseline. Special observations at Jodrell Bank showed that 3C 295, like Cygnus A, was a double radio source, with a lobe spacing of only 4 arcseconds, or about 25 times smaller than for Cygnus A, roughly consistent with its 10 times larger redshift.[46]

Observations of other radio galaxies with interferometers and multi-element arrays showed that, as in Cygnus A and 3C 295, the radio emission from radio galaxies comes primarily from extended radio lobes that are well removed from the parent galaxy.[47] At the 1958 Paris Symposium on Radio Astronomy, the British astrophysicist Geoffrey Burbidge explained that if the radio emission is due to synchrotron radiation, then the energy contained in relativistic particles and magnetic fields had to be at least 10^{60} ergs – sometimes more, and the corresponding magnetic fields typically ~10^{-4} Gauss.[48]

A key breakthrough in understanding the energetics of radio galaxies came in early 1973, when Martin Ryle and his student Peter Hargrave found a small compact radio source, less than a few light-years across, located between the much larger radio lobes of Cygnus A and coincident with the optical counterpart.[49] Later, more sensitive observations revealed radio jets that apparently carry the energy from the galaxy to the extended radio lobes several hundred thousand light-years away.[50] The source of this tremendous energy requirement and how this energy is transformed into relativistic particles remained challenging questions surrounding mid-twentieth century astrophysics.

The Great Polarization Controversy

A key property of synchrotron emission is that the radiation should be linearly polarized. Following a 1953 suggestion by Shklovsky that both the optical and radio emission from the Crab Nebula should be linearly polarized, a number of astronomers in the USSR, the Netherlands, and the United States showed that the observed optical radiation from the Crab Nebula was indeed linearly polarized, and in 1957 scientists at the US Naval Research Laboratory succeeded in detecting 7.5 percent linearly polarized radio emission at 3.15 cm (9.5 GHz).[51] But measuring polarization from radio galaxies would be more challenging. Concentrating on the strong radio sources Taurus A, Cygnus A, and Cen A, the NRL group observed polarization in all three sources, but suspiciously found that the polarization had nearly the same position angle in each source, suggesting a possible hidden instrumental effect, and no mention was made of the Cen A results in their published paper.[52] As was characteristic of the friendships within the global radio astronomy community, largely established by the 1955 and 1958 IAU radio astronomy symposia, NRL scientist Cornell Mayer (Figure 3.9) circulated unpublished preliminary results to colleagues at other observatories in the hope that someone could confirm the discovery of polarization in radio galaxies.

Ronald (Ron) Bracewell (Figure 3.10) had been an early member of the Radiophysics staff but, in 1955, he joined Stanford University as Professor of Electrical Engineering where he initiated an aggressive program in radio astronomy. He was an authority on image reconstruction in radio astronomy as well as medical imaging, was the author of an authoritative book on Fourier transforms that went through many editions, and was also the co-author with Joe Pawsey of a widely used textbook on radio astronomy.[53] Using the Stanford microwave spectroheliograph, Bracewell, together with Alex Little, discovered the double nature of the inner part of the Cen A radio source (NGC 5128).[54] In April

Figure 3.9 Cornell Mayer. Credit: AAS.

Figure 3.10 Ronald Bracewell. Credit: NRAO/AUI Archives, Papers of R. Bracewell.

1962, during a visit with the Sydney University Physics Department, Bracewell traveled to Parkes to see the newly-completed 210 foot radio telescope. The first polarization observations at Parkes began in March 1962 using the dual 20/75 cm feed system to observe strong radio sources, but the results were inconclusive.[55] To extend the polarization work to Jupiter's radiation belts (Chapter 7) as well as the galactic background and radio galaxies, the Parkes staff installed a rotatable feed for use with the recently completed 10 cm receiver and planned an extensive polarization observing program. According to Bracewell, Radiophysics Chief Taffy Bowen had given him permission to use the Parkes telescope to observe the structure of Cen A, although John Bolton, the Parkes Director, had apparently not been consulted.[56] At the time of Bracewell's visit to Parkes, Brian Cooper and Australian National University PhD student Marcus Price were scheduled to observe the Galactic Center but, since Cen A rose about 4.5 hours earlier than the Galactic Center, there was no problem in squeezing Cen A into the observing schedule. During his observing session, Bracewell rotated the feed and was easily able to detect the polarization of Cen A. He quickly submitted a letter to *Nature* reporting the discovery of 15 percent polarization in Cen A, that, "at that time," according to Bracewell, was "an unprecedented degree of extragalactic polarization."[57] Following well-established Radiophysics custom, Bracewell included Brian Cooper, who had designed and built the 10 cm receiver, as a co-author. However, Bolton was reportedly "furious" when he learned that "Bracewell had apparently hijacked his telescope without proper authorization and that [considering that the Parkes team had spent over a year in getting the telescope ready] his behavior had been unethical and deceitful."[58]

That weekend, as a student far from his home in Colorado, Marc Price found himself as the lone scientist in Parkes during the Easter holiday period. Although the telescope was closed for the holiday weekend, he wondered, "Well gee, it would be interesting to see if Cen A is polarized at different frequencies too."[59] As one of only a few scientists to be certified by Bolton to run the telescope, and being familiar with the telescope instrumentation, Price installed a 21 cm receiver and was able to confirm Bracewell's detection of polarization. But there was a problem. The position angle of the polarization differed from that reported by Bracewell. As Price later recalled, he was sure that Bracewell "had got his feed angle wrong" and, on the following Monday, he telephoned the Sydney Radiophysics headquarters. After being chided by Bolton for using the telescope when it was shut down for Easter, Price told Bracewell, "Hey, before you send off your paper you'd better check the feed angle, because I've done it at 21 cm and it's 90° out." Bracewell expressed his usual self-confidence, or what Price later described as "arrogance," and told him that he had not made any mistake with the position of the feed angle.

Price recalled learning about Faraday rotation in his physics classes and consulted Bolton and Frank Kerr to see if that might explain the apparent polarization position angle discrepancy. But Price also noted that, in his well-known book on *Cosmic Radio Waves*, Shklovsky had indicated that, with the bandwidths normally used in radio astronomy, differential Faraday rotation across the bandpass would smear out any rotation.[60] Undaunted, Price teamed up with Cooper and showed from observations at eight frequencies between 970 and 3,000 MHz that the Cen A polarization angle rotated following the classical λ^2 law expected from Faraday rotation. Cooper and Price made the important conclusion that the rotation must take place either within the galactic interstellar medium or in the outer reaches of Cen A. Like Bracewell, Cooper and Price hurried off a letter to *Nature* reporting their discovery of Faraday rotation in the Cen A radio galaxy.[61] Their paper appeared in the 15 September 1984 issue of *Nature.* Meanwhile, Frank Gardner and John Whiteoak used the Parkes 20 cm receiver to detect polarization in seven out of nine different radio galaxies that they studied, which, as they wrote, "considerably strengthens the hypothesis that the synchrotron mechanism is responsible for the radiation from nonthermal sources."[62] Their quickly written paper, which was sent to *Physical Review Letters*, was received by the journal on 11 July and published in the 1 September issue.

Meanwhile, Bracewell's original discovery paper was languishing at *Nature*, and was not published until the 29 September issue, after the two follow-up papers by Gardner and Whiteoak and by Cooper and Price had already appeared. According to Bracewell, he was surprised to learn that Tom Cousins' name had been added to the author list and that someone had included a "Note added in Proof" that C. H. Mayer "has independently detected [similar] linear polarization in the central component of Cen A."[63] Bracewell claimed to have not known about the NRL result until after he had returned to Stanford, hearing about it first in a letter from Cooper, and then at the August 1962 meeting of the American Astronomical Society (AAS) in New Haven, Connecticut, where Mayer presented the new NRL Cen A polarization work.[64] Bracewell was annoyed by the implication that he knew about the NRL result before his Parkes observations, and hinted that the inclusion of the NRL Cen A polarization at the AAS meeting was motivated by their knowledge of his Parkes results. It is unclear why and by whom publication of the original Bracewell discovery paper was delayed until after the two follow-up papers had already appeared and the NRL result had been announced at the AAS meeting. Haynes et al. blamed John Bolton for delaying publication, which Bolton denied, but suggested that Taffy Bowen had intervened with the editor of *Nature*.[65] The story was further exacerbated by an article that appeared in the 15 September 1962 *Sydney Morning Herald* that reported on the Cooper and Price discovery that radio waves were linearly polarized, and that the plane of polarization changed with frequency. No mention was made of Bracewell's discovery that led to the Cooper and Price investigation. It remains a mystery whether or not Bracewell knew of the NRL result before his Parkes observations and if he really had permission from Bowen to use the telescope or was just a visitor who took advantage of the opportunity.

Probably Bowen had told Bracewell he could use the Parkes telescope at 10 cm to exploit the better Parkes resolution needed to map the inner radio lobes of Cen A that he and Little had previously discovered. But when he got to Parkes, Bracewell found the rotatable feed and followed his curiosity to discover the strong polarization in the northern lobe of the radio galaxy. Or, perhaps Bowen had told Bracewell only that he could go to Parkes to "observe," meaning to "watch," and not to "observe," meaning to use the telescope. Whatever the correct background, the early 1960s polarization work at NRL and Parkes confirmed the synchrotron radiation interpretation of radio galaxies and opened the door to the full appreciation of the important role of cosmic magnetic fields in astrophysics.[66] Latter polarization observations of radio galaxies with the Cambridge, Westerbork, VLA, VLBA, and ATCA synthesis telescopes (Chapter 11) led to a better understanding of the complex emission processes in radio galaxies, quasars, and active galactic nuclei and, as discussed in Chapter 5, their use as probes of the interstellar and intergalactic magnetic fields.

4
Quasars and AGN

> I had a reputation for being a radical and I was afraid to go out on a limb with such an extreme idea.[1]

The discovery of quasars as the ultra-luminous nuclei of active galaxies, with their unprecedented large redshifts, unveiled a new class of previously unrecognized objects and opened the door to exploring the most distant parts of the Universe. Twenty years earlier, during the middle of the Second World War, Vanderbilt University astronomer Carl Seyfert (Figure 4.1) reported on his observations of enhanced activity in the center of six galaxies or, as he called them, "extragalactic nebulae."[2] Seyfert's now famous paper was essentially ignored until 1962 when the Russian astrophysicist Iosef Shklovsky drew attention to the possible connection between what are now called Seyfert galaxies and radio galaxies.[3] Even earlier, the British scientist, Sir James Jeans speculated that[4]

> The centres of the nebulae are of the nature of singular points, at which matter is poured into our universe from some other and entirely spatial dimension, so that to a denizen of our universe, they appear as points at which matter is being continually created.

In 1958, the Armenian scientist Viktor Ambartsumian (Figure 4.2) proposed what he called "a radical change in the conception of the nuclei of galaxies," and argued that, "Apparently we must reject the idea that the nucleus of a galaxy is composed of stars alone."[5] However, these speculations played no role in the actual discovery of active galactic nuclei (AGN) or quasars, which resulted from the quest to find more distant radio galaxies and was aided by a fortuitous series of lunar occultations of the bright radio source 3C 273. The route to the recognition of quasars as the most powerful and most distant constituents of the Universe was a convoluted path fraught with mis-directions, ignored clues, and the rejection of data that did not appear to agree with preconceived notions.

Figure 4.1 Carl Seyfert. Credit: Vanderbilt University.

The Tortuous Road to the Discovery of Quasars

As discussed in Chapter 3, by 1960, it was widely accepted that most of the discrete radio sources located away from the Galactic Plane were powerful distant radio galaxies. The idea of radio stars which, until the identification of Cygnus A, was popular, particularly in Cambridge, seemed dead. Caltech astronomer Jesse Greenstein even offered a case of the best Scotch whiskey to whoever found the first true galactic radio star. In the quest to find more distant galaxies, the Caltech and Mount Wilson and Palomar Observatories (MWPO) identification programs concentrated on small diameter radio sources under the assumption that the most distant sources would appear smaller. Moreover, John Bolton wanted to find the optical counterpart of a small diameter radio source to align his new Owens Valley Radio Observatory (OVRO) interferometer with the optical reference frame.

One of the best candidates was the radio source 3C 48. Interferometer measurements at Jodrell Bank had shown that 3C 48 was less than about 1 second of arc in diameter.[6] The Caltech interferometer position was good to a few arcseconds accuracy, which was typically sufficient to identify a radio source with a distant galaxy. But when Caltech radio astronomers Tom Matthews and John Bolton

Figure 4.2 Viktor Ambartsumian. Credit: AIP Emilo Segrè Visual Archives, Physics Today Collection.

inspected the 48 inch Schmidt survey plates, they were surprised to find only a 16th magnitude star within the small position error box of the 3C 48 radio position.

Bolton asked Jesse Greenstein to observe 3C 48 with the Palomar 200 inch telescope. However, Greenstein's next 200 inch session was clouded out, so Matthews passed his finding chart on to Mount Wilson and Palomar Observatories' Allan Sandage who was scheduled to observe on 4 September 1960. When Matthews inspected Sandage's 200 inch photograph he found what he described as a faint red "wisp" 3 × 8 arcseconds in size alongside the star. Over the next months, spectroscopic observations by Sandage, Greenstein, and Guido Münch showed multiple emission and absorption lines as well as a strong continuum ultra-violet excess. Attempts to identify the spectral lines were inconclusive, and further investigations appeared to confirm the nature of 3C 48 as a real radio star. Sandage showed that, except for the faint "wisp," most of the optical light came from the stellar-like object, that the color was "peculiar," and that the optical counterpart varied by at least 0.4 magnitude (20 percent) over a time scale of months, thus supporting the notion that 3C 48 was just a galactic star and not a distant radio galaxy (Figure 4.3).

Figure 4.3 From left to right, Jesse Greenstein (Credit: Caltech Archives), Allan Sandage (Credit: Carnegie Institution for Science), and Tom Matthews (Credit: Caltech Archives).

John Bolton was not convinced. Although he had no background in optical spectroscopy, Bolton realized that he could interpret the spectrum if the lines had a redshift of 0.37. He wrote to Joe Pawsey back in Australia,[7]

> I thought we had a star. It is not a star. Measurements on a high dispersion spectrum suggest the lines are those of Neon [V], Argon [III] and [IV] and that the redshift is 0.367. The absolute photographic magnitude is –24 which is two magnitudes greater than anything known . . . I don't know how rare these things are going to be, but one thing is quite clear – we can't afford to dismiss a position in the future because there is nothing but stars.

Bolton later claimed that he could not convince Greenstein or Ira Bowen, Director of the MWPO, that 3C 48 had such a large redshift. Both Bowen and Greenstein were experts in stellar spectroscopy. As Bolton later related, they both argued that if 3C 48 had a redshift of 0.367, there were small 3 or 4 Angstrom discrepancies in the relative positions of different lines that were too large to accept as measurement error or due to Doppler shifts resulting from small internal velocities.[8] Probably unspoken at the time, Bowen and Greenstein also recognized that, with a redshift of 0.37, 3C 48 would have an extraordinary absolute magnitude of –24, or brighter than any then known object in the Universe. Uncharacteristically, Bolton apparently deferred to his distinguished colleagues, probably also skeptical about the inferred unprecedented optical luminosity. Bolton left Caltech in December 1960 to return to Australia to take charge of the new 210 foot radio telescope then under construction at Parkes. While sailing to Australia on the SS *Orcades*, he wrote again to Pawsey, "The latest news on 3C 48 as I left Caltech was – It is most likely a star," and he mailed his letter during a stopover in Honolulu.[9]

With Bolton gone and Allan Sandage having already departed Pasadena to attend the meeting of the American Astronomical Society (AAS) in New York City, Matthews was excited by the thought that they had discovered the first real radio star and sent a telegram to Sandage, who presented a late paper at the AAS meeting reporting on the discovery of the first radio star.[10] *Sky and Telescope* later reported the discovery of the "First True Radio Star," saying, "there is general agreement among the astronomers concerned that it is a relatively nearby star with most peculiar properties."[11] A few months later, Greenstein, apparently having been convinced that 3C 48 was his long sought "first radio star," wrote an article in the Caltech *Engineering and Science* publication announcing "The First True Radio Star."[12]

Encouraged by the apparent discovery of the first radio star, Matthews and Sandage went on to find two other small radio sources apparently associated with stellar objects. In their paper, "Optical Identification of 3C 48, 3C 196, and 3C 286 with Stellar Objects," they wrote that they were not able to find "any plausible combination of red-shifted emission lines."[13] Meanwhile, Greenstein went on to make an exhaustive study of the 3C 48 optical spectrum. After two years of analysis, he submitted a paper to *The Astrophysical Journal* with the title "The Radio Star 3C 48."[14] In this paper, Greenstein concluded that 3C 48 was the stellar remains of a supernova, and that the unfamiliar spectrum represented highly ionized rare earth elements. The abstract stated, "The possibility that the spectral lines might be greatly redshifted forbidden emissions in a very distant galaxy is explored with negative results," and the first sentence of the paper argued that "The first spectra of the 16th magnitude stellar radio source 3C48 obtained in October 1960 by Sandage were sufficient to show that this object was not an extragalactic nebula of moderate redshift." Greenstein discussed possible line identifications with several possible redshifts. Although he commented that "except for 0.367 no redshift explains the strongest lines of any single ionization," he continued, "the case for a large redshift is definitely not proven." Nevertheless, with great prescience, he then went on to point out that if the 3C 48 spectrum is really the red-shifted emission spectrum of a galaxy, then for $\Delta\lambda/\lambda > 1$, Ly-α and other strong UV lines would be shifted into the visible spectrum. When asked years later why he rejected what appeared to be a very important and satisfactory interpretation in terms of a large redshift, Greenstein explained, "I had a reputation for being a radical and I was afraid to go out on a limb with such an extreme idea."[15]

3C 273 Unlocks the Puzzle

3C 273 is the seventh brightest radio source in the northern sky. Its optical counterpart – the quasar, is 13th magnitude, or more than 100 times brighter than many of the radio galaxies that Caltech astronomers had already identified based on radio positions measured with the OVRO interferometer. The typical position accuracy of the OVRO interferometer was better than 10 arcseconds for the stronger sources, which was normally adequate to identify an optical counterpart, especially one as bright as 3C 273. But the Caltech radio astronomers knew from the interferometer observations at Jodrell Bank[16] and at the French Nançay Observatory[17] that 3C 273 was not one of the smaller radio sources, so they made no special effort to find an optical counterpart. By 1961 the OVRO positions of 3C 273 and other sources had been measured by Caltech graduate students Ed Fomalont and Richard Read working together with Matthews.[18] They were able to locate the position of the 3C 273 radio source within 2 arcseconds of the currently measured quasar position. In his 1962 PhD dissertation, Read discussed likely optical counterparts for four radio sources, including 3C 273, all with faint galaxies seen only on 200 inch photographic plates. Interestingly, in his later published paper, Read again discussed four identified sources, three of which were the same sources discussed in his thesis, but 3C 273 was not included and had been replaced by 3C 286.[19] When observing Read's proposed optical identifications with the Palomar 200 inch telescope in May 1962, inexplicably, Maarten Schmidt looked at a galaxy that was located about an arcminute west of the measured radio position of 3C 273.[20]

The breakthrough occurred in 1962 when 3C 273 was occulted by the Moon. As early as 1960, the British radio astronomer Cyril Hazard (Figure 4.4) had used the Jodrell Bank 250 foot radio telescope to observe a lunar occultation of the radio source 3C 212.[21] By measuring the time of immersion and emersion from the limb of the Moon, Hazard was able to determine the position of 3C 212 with an accuracy of 3 arcseconds, but his inspection of Palomar Sky Survey plates failed to recognize the 19th magnitude stellar identification which later turned out to be a quasar.

Hazard had another opportunity to discover quasars in 1962 when a series of occultations of the strong radio source 3C 273 was predicted to occur at the Parkes radio telescope on 15 May,[22] 5 August, and 26 October. By this time Hazard had moved to the University of Sydney, where he joined his old mentor, Robert Hanbury Brown, who was building an optical intensity interferometer at Narrabri in New South Wales. Tensions between the CSIRO Radiophysics Lab and the University of Sydney were already strained, in part due to the aggressive leadership of the University School of Physics Head, Harry Messel, and the earlier abrupt transfer of Bernie Mills and Chris Christiansen from CSIRO to the University under less than agreeable circumstances.[23] Nevertheless, Joe Pawsey and Parkes Director John Bolton invited Hazard to use the Parkes radio telescope (Figure 4.5) to observe the three occultations of 3C 273 expected to take place in 1962.[24]

In May, only the emersion (reappearance) of 3C 273 from behind the rising Moon was visible from Parkes.[25] Hazard got lost in driving to Parkes and arrived late at the Observatory. Due to the strong radio emission from the Moon itself, the

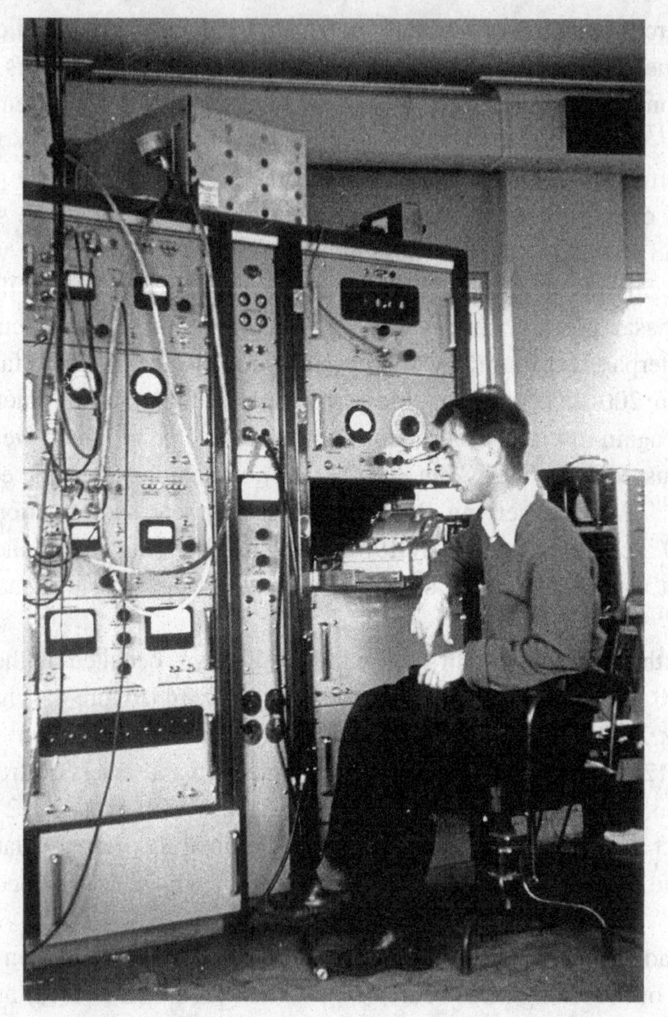

Figure 4.4 Cyril Hazard. Courtesy of Cyril Hazard.

data from the May occultation of 3C 273 were of limited value but did show a classical diffraction pattern, demonstrating that at least some of the radio emission came from a very small region. This generated great interest in the upcoming August occultation when both immersion and emersion would be visible.

But there was a problem. The expected emersion of the radio source from behind the Moon on 5 August was predicted to occur slightly below the 30 degree elevation angle limit of the Parkes antenna. To make sure that the event would not be missed, Bolton deactivated the elevation limit switches, removed a ladder from the antenna structure, and ground down part of the antenna gear box in order to extend the length of time the telescope could track 3C 273 below the nominal 30 degrees elevation

Figure 4.5 Parkes 210 foot radio telescope. Author photograph.

limit. He also had a trench dug next to the antenna, in case the emersion occurred when the edge of the antenna was below the ground. As it turned out, due to an incorrect position of the telescope that had been used to predict the time of the occultation, 3C 273 emerged from behind the Moon well before the telescope reached the nominal 30 degree limit, and so Bolton's rather drastic attack on the antenna proved unnecessary.[26] But it made a good story that Bolton enjoyed telling and retelling. To avoid any potential radio frequency interference to the occultation record, Bolton blocked all the roads leading to the Observatory, disabled all unneeded electrical equipment at the site, and even arranged for the local Parkes radio station to cease broadcasting during the critical period of the occultation. As insurance against the loss of data due to any malfunction of the chart recorder, he hooked two separate recorders to the receiver output.

The occultation was successfully observed on 5 August at two frequencies, 136 MHz (2.2 m) and 410 MHz (73 cm). Bolton and Hazard carried the chart records back to Sydney on separate airplanes. Analysis of the August and later October occultation data[27] showed that 3C 273 was a complex source consisting of a small flat spectrum component (B) less than an arcsecond wide and an elongated steep spectrum component (A) coincident with the optical jet.[28] At the time, Rudolph

Figure 4.6 Maarten Schmidt. Credit: Caltech Archives.

Minkowski was visiting the Mount Stromlo Observatory in Canberra, and had brought a Polaroid copy of a Palomar 200 inch image of the 3C 273 field to Australia.[29] The position of the incorrect faint galaxy identification that Schmidt had tried to observe in May of that year was marked on the print. Hazard and Minkowski noted that the occultation indicated a position that was close to the 13th magnitude star and a faint elongated feature that Bolton had previously noted based on a rough position that he had determined using the Parkes telescope. They assumed that this elongated feature was an edge-on galaxy that was the radio source.

On 20 August, just two weeks after the occultation and unknown to Hazard, Bolton wrote to Maarten Schmidt at Caltech discussing the Parkes program of radio source identifications and requesting 200 inch follow-up with Palomar on the radio galaxy PKS 2216-28 (Figure 4.6). Somewhat parenthetically, or as an afterthought, at the end of his letter, Bolton gave the preliminary occultation positions of 3C 273 components A and B and asked that Schmidt pass the coordinates along to Tom Matthews. In his letter, Bolton indicated no special interest in 3C 273 and made no mention of the obvious optical counterparts.[30] However, due to an error by Hazard in analyzing the time of immersion and emersion, the positions communicated by Bolton on 20 August were in error by about 15 arcseconds and were further away from the true optical counterpart than the positions that had been previously determined with the OVRO and Parkes antennas (Figure 4.7).

3C 273 was too close to the Sun for Schmidt to attempt any optical spectroscopy until December 1962, when Schmidt was scheduled to use the 200 inch telescope to observe a list of radio galaxies supplied by Matthews. Just before dawn on 27 December, Schmidt turned the telescope to observe the spectrum of the 3C 273 optical counterpart. Based on the faint optical galaxies that were being

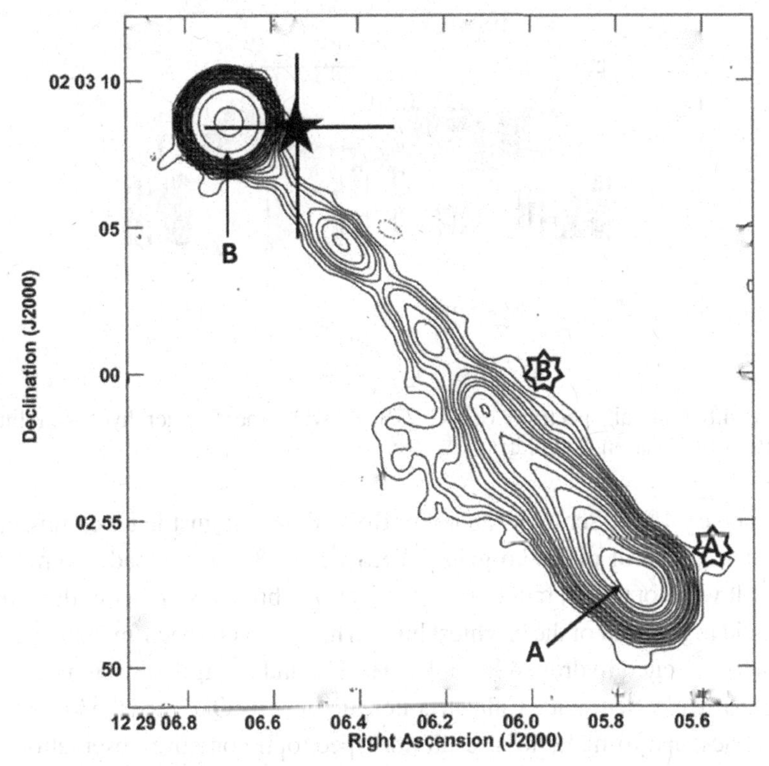

Figure 4.7 Six centimeter contour map of 3C 273 showing the preliminary occultation positions (open eight-sided stars) of components A and B communicated to Schmidt on 20 August 1962. The arrows point to the revised 26 January 1963 positions. The filled star indicates the OVRO interferometer position and its uncertainty. Credit: Contour map from Perley and Meisenheimer (2017).

associated with other radio sources, Schmidt, like Minkowski, assumed that the correct optical counterpart of 3C 273 was the faint nebulosity located next to the bright stellar object. Although at the time there were already three other radio sources, 3C 48, 3C 196, and 3C 286, that had been identified with apparent stellar counterparts, Schmidt assumed that the magnitude 13 stellar object was far too bright to be associated with the 3C 273 radio source.[31] Nevertheless, in order to eliminate the stellar object from further consideration, Schmidt decided to first obtain a spectrum of the "star." Having spent most of his 200 inch experience observing very faint objects using the world's largest optical telescope, he overexposed the spectrum on 27 December. The following night, Schmidt concentrated on his other radio galaxies, and ignored 3C 273. On 29 December, he returned to 3C 273 and obtained a properly exposed spectrum, and did so again on 23 January 1963 during his next session on the 200 inch telescope (Figure 4.8).

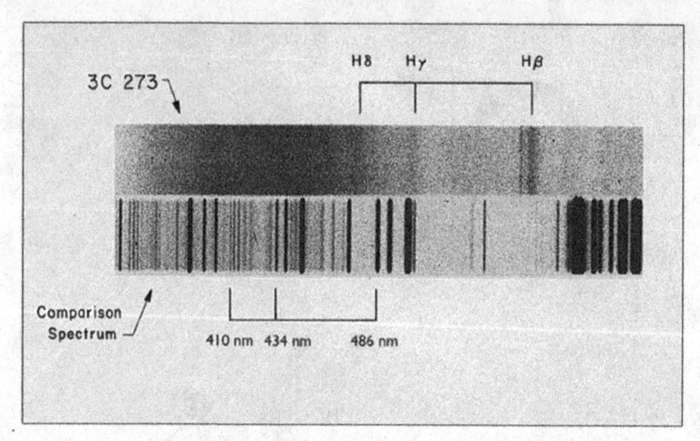

Figure 4.8 Optical spectrum of 3C 273 showing the Balmer hydrogen lines. Courtesy of Maarten Schmidt.

On 8 January 1963, Schmidt replied to Bolton's 20 August letter commenting on his 200 inch image of the radio galaxy PKS 2216-28, but he made no mention of 3C 273.[32] It was not until a month later, on 5 or 6 February, that Schmidt recognized that he could match four of the brightest lines in his 200 inch spectrum with the Balmer series lines of excited hydrogen lines, H_β, H_γ, H_δ, and H_ε, if they were redshifted by 0.16.[33] Meanwhile, Schmidt's Caltech colleague Beverly Oke used the Mount Wilson 100 inch telescope to make low resolution spectrophotometric observations of 3C 273 that extended into the infrared. Oke's spectrum included the Balmer Hα line at 7,599 Å, and supported Schmidt's redshift of 0.158. Applying this redshift, Schmidt was able to associate the strong line at 3,239 Å, near the short violet limit of his spectrum, with the 2,798 Å ionized Mg II line. Since the Earth's atmosphere is opaque at this ultra-violet wavelength, the Mg II line had been previously seen only in the solar chromosphere from rocket borne instruments. Schmidt considered two possible interpretations of the 0.16 redshift. Either 3C 273 was a 10 km diameter star with a strong gravitational redshift or, more likely, the redshift was cosmological and that with an apparent magnitude of 13, the optical luminosity of 3C 273 was some hundred times greater than typical galaxies previously identified with radio sources.

Still believing that 3C 273 was a galactic star, on 26 January 1963, Bolton again wrote to Schmidt with the corrected occultation position.[34] Matthews measured the coordinates of the optical counterparts from Sandage's 200 inch plate and showed that the magnitude 13 stellar object was coincident with the radio component B, and that the faint jet was coincident with component A.[35] But this was only after Schmidt had already observed the spectrum of the stellar object, although before he understood that it had such a large redshift.[36] Interestingly, it was possible to recognize the "large" redshift of 3C 273, not because it was large, but because it was so "small" that the Balmer series of lines, H_β, H_γ, H_δ, and H_ε, were still seen within the small classical optical window. 3C 273 is probably unique in being the only quasar whose spectrum can be so easily determined without prior knowledge of the line identification. In 3C 48, by contrast, the redshift is so large that all the spectral lines familiar to astronomers were shifted into the infrared.

Excited by his remarkable result, Schmidt called Jesse Greenstein to his office to discuss his discovery. Together they re-inspected Greenstein's three year old spectrum of 3C 48, and realized that if the broad 3C 48 line at 3,832 Å is the same Mg II line that Schmidt found in the 3C 273 spectrum, the redshift of 3C 48 would be 0.3679, the value that John Bolton had claimed three years earlier and that Bowen and Greenstein had rejected. Using this value of the redshift, the other lines in the 3C 48 spectrum fell into place as [Ne V], [O II], and [Ne III]. Greenstein then hurriedly withdrew his paper, "The Radio Star 3C 48"[37] that he had submitted some months earlier to *The Astrophysical Journal*, while Matthews and Sandage added a section to their paper based on the newly-determined redshift of 0.37 after the paper had already been accepted for publication by the same journal. This was the second largest redshift ever measured, next to 3C 295 at $z = 0.45$. Two days later, Schmidt wrote to Hazard that he was still unclear whether the 3C 48 and 3C 273 were extragalactic with cosmological redshifts or were dense neutron stars of only a few tens of kilometers in extent.[38]

The four classic papers, by Hazard et al. describing the occultation measurements,[39] the determination by Schmidt of the redshift,[40] the confirmation based on Oke's infrared observations,[41] and the redshift of 3C 48 by Greenstein and Matthews,[42] were published as consecutive papers in the 16 March 1963 issue of *Nature*. Further exacerbating already existing tensions between CSIRO and the University of Sydney radio astronomers, Hazard's name appeared with a CSIRO Radiophysics Laboratory affiliation, although Hazard was on the University of Sydney staff. Following standard practice for non-Radiophysics staff members, Bolton had determined that, since Hazard was not sufficiently familiar with the operation of the telescope or the instrumentation to use the telescope himself, he added John Shimmins and Brian Mackey to the observing team to provide telescope and instrumental support, respectively. Characteristically Bolton, who prepared the telescope, carried out most of the observations, and was the key link to Tom Matthews and Maarten Schmidt, declined to include his name on the paper, claiming he was only doing his job as the director. But Hazard's incorrect affiliation has remained controversial. Although CSIRO blamed it on an editorial error by *Nature*, someone had written the word "delete" next to Hazard's University of Sydney address before CSIRO submitted the manuscript for publication in *Nature*.[43]

The discovery of quasars with their large redshifts and correspondingly unprecedented large radio and optical luminosities initiated a multitude of new observational and theoretical investigations, as well as a plethora of conferences, particularly the series of *Texas Symposia on Relativistic Astrophysics and Cosmology*. The following year, Greenstein and Schmidt published a paper extensively discussing the physical conditions in 3C 48 and 3C 273. They reconsidered the possibility that the observed redshifts were non-cosmological and might be the gravitational redshift from collapsed neutron stars or from "very massive objects," but confirmed Schmidt's earlier calculation that a cosmological redshift was more likely.[44] Motivated by the possibility of extending the Hubble relation to higher redshifts and determining the value of the deceleration constant, q_o, a race began to find the largest redshifts. Within two years of the 3C 273 breakthrough, redshifts as high as two were reported by Schmidt and others, but it would take a decade of searching before quasars with redshifts greater than three were observed. Subsequent work has extended quasar redshifts to values greater than seven. Ironically, some luminous galaxies have now been recorded with even larger redshifts than any quasar and, unlike for galaxies, the dispersion in the intrinsic quasar luminosities is too great to use quasars as standard candles in deriving the Hubble relation.

Nomenclature

The discovery of quasars, with their large redshifts and their unprecedented radio and optical luminosity, was certainly one of the outstanding discoveries of twentieth century astronomy and was recognized by Maarten Schmidt's picture on the cover of *Time* magazine. Unable to come up with a clever sounding name for these stellar-like objects that were not stars but the powerful nuclei of galaxies, Schmidt called them "*quasi-stellar radio sources*." This was quite a mouthful and, in a *Physics Today* review article, Hong-Yee Chiu coined the shorter name "quasar."[45] "Quasar" became widely used in oral discussion and in the popular media, but was not accepted by the *Astrophysical Journal (ApJ)* until, in his 1970 paper on the "Space Distribution and Luminosity Function of Quasars," Schmidt wrote,[46]

> We use the term "quasar" for the class of objects of starlike appearance (or those containing a dominant starlike component) that exhibit redshifts much larger than those of ordinary stars in the Galaxy. QSOs are quasars selected on the basis of purely optical criteria, while QSSs are quasars selected on both the optical and radio criteria.

The *Astrophysical Journal* editor, Subrahmanyan Chandrasekhar, responded with a footnote saying, "*The Astrophysical Journal* has until now not recognized the term 'quasar'; and it regrets that it must now concede: Dr. Schmidt feels that, with his precise definition, the term can no longer be ignored." The word "quasar" has caught on and is now commonly used in both the popular and professional literature. "QSO" refers to the optical counterpart, the Quasi Stellar Object, but increasingly "quasar" is used independent of any radio emission. As observations have been extended to cover the entire electromagnetic spectrum, and as improvements in technology have resulted in increasingly detailed descriptions of both continuum and line spectra, variability, and morphology, quasars have become classified and sub-classified based on their line spectra, as well as their radio, optical, and high energy spectral distribution. Optical spectroscopy and photometry have defined QSO1s, QSO2s, Broad Absorption Line quasars (BALs), Low-Ionization Nuclear Emission-Line Regions (LINERS), and BL Lacs, all collectively referred to as QSOs or quasars. Radio astronomers have defined Flat Spectrum and Steep Spectrum Radio Quasars and Radio Loud and Radio Quiet Quasars. Radio Quiet quasars have been referred to as Interlopers, Quasi-Stellar Galaxies (QSGs), and Blue Stellar Objects (BSOs). Relativistically beamed quasars are known as blazars; X-ray and gamma-ray observations have defined High and Low Spectral Peaked Quasars. Collectively they are all referred to as Active Galactic Nuclei (AGN), although the term AGN was originally used to describe the low luminosity counterpart to quasars, such as originally discussed by Carl Seyfert.

Where, When, and by Whom Was the First Quasar Discovered?

Although both radio and optical astronomers were concentrating on small diameter radio sources in their quest to locate high redshift galaxies, the occultation observations of 3C 273 in 1962 were unrelated to the quest for distant galaxies. Because it was known to be well resolved, 3C 273 was not on the Caltech list of high priority sources that might be at high redshift, so it would not have been observed by Schmidt in December were it not for John Bolton's August letter that motivated Schmidt to take the spectrum of 3C 273 as it rose just before the December morning twilight. Although the crucial observations were made with the Palomar 200 inch telescope, at the time the largest in the world, optical measurements of 3C 273 were recorded as much as 75 years earlier,[47] and the defining spectral observations could have been made much earlier and with a much smaller telescope had anyone thought to look.

As early as 1960 Caltech was measuring sufficiently accurate radio source positions and had a successful program of identifying weaker radio sources with much fainter optical counterparts than 3C 273. Although the possibility of identification with apparent stellar objects was already well established with the identifications of the fainter apparently stellar objects 3C 48, 3C 196, and 3C 286, the association of an extended radio source with a magnitude 13 star apparently appeared too extreme to seriously consider, although the position of 3C 273 was known for several years, and was certainly known prior to Schmidt's May and December 1962 observations with the 200 inch Palomar telescope. The 19.5

magnitude galaxy first thought to be the optical counterpart of 3C 273 was observed by Schmidt in May 1962, although it was about a minute of arc west of the true position. Schmidt and the Caltech radio astronomers must have believed that their radio position was accurate to within a few arc seconds, as that kind of accuracy would be required to accept an identification with such a faint barely detectable galaxy.

Possibly, somewhere along the line, there may have been an error in conveying the OVRO radio position to the Caltech optical astronomers, perhaps a misunderstanding as to whether the position conveyed to Schmidt had been precessed to 1950 coordinates. Ironically, the occultation positions that Bolton sent to Schmidt on 20 August 1962, which was the basis of the spectrum taken by Schmidt in December, were about 15 arcseconds away from the correct positions. The correct positions were not known by Schmidt until he received Bolton's 26 January 1963 letter, more than a month after he had obtained the spectrum of the quasar.

Thus, it appears that the sub-arcsecond Parkes occultation position did not play a direct role in the identification of 3C 273 as a quasar, that was only recognized as the optical counterpart after Schmidt's 200 inch observations showed it to have a peculiar spectrum, and was confirmed by Matthews in January after Bolton had communicated the correct position to Schmidt. Had he not received the occultation positions from Bolton, in August 1962, even though incorrect, Schmidt probably would not have tried to observe the spectrum of the stellar counterpart. Sooner or later, an object as bright as 3C 273 would have been identified on the basis of the OVRO interferometer position and should have been identified at least a year or two earlier. But, as Schmidt later explained, "There was, of course, a mental barrier against considering the possibility of a star-like object having a redshift."[48]

Considering that two years earlier, Bolton had argued that 3C 48 was not a galactic star but had a redshift of 0.37 and that he, together with Matthews, made the original identification of 3C 48 with a star-like object, it is surprising that his name was omitted from the Greenstein and Matthews (1963) *Nature* paper. When asked, Greenstein later claimed that Bolton played no role in the 3C 48 story, although Bolton's name appears second on the paper that Sandage presented at the December 1962 AAS meeting and, Greenstein claimed, that it was Tom Matthews who first suggested that 3C 48 might have a high redshift. So the 3C 48 paper, written more than two years after Bolton had left Caltech, was authored by only Greenstein and Matthews.[49] Later, Matthews explained that his early suggestion of a large redshift was based simply on the small angular size of the radio source, not on the optical spectrum, but that, once they had the optical identification, he too assumed 3C 48 was a galactic star.[50] Strangely, Bolton had kept quiet about his 1960 determination of the 3C 48 redshift until 1989, when he gave a talk at a meeting of the Astronomical Society of Australia in which he reminisced about his work at Caltech during the period from 1955 to 1960 while he was away from Australia to start the radio astronomy program at Caltech. In his talk, Bolton recalled that, back in 1960, more than two years before the redshifts of 3C 273 and 3C 48 were announced, he had claimed that 3C 48 had a redshift of 0.37 but that he could not convince Greenstein or Bowen.[51] Why did Bolton finally speak out after nearly 30 years? Later, he explained, "I had got more and more irritated as the years went by Australians claiming that QSOs were an Australian discovery."[52] They were, in a way, but not at Parkes. Curiously, when Schmidt visited Bolton in Australia in 1963, just a month after the publication of the *Nature* papers, Bolton made no mention of his earlier claim that 3C 48 had a large redshift.

The search for ever larger redshifts following the 1963 understanding of the 3C 48 and 3C 273 spectra was highly competitive between Pasadena astronomers and others, and among Pasadena astronomers themselves. Along with the complex events, multiple personalities (Bolton, Greenstein, Hazard, Matthews, Minkowski, Oke, Sandage, and Schmidt) surrounding the discovery of quasars, and their competing claims of credit, it led to increased tensions between Caltech and Carnegie astronomers in Pasadena. Sandage and Greenstein were particularly upset with each other, both were upset with Bolton, and everyone was annoyed with Fritz Zwicky (see next section). Their colleague Bev Oke was apparently unhappy that he did not receive his deserved recognition for his contribution in determining the redshift of 3C 273.[53] These tensions later contributed to the separation of the MWPO by Schmidt when he later became Director of the Hale Observatories and Chairman of the Caltech Division of Physics, Mathematics, and Astronomy.[54]

Interlopers, Blue Stellar Objects, Quasi-Stellar Galaxies, and AGN

Soon after their discovery, in the quest for high redshift quasars, many astronomers realized that, compared with stars, quasars appeared unusually blue. This suggested that quasars might be identified by their blue color only, without the need for very precise radio positions to distinguish them from stars. In pursuing radio source identifications with "blue stellar objects" (BSOs), Allan Sandage noticed that there were many BSOs in the sky that were not coincident with known radio sources. He claimed that they had a sky density about a thousand times greater than that of 3C radio sources. Excited by his new discovery, he quickly wrote a paper with the provocative title, "The Existence of a Major New Constituent of Universe: The Quasi-Stellar Galaxies," that was received at the *Astrophysical Journal* editorial offices on 15 May 1965.[55] *The Astrophysical*

Figure 4.9 Fritz Zwicky. Credit: Caltech Archives.

Journal editor, Chandrasekhar, was apparently so impressed by Sandage's claimed discovery that he did not send the paper to any referee and delayed publication so that Sandage's paper could appear in the 15 May issue of the *Journal.*

Sandage's claim was received with skepticism, no doubt in part generated by what was perceived as privileged treatment of his paper. Caltech physics professor, Fritz Zwicky (Figure 4.9), who did not get along well with other Caltech faculty or MWPO astronomers, was particularly incensed. Two years earlier, at a meeting of the American Astronomical Society, Zwicky himself had argued that what others were calling "radio stars" he thought lay at the most luminous end of the sequence of compact galaxies.[56] Following the publication of Sandage's paper, Zwicky pointed out that, "All of the five quasi-stellar galaxies described individually by Sandage evidently belong to the subclass of compact galaxies with pure emission spectra previously discovered and described by the present writer."[57] A few years later, he was characteristically more direct. In a paper sent to the *Astrophysical Journal*, he wrote,

> In spite of all these facts being known to him in 1964, Sandage attempted one of the most astounding feats of plagiarism by announcing the existence of a major new component of the Universe: the quasi-stellar galaxies ... Sandage's earthshaking discovery consisted in nothing more than renaming compact galaxies, calling them "interlopers" and quasi-stellar galaxies, thus playing the interloper himself.

But Chandrasekhar rejected Zwicky's paper writing, "Communications of this character are outside the scope of this journal."[58]

Other astronomers were more circumspect than Zwicky in attacking Sandage, but still argued that most of Sandage's BSOs were just that, blue stellar objects, located in the Galaxy, and not very compact luminous external galaxies.[59] Today astronomers recognize that there is indeed a population of so-called radio quiet quasars, but they are only about an order of magnitude more numerous than the classical "radio loud" quasars, or a factor of about a hundred less than argued by Sandage. The radio quiet quasars are much weaker than the radio loud quasars, but they are not radio silent. Their radio emission is probably not from the quasar itself, but is due to ongoing star formation and supernovae activity that accelerates cosmic ray electrons in the host galaxy.[60] Other previously cataloged visual "stars," including W Comae, AP Libre, and BL Lacerte, were later identified with radio sources and are now recognized as part of the broad category of active galactic nuclei (AGN), with quasars at the upper end of the AGN luminosity range. Seyfert and other galaxies with AGN such as NGC 1275 with intense nuclear emission are about a thousand times weaker than quasars.

At the high end of the luminosity range, quasars are the extremely bright nuclei of galaxies, and are so luminous that they outshine the rest of the galaxy by so much that the surrounding galaxy is only visible in images of a few relatively nearby quasars. At the other extreme are the compact radio sources found in the nuclei of nearby spiral galaxies such as M31 and M81, as well as the radio source at the center of the Milky Way, Sgr A.*

Sgr A*: The AGN in the Milky Way

Although the 5 May 1933 *New York Times* proudly declared "Radio Waves Traced to the Center of the Milky Way," as seen in Figure 1.9, Karl Jansky's observations did not quite reach the center of the Galaxy. The first clear evidence for radio emission from the Galactic Center can be seen on Grote Reber's 480 MHz (62 cm) map of the Milky Way radio emission, which Reber commented coincided with the infrared intensity maximum at 1 micron that Joel Stebbins and Albert Whitford had previously noted occurred in the region of the center of the Galaxy.[61] Two groups of Australian radio astronomers were probably the first to specifically discuss the radio emission from the Galactic Center.[62] Due to the heavy obscuration in the Galactic Center region, the actual nucleus is obscured to optical astronomers. But the radio observations, particularly those using an innovative hole in the ground antenna[63] (Figure 4.10) and later 21 cm H I observations of galactic rotation,[64] later defined the location of the Galactic Center, which was about 32 degrees away from the previously assumed position. In 1958 the International Astronomical

Figure 4.10 CSIRO hole-in-the-ground used to discover the Sgr A radio source at the Galactic Center. Credit: CRAIA.

Union decided to adopt a new galactic coordinate system based on the radio data,[65] but the Sagittarius A (Sgr A) radio source at the center of the Galaxy is a complex mixture of thermal and nonthermal sources, so the precise location of the Galactic Center remained uncertain.

The Green Bank Interferometer (GBI) (Figure 4.11) was built by NRAO, primarily as a test instrument to verify that it was possible to do interferometry over the 35 km long baselines planned for the Very Large Array. The GBI consisted of three 85 foot diameter antennas located on the NRAO site in Green Bank, WV, and a 45 foot antenna located on a mountaintop 35 km from Green Bank that was connected to the other antennas by means of a radio link. While using the GBI in February 1974 to search for bright thermal knots in the ionized hydrogen (H II) regions, including the one surrounding the Galactic Center, Bruce Balick and Robert Brown discovered a very compact radio source, that they determined was less than 0.1 arcseconds (1,000 AU) in size and was coincident with an unresolved strong infrared source located within one parsec of the nucleus

Figure 4.11 The Green Bank Interferometer. Credit NRAO/AUI/NSF.

of the Milky Way Galaxy.[66] The corresponding radio brightness temperature was greater than 2 million degrees K, so the emission clearly was not due to thermal emission from an ionized hydrogen region that has a typical temperature of 10,000 K. Nearly a decade later, when Brown recalled from his PhD work that atomic spectroscopists used the asterisk symbol to indicate metastable excited states, he decided to name their Galactic Center radio source Sagittarius A* or Sgr A* on the grounds that "it was pretty excited."[67]

In their paper, Balick and Brown noted the "morphological similarities between Sgr A* and the more energetic nuclei of other galaxies in terms of their size and brightness," which "strongly suggests that this structure is physically associated with the Galactic Center (in fact, defines the Galactic Center)." Nearly half a century later, Balick recalled that, at the time, he had speculated that Sgr A* might be the result of a massive black hole at the Galactic Center. But being only a postdoc on the job market, and aware that at the time black holes were still in the realm of science fiction, Balick avoided any mention of a black hole in their paper.[68] Later observations with interferometer baselines up to nearly 4,000 km (2,500 miles) showed that Sgr A* contained a radio source less than 0.001 arcseconds (10 Astronomical Units) across.[69] With a luminosity of only 10^{25} Watts, Sgr A* was 10,000 times weaker than the radio sources found in the nuclei of nearby spiral galaxies such as the Andromeda

Nebula (M31) and more than 10^{12} times weaker than the nuclei of strong radio galaxies and quasars. Sgr A* was at the low end of a sequence running through spiral and elliptical galaxies, radio galaxies, and quasars, all apparently powered by supermassive black holes of up to several billion times the mass of the Sun. Indeed, the measured small size of Sgr A* appeared to confirm the earlier suggestion of the British astrophysicists Donald Lynden-Bell and Martin Rees[70] that the detection of a radio source with milli-arcsecond dimensions would be a test of the presence of a black hole at the Galactic Center. Firm confirmation came only decades later with the Nobel Prize winning observations of the orbits of stars orbiting Sgr A* within a few hundred Astronomical Units from the four million solar mass black hole at the Galactic Center.[71] When the 2020 Nobel Prize was announced, Brown's granddaughter, Piper, noted "Gramps gets the assist!"[72] More recently, the international Event Horizon Telescope Collaboration published a VLBI image of the shadow of the black hole with an effective angular resolution of only 20 microarcsec.[73]

Balick and Brown found no evidence for small scale structure in any of the H II regions that they observed, which was the goal of their original proposal. Later, they learned that Dennis Downes and Miller Goss, both working in Europe, had proposed to use the NRAO GBI specifically to search for evidence of a black hole in the Galactic Center. Although offered observing time before the Balick and Brown observations, they had to decline due to their other responsibilities.[74]

Understanding Quasars and AGN

As shown in Figure 4.12, for nearly all the radio sources in the 3C Catalogue the flux density decreases with increasing frequency, but there were a few exceptions that went unrecognized until a Caltech graduate student accidently uncovered a whole new population of radio sources. Daniel (Dan) Harris was the first graduate student in radio astronomy at John Bolton's new Owens Valley Radio Observatory (OVRO) and was assigned by Bolton to use the new OVRO 90 foot steerable antenna to more accurately locate the position of 3C radio sources. Working at what was then considered a high frequency of 960 MHz ($\lambda = 30$ cm wavelength), Gordon Stanley built a very sensitive receiver. Because there were issues with receiver stability when the antenna was moved to track a radio source, Bolton instructed his students to set the antenna at the declination of the source about 12 minutes ahead in right ascension and then stop the antenna to let the radio source drift through the antenna beam as the Earth rotated. The telescope output was recorded on a moving chart pen recorder. Like graduate students everywhere, Harris had to juggle the demands of studying for his exams, building and testing new instrumentation, observing, and when possible eating, drinking, and grabbing some sleep. During several nighttime observing sessions, Harris set the telescope

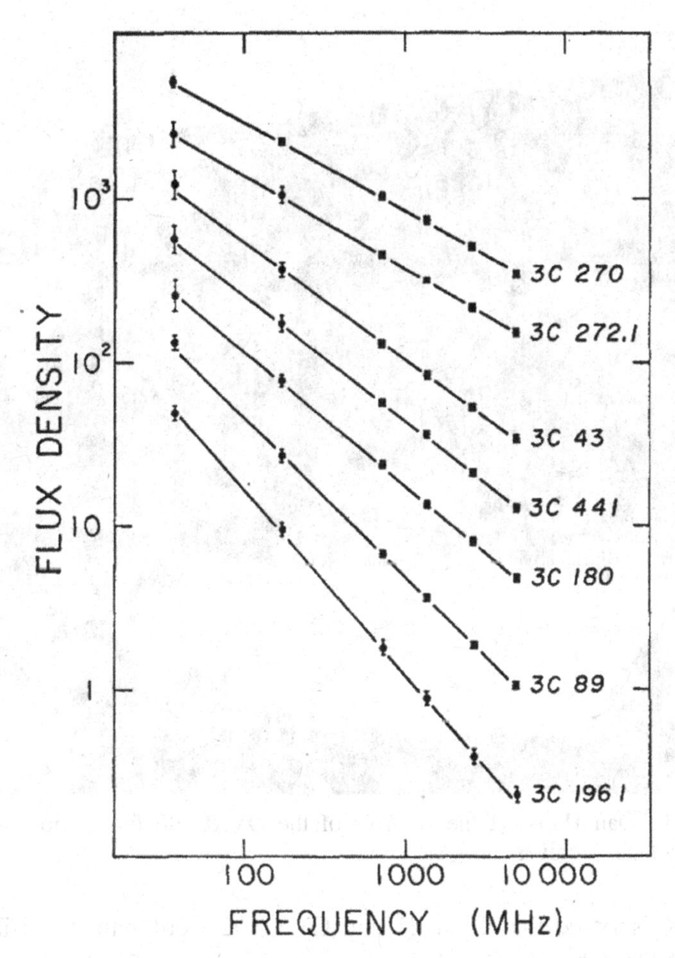

Figure 4.12 Typical extragalactic radio source spectra showing flux density versus frequency. Displaced zeros. Credit: Kellermann, Pauliny-Toth, and Williams (1969).

to observe a 3C source, but tired, and apparently dulled by a few beers, he fell asleep at the controls. When he woke up, his 3C sources were visible on the chart record, but other sources also appeared on the long chart record that had not been previously seen in any previous radio source survey. This was effectively the first radio source survey made below 75 cm wavelength (Figure 4.13).

In their published catalog describing the results of their study, Harris and Roberts listed not only the 3C or MSH names[75] of their sources, but also a running number from 1 to 106, and suggested that the not-previously cataloged sources be referred to as CTA followed by a running serial number, where CT stood for Caltech and the letter A for the first list of radio sources published by the Observatory.[76] Subsequent inspection of the original Cambridge 178 MHz 4C survey data by one of the present

Figure 4.13 Dan Harris at the controls of the OVRO 90 foot radio telescope. Credit: Caltech Archives.

authors (KIK) showed that CTA 21 and CTA 102 were actually visible on the Cambridge 178 MHz records, although they were below the 4C Catalogue flux density limit. Further observations at OVRO at other wavelengths showed that the flux density of both CTA 21 and CTA 102 peaked in the general region of 1 GHz and dropped off at both higher and lower frequencies.[77] CTA 21 and CTA 102 in particular were the prominent prototypes of a new class of radio sources that later became known as Gigahertz Peaked Spectrum (GPS) sources (Figure 4.14).

At the end of 1960, John Bolton returned to Australia to take charge of the completion and operation of the new 210 foot Australian radio telescope at Parkes. The discovery of the unusual spectra of CTA 21 and CTA 102 alerted Bolton to be on the lookout for other radio sources with similar spectra. One of the first programs that Bolton initiated with the Parkes telescope was a decimeter wavelength survey of the relatively unexplored southern sky. Each cataloged source was labelled by a number describing its approximate position in the sky using the hours and minutes of right ascension followed by the degrees of declination, a system later adopted by the IAU for all radio sources. In order to determine their radio spectra, the Parkes survey was

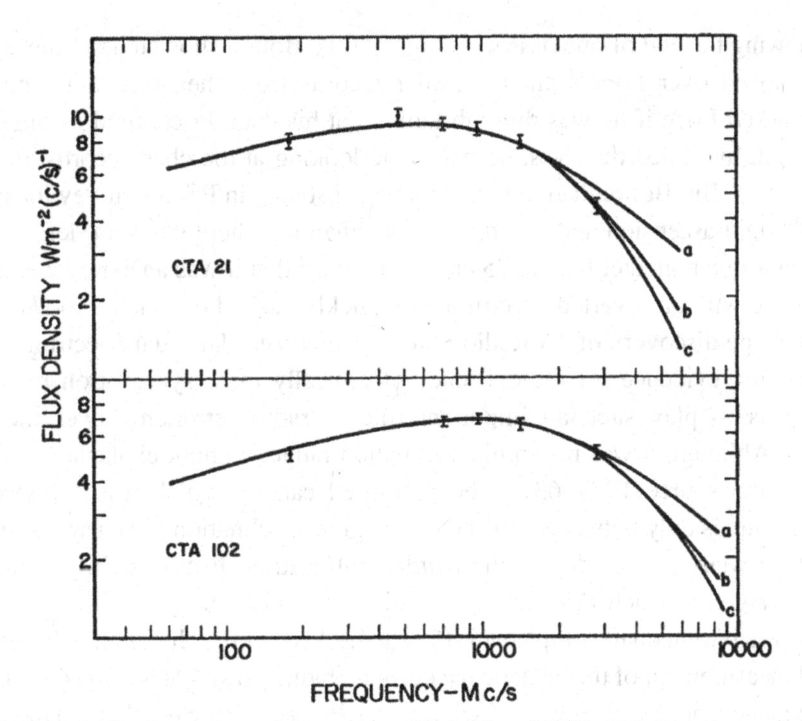

Figure 4.14 Radio spectra of the quasars CTA 21 and CTA 102. The three curves represent three theoretical models that fit the data. Adapted from Kellermann (1964).

conducted with simultaneous observations at both 408 MHz (75 cm) and 1,410 MHz (21 cm) and was followed up with 2,650 MHz (11 cm) observations for all sources that were detected at the lower frequencies. In order to quickly spot sources that deviated from the traditional power law, Bolton set the gain on the two-pen chart recorder to show equal deflection for a "normal" power law source.

The Parkes survey was divided into separate declination zones. Bolton himself, assisted by staff members Frank Gardner and Brian Mackey, took the zone from declination –20 degrees to –60 degrees, and he assigned the zone from –60 degrees to –90 degrees to the young American graduate student, Marcus (Marc) Price who was assisted by Doug Milne. Price had done his undergraduate work at Colorado State University supported by an ROTC scholarship, and was able to defer his military service obligation in order to pursue graduate work at the Australian National University (ANU) in Canberra under a Fulbright Fellowship. Bolton insisted that all observations be made at night in order to avoid interference from the Sun and to allow the daytime to be used for telescope and instrument maintenance and upgrades. Price, being only a graduate student and low in the pecking order, was assigned the graveyard shift from midnight to dawn.

Following the end of one of Price's night shifts, Bolton showed up in the control room, looked over Price's chart recorder records from the preceding night and calmly asked Price if he was through looking at his data. Price, tired from his all-night vigil, mumbled that, yes, he was done looking at the chart records and was going to bed. But Bolton had noted one unusual source in Price's survey located at $19^h\ 34^m$ right ascension and –63 degrees declination, where the chart deflection at 21 cm was much greater than at 75 cm. He realized that it was an extreme example of a source with an inverted spectrum and quickly dashed off a letter to *Nature* to announce the discovery of "A Radio Source with a Very Unusual Spectrum." This was the first evidence for the existence of optically thick synchrotron radiation, which was to play such an important role in radio astronomy over the next decades. Although it was not in his declination range, without explanation Bolton included the source 1934–63 in the published catalog which was supposed to contain sources only between –20 and –60 degrees declination.[78] When asked later why Price was not included in the *Nature* publication, Bolton simply explained that he wanted to teach Price a lesson – to look at his data.

Marc Price went on to complete his PhD at the ANU with a dissertation based on a careful measurement of the galactic background radiation at 408 MHz (74 cm).[79] At that frequency the sky's brightness is dominated by the galactic nonthermal radiation. Had he chosen a higher frequency for his project, he might well have discovered the 2.7 degree cosmic microwave background later reported by Arno Penzias and Robert Wilson at Bell Laboratories.[80] After receiving his PhD in 1966, Price put his missed discoveries behind him and went on to a distinguished career. After satisfying his military obligation with the US Signal Corp, he joined MIT as an Assistant Professor and was later a Visiting Scientist at the Max-Planck-Institut für Radio Astronomie in Bonn, Germany and at NRAO. He then joined the National Science Foundation as the Radio Frequency Spectrum Manager and rose to become the Head of the Astronomy Section and Acting Director of the Division of Astronomy. Price than returned to Australia for four years as the Officer in Charge of the Parkes Observatory. He closed out his career as Head of the Department of Physics and Astronomy at the University of New Mexico.

Most of the radio sources in the *Revised 3C Catalogue* have the traditional power law spectrum where the flux density monotonically decreases with increasing frequency. In many cases the spectrum becomes steeper at high frequencies consistent with expected electron energy losses due to radiation.[81] A few 3C sources, however, markedly deviated from this simple spectral picture, as do the spectra of CTA 21, CTA 102, and Parkes 1934–63. Instead, their flux density changes only slightly with frequency. On a log–log plot of flux density versus frequency their radio spectra are flat or showed small multiple increases and decreases in flux density. The two radio source populations are referred to as "flat spectrum" and "steep spectrum" sources, respectively. Radio source surveys made at shorter wavelengths in Australia at 11 cm wavelength[82] and at NRAO at 6 cm[83] discovered large numbers of these flat, inverted, or peaked spectrum radio sources. At 6 cm, radio surveys contained flat spectrum sources in about equal numbers as the traditional steep spectrum sources first found in the low frequency surveys. It was these same few flat spectrum 3C sources that appeared unresolved in the longest 61,000 wavelength spacing of Jodrell Bank long baseline interferometer observations.[84] These flat spectrum sources all turned out to have very small angular dimensions with corresponding high surface brightness.[85] The reason their spectra are flat is because they are so small that the low frequency emission is absorbed by the relativistic electrons in the path of the radiation, or what optical astronomers call "optically thick," and the low frequency emission is suppressed. In the so-called flat spectrum radio sources, different parts of the source become optically thick at different frequencies, and the superposition of these different spectral components leads to the appearance of "flat spectra."

The flat, inverted, and peaked spectrum radio sources are identified primarily, but not entirely, with quasars rather than radio galaxies. Radio galaxies primarily have power law or quasi power law spectra, although in a few cases, the spectrum is dominated by a compact AGN and the spectra are flat (Figure 4.14).

Self-absorption of synchrotron radiation becomes important only when the brightness temperature of the source is comparable to the equivalent electron energy or about 10^{11} K. Prompted by the unusual spectrum of CTA 21 and CTA 102, Vyacheslav Slysh, a former student of Shklovsky, showed that the angular size is related to the peak frequency of the source, depends only slightly on the magnetic field strength, and can be written as

$$\theta = 0.015 B^{1/4} S^{1/2} \nu_{\mathrm{m}}^{-5/4} \text{ arcseconds},$$

where B is the magnetic field strength in Gauss, S the flux density in Janskys, and ν_m the frequency in GHz where the flux density is maximum.[86] So, for a magnetic field of 10^{-4} G and a peak flux density of 1 Jy at 1 GHz, the angular size is only about one thousand of a second of arc.

In the USSR, Shklovsky was quick to realize that, due to expansion, the flux density and peak frequency of such small radio sources such as CTA 21 and CTA 102 or Parkes 1934–63 might change on timescales of a year or less.[87] At the time, there were no radio telescopes in the Soviet Union capable of testing Shklovsky's prediction. As a result of his role in the Soviet space program, Shklovsky was aware of a classified military facility located near Yevpatoria in Crimea that was used to track Soviet satellites (Figure 4.15). He was able to obtain permission to send his student, Gennady Sholomitsky, to Crimea to use the tracking antenna to observe CTA 21 and CTA 102 as well as other sources.

Sholomitsky observed at 921 MHz (32.5 cm) between August 1964 and February 1965 and, as shown in Figure 4.16, he found a remarkable variation in the flux density of CTA 102.[88] Although this was the first reported variation in any extragalactic radio source, for several reasons Sholomitsky's exciting result was received by Western scientists with great skepticism:

(1) The published papers were exceedingly brief and gave no information on the radio telescope or instrumentation used to obtain this dramatic result.
(2) Observations of CTA 102 at Caltech and Arecibo showed no evidence of variability.[89]
(3) The pronounced variability meant that the size of the radio source could not be more than a few light-months across. Using Slysh's formulation of the relationship between angular size and peak frequency, Sholomitsky calculated that the angular size should be about 0.01 arcsecs. The linear size of not more than a few tenths of a light year deduced from the variability time scale suggested that this meant that in order for such an intrinsically small source to appear so large, CTA 102 had to be located within the Galaxy. By the time Sholomitsky's paper was published, CTA 102 had been identified with a quasar at a redshift of 1.04, in apparent contradiction to the distance derived by Sholomitsky from the timescale of the flux density variations. The large redshift of CTA 102 also created an additional problem since at such a great distance, the radio luminosity and surface brightness would be so great that the relativistic electrons would lose energy not by synchrotron radiation but by inverse Compton radiation caused by interaction with the dense radiation field, and the radiation would appear as X-rays and not as radio waves.[90] This meant that the relativistic electrons would then encounter the even stronger X-ray radiation field, and the source would self-destruct by what is called "the inverse Compton catastrophe." The apparent paradox could only be understood if CTA 102 was closer than indicated by the measured redshift or if the observed variable radiation was not due to conventional synchrotron radiation.
(4) Shklovsky and Nikolai Kardashev understood the theoretical implications and, noting that the oscillation period was close to 102 days, whimsically suggested to a TASS reporter that CTA 102 might be a signal from an extraterrestrial civilization. The subsequent report of alien signals in the 14 April 1965 edition of *Pravda* received worldwide attention and further detracted from the credibility of the claimed variability.

It was later understood that the lack of any technical details in Sholomitsky's published reports was due to restrictions imposed by the Soviet military in return for permission to use their antenna for radio astronomy. As shown in Figure 4.15, the antenna used by Sholomitsky consisted of an array of eight 16 m diameter

Figure 4.15 Soviet antenna system used to discover radio variability in CTA 102. Credit: Courtesy of Leonid Gurvits.

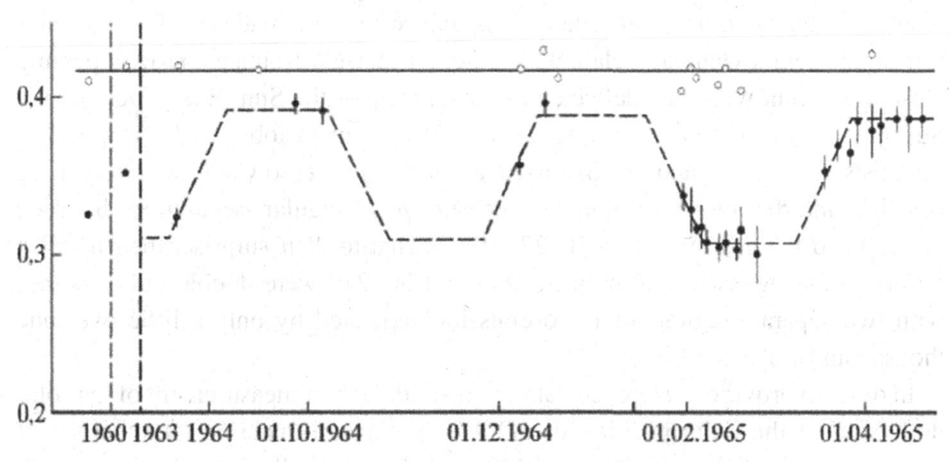

Figure 4.16 Radio variability of CTA 102 (closed circles) and CTA 21 (open circles) as shown in *Pravda*. Credit: Courtesy of Sholomitsky Family.

antennas mounted on part of an abandoned military railway bridge. The elevation axis was built around two Italian submarine hulls taken as WWII reparations in 1945 and the azimuth bearing came from a discarded naval artillery weapon.[91] The Caltech observations of CTA 102 were apparently made at different times when there were no observed changes in flux density. Later observations showed that the flux density of CTA 102 does change on the time scales reported by Sholomitsky.

Around the same time, an America student, William (Bill) Dent, at the University of Michigan, discovered that the high frequency (short wavelength) flux density of several flat spectrum quasars, including 3C 273, as well as the compact core of the radio galaxy NGC 1275 (3C 84) varied on time scales of only a few months.[92] Dent's observations covered a period of nearly three years dating back to mid-1962 but were not published until after Sholomitsky's papers. Unlike Sholomitsky's work, there was now no doubt that the flux density of quasars and AGN do change with time. Subsequent observations at many observatories around the world suggested that enormous energies, apparently up to 10^{60} ergs, can be released on surprisingly short time scales and that the acceleration of electrons to ultra-relativistic energies takes place in a few years and not, as previously thought, over billions of years.[93] Unless the quasars were not as far away as indicted by their redshift, the theoretical problems first raised by Sholomitsky's observations of CTA 102 became a serious challenge.

Faster than the Speed of Light

The answer to the puzzle of rapid radio source variability came in an unexpected way. As we discuss in Chapter 11, the development of very long baseline interferometry, with its unprecedented angular resolution, made possible a precise test of Einstein's General Relativity prediction that electromagnetic rays passing close to the Sun would be deflected by the gravity of the Sun. Every October the Sun passes very close to the quasar 3C 279 and, in October 1970, a group of scientists from MIT and NASA used a Massachusetts-to-California very long baseline interferometer system to measure the angular separation between 3C 279 and the nearby quasar 3C 273. Somewhat to their surprise, their detailed observations suggested that both 3C 273 and 3C 279 were double radio sources with two separate regions of radio emission separated by only a little over one thousandth of a second of arc.[94]

In order to provide a reference data set to validate their measurement of the solar deflection of the radio position of 3C 279, the group made an identical set of observations in February 1971 when 3C 279 was well away from the Sun. Meanwhile, another group of scientists from Caltech and NRAO, who were studying the small scale radio structure of quasars, were also scheduled to observe in February using the same pair of antennas – the 120 foot Haystack antenna in Massachusetts and the JPL Goldstone 210 foot antenna. With its large collecting area and low noise maser receiver, the JPL 210 foot Goldstone antenna was probably the most sensitive radio telescope in the world. As part of the JPL Deep Space Network (DSN), the Goldstone antenna was in high demand to track interplanetary space craft, and the two radio astronomy groups were both fortunate to get approval for the February time to observe quasars.

The MIT–NASA group shared their preliminary results with the Caltech–NRAO group. Knowing that both 3C 273 and 3C 279 were rapidly varying radio sources, the Caltech–NRAO group planned to use part of their valuable observing time to see if there had been any changes in the structure of the two sources during the four-month period between the two observing sessions. Caltech Professor Marshall Cohen assigned graduate student David Shaffer the responsibility of analyzing the data. When Shaffer plotted the data from the two epochs together, he realized that the components of 3C 273 and 3C 279 had separated during the four-month period by 0.00035 and 0.00024 arcseconds, respectively. Using the known redshift of the two quasars, he converted the angular velocities to linear velocities and was surprised to see that the components of both sources appeared to be separating faster than the speed of light.[95] The MIT–NASA group reached the same conclusion from their data[96] and, a few months later, both groups reported their discovery of what is now called superluminal motion at a conference held at the American Academy of Arts and Sciences in Cambridge, Massachusetts.[97]

Although seemingly violating the well-known restriction of special relativity that nothing can travel faster than the speed of light, the apparent superluminal motion found in 3C 273 and 3C 279 has a simple explanation. If one of the components is moving with nearly the speed of light along a trajectory close to the line of sight, it nearly catches up with its own radiation; time scales seen by a distant observer thus appear to be compressed, and so the apparent velocity, calculated from dividing the distance moved by the elapse time, can be arbitrarily high. Moreover, due to the effects of relativistic beaming, the radio emission is concentrated in a narrow cone so the apparent luminosity is enhanced compared to what would be observed if the radiation is isotropically distributed. The combined effects of relativistic beaming and time compression explain the apparent rapid time variability in quasars and avoids the effects of the Compton catastrophe that would occur if the radiating source is at rest and the radio emission isotropic.[98] Quasars that contain relativistically beamed ejections are known as "blazars."

As it turned out, a year before the discovery of superluminal motion, Martin Rees, a student at Cambridge University, had pointed out that the inverse Compton catastrophe could be avoided if the source were expanding at nearly the speed of light.[99] Rees's expanding sphere model was similar to the one we have just discussed for 3C 273 and 3C 279, but more complex than the simple two component picture found from observation. Even earlier, the young Russian scientist Leonid Ozernoy, working together with V. Sazonov, had considered a model describing a radio source "with components flying apart at relativistic velocities" that corresponded closely to what radio astronomers observed in 3C 273 and 3C 279.[100] Ironically, both the Caltech–NRAO and MIT–NASA groups overinterpreted their data. Although the simplest interpretation of the limited one baseline interferometer data was, indeed, a simple separation of two components, more complex models could also fit the limited data, including stationary models with variable component flux density. As more detailed imaging observations have shown, 3C 273 and 3C 279 (Figure 4.17) are indeed more complex. The interpretation of the original single baseline interferometer data was naïve and unjustified, but the new observations with more sophisticated multi-element interferometer systems have verified the widespread occurrence of superluminal expansion in blazars.[101]

Martin Rees went on to a distinguished career in astrophysics. In 1995 he became the UK Astronomer Royal and later President of the prestigious Royal Society of London. In 2005, he was elevated to the House of Lords as Baron Rees of Ludlow. Leonid Ozernoy faced a less pleasant future. After being denied permission to emigrate to the United States, he became a vocal activist and dissenter of the Soviet regime, voicing support for Andrei Sakarov, and lost his position at Moscow University. Following several hunger strikes, he received support from the International Astronomical Union and US Senator Edward

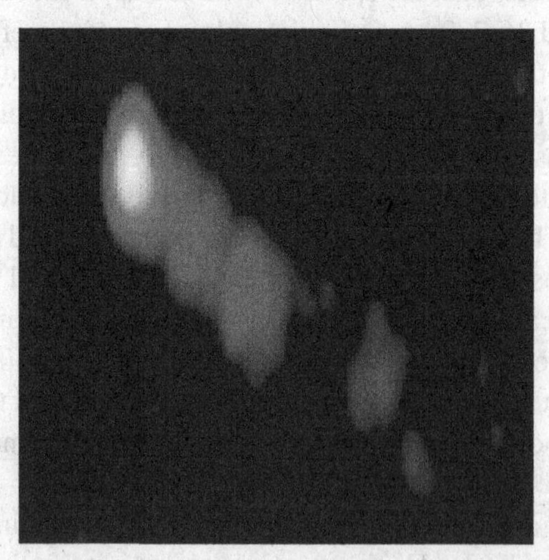

Figure 4.17 Very Long Baseline Array 2 cm image of the quasar 3C 279. Image is 0.015 arcseconds on a side. Credit: MOJAVE image, Lister et al. (2018).

Kennedy and was finally allowed to leave for the United States in 1986 but died in 2002 at the age of 62.

Now, more than half a century after their discovery, blazars, quasars, and AGN have become part of mainstream astronomy, with numerous AGN and quasar conferences held each year, along with publication of many books and thousands of papers. As currently understood, quasars are the ultra-luminous nuclei of galaxies, building on the ideas introduced much earlier by Jeans, Seyfert, Ambartsumian, and others. Based on earlier arguments by the Russian scientist Vitaly Ginzburg, Fred Hoyle and William Fowler were quick to suggest that gravitational collapse with masses up to 10^9 times the mass of the Sun could supply the energy needed to power radio galaxies and quasars.[102] Following Donald Lynden-Bell's classic paper suggesting that quasars are powered by supermassive black holes located at the center of galaxies, quasars, AGN, and black holes now play a major role in discussions of galaxy formation and evolution.[103] "For their seminal contributions to understanding the nature of quasars," Maarten Schmidt and Donald Lynden-Bell shared the 2008 Kavli Prize in Astrophysics.

5

Radio Astronomy, Cosmology, and Cosmic Evolution

> There seems to be no way in which the observations can be explained in terms of a Steady-State theory.[1]

During the middle of the twentieth century, there were two different approaches to cosmology. The widely accepted picture discussed by most astronomers postulated that the Universe began at some time in the distant past through an explosive event, now thought to have occurred about 14 billion years ago, and that it continues to expand, as evidenced by the observed redshift of galaxies, with the most distant galaxies moving at the highest speed.[2] As described by Allan Sandage in a 1970 article in *Physics Today*,[3] cosmology of the 1950s was the search for two numbers: the local rate of expansion per Megaparsec of distance, known as the Hubble constant H_0, and the rate of deceleration due to the gravitational attraction from all the matter in the Universe. Depending on the total amount of matter in the Universe, the rate of slowing down, or the deceleration constant, q_0, was thought to be in the range from 0 to 1. With $q_0 = 0.5$, the so-called closed Universe would continue to slow down until it stopped, and then gravity would cause the Universe to begin to contract. In an open Universe, with $q_0 = 0$, the Universe would continue to expand forever. This "big-bang" Universe, as it was mockingly referred to by Fred Hoyle, predicted that galaxies in the most distant parts of the Universe were older, and so the high-redshift Universe should look different from the local Universe. It was attractive in that, at least in broad terms, it was consistent with the biblical description of creation, although the time scales differed by a factor of a million. But the big-bang cosmology left unanswered the physics of "creation" and what happened before the big-bang.[4]

However, the philosophically attractive and competing "steady-state" theory promoted by Hermann Bondi, Thomas Gold, and Fred Hoyle was not widely accepted by other astronomers[5] (Figure 5.1). In the steady-state cosmology, the Universe is, and always was, expanding and so everywhere looks the same. In

Figure 5.1 Left to right: Thomas Gold, Herman Bondi, and Fred Hoyle in the front row of a 1960 conference. Credit: NRAO/AUI Archives, Papers of R. Bracewell.

order to maintain the requirement of an unchanging but expanding Universe, the steady-state theory required the controversial continuous creation of new matter by an unknown and unspecified process, but, argued its proponents, a process no more mysterious than the one that created the big-bang. The steady-state model explained "the expansion of the Universe in physical terms instead of assuming it *ad hoc*, as was done in other theories." To keep a constant density of matter with time, the steady-state theory required that the rate of expansion of the Universe increase with time or accelerate rather than decelerate, with a value of $q_0 = -1$.[6]

The distinction between these various cosmological models is only apparent at high redshifts ($z \sim 0.5$) corresponding to large distances and early lookback times in the age of the Universe. But at these large distances even the largest optical telescopes were unable to accurately detect, let alone determine, the brightness or size of galaxies, or indeed any other parameter that might give clues to the relevant cosmological model. As discussed in Chapter 4, with identification of Cygnus A and other radio galaxies, it became clear that radio astronomers were able to study the Universe at far greater distances (early epochs) than optical astronomers. However, without knowing the redshifts and therefore the distances, it would turn out to be more difficult in practice to fully exploit the power of radio telescopes. There was a long and bitter debate over whether the spatial distribution of radio galaxies indicated an evolving big-bang universe or if the data were consistent with the controversial steady-state cosmology that required the Universe to look the same everywhere. The competition between the Cambridge and Sydney radio astronomy groups and the controversy that developed around their claims became a black mark on radio astronomy. As discussed in Chapter 6, all that changed in 1965, with the serendipitous discovery of the cosmic microwave background, which was arguably the most important astronomical discovery of the mid-twentieth century.

The Cambridge–Sydney Source Count Controversy

Early Australian and Cambridge radio source surveys showed that there were two classes of radio sources: Class I, that seemed to be distributed along the Galactic Plane and was presumably a population of sources located within the Milky Way Galaxy, and Class II, that was randomly distributed around the sky.[7] The Class II sources could be very local objects located within the thin galactic disk or very distant galaxies outside the Milky Way.

While some of the radio sources near the Galactic Plane were recognized as supernova remnants or ionized gas regions, it was unclear whether the second class consisted of relatively nearby stars or were extragalactic. Since there appeared to be no statistical association of these radio sources with the 1,249 galaxies in the Shapley Ames catalog of galaxies, and because observations of several nearby galaxies showed, at best, only weak radio emission,[8] it was argued, especially by Martin Ryle, that these sources were nearby radio stars of an unusual type and were distributed over the same region of space as the visible stars.

As the radio position measurements improved, a number of the stronger Class II sources were identified with moderately distant galaxies, and some were shown to have non-stellar sizes. Nevertheless, until the mid-1950s, Cambridge radio astronomers continued to call them "radio stars." When it had become clear, even to Ryle, that most of the discrete radio sources away from the plane of the Milky Way were actually distant powerful radio sources, they were then called "radio galaxies." Radio stars were dead, and Ryle turned his attention to using radio galaxies to explore the distant early Universe.[9] As he later explained, the identification of Cygnus A, and later 3C 295, "indicated the possibility that observations of similar sources at greater distance might prove a powerful method of exploring the Universe on the largest scale" and that "this hope has not proved unfounded."[10] Because the radio sky is cold (dark in optical terms) and radio galaxies and quasars are very bright, unlike at optical wavelengths where faint distant galaxies blend into the night sky, radio astronomers are only limited by the sensitivity and resolution of their instruments. Also, at radio wavelengths galactic absorption is negligible, and corrections for uncertain and varying extinction (dimming) due to intervening matter, necessary at visible wavelengths, is avoided. Cosmology was no longer an abstract mathematical exercise. Radio astronomers thought that, by building the right instruments, they could investigate the geometry of the Universe.[11]

Although observations of distant radio galaxies corresponded to the early Universe where one would expect to see differences between the steady-state and big-bang universes, there was still a problem. Radio observations alone give no indication of distance. Any individual radio source might be a nearby galactic supernova remnant or other galactic source, a relatively close low luminosity radio galaxy, or a much more powerful distant radio galaxy. Without optical identifications of radio sources and measurement of their redshift (distance), radio astronomers were unable to determine the nature of their radio sources. Many of the brighter radio sources were identified with galaxies, but most of the fainter, and hence more distant, radio sources had no visible counterparts. Even the most powerful optical telescopes were unable to study the faintest, and presumably more distant galaxies, where differences between a steady-state and big-bang Universe were expected. The optical astronomers couldn't see far enough to distinguish between the two world models, and the radio astronomers didn't know how far away their radio sources were.

Soon after they started to catalog significant numbers of radio sources, radio astronomers in both Cambridge and Sydney tried to build on a simple technique developed earlier by Edwin Hubble to study the spatial distribution of galaxies.[12] In a static Universe with conventional Euclidean geometry, the measured flux density is expected to fall off as the inverse square of the distance, while the number of sources of a given luminosity should increase as the volume or cube of the distance. The number, N, of detected radio sources above a given flux density, S, then becomes proportional to $S^{-3/2}$. Or equivalently,

$$\log N = -1.5 \ \log S + \text{constant}, \tag{5.1}$$

which became known as the $\log N - \log S$ relation. The slope of the number of detected sources when plotted against flux density on a logarithmic scale was therefore expected to be −1.5. Apparently, it was not fully understood by radio astronomers that, in the real Universe, a more complex four-dimensional non-Euclidean geometry was appropriate. Also, as discussed decades earlier by Hubble,[13] due to the redshift the flux density of a receding source falls off faster than the square of the distance. Nevertheless, Equation 5.1 became a convenient standard for comparison with the radio source counts that became available starting in the 1950s, although it actually held no meaning in the real world.

Probably the first survey to show a deviation from a uniform distribution was the 100 MHz nearly all sky sea interferometer survey by John Bolton and colleagues that contained 104 discrete sources[14] (Figure 5.2). The 83 sources that

Figure 5.2 Twelve element Yagi array at Dover Heights used by John Bolton to catalog 104 radio sources. Credit: CRAIA.

were located away from the Galactic Plane appeared to be randomly distributed in the sky and had a source count that rose more steeply than $S^{-1.5}$ for the weaker sources. While the excess of weak sources found by Bolton et al. was probably the result of instrument noise and confusion and was not statistically significant, the authors dismissed their evidence for non-uniformity by incorrectly claiming that "it could also be produced by a large dispersion in absolute magnitude among the sources of the survey."[15]

The first really large radio source survey where this test could be applied was the 81 MHz Cambridge 2C survey which was the PhD project of John Shakeshaft. The published source list contained 1,906 small diameter sources that were isotropically distributed in the sky.[16] From inspection of Palomar 48 inch survey plates, Rudolph Minkowski could not find more than a few possible optical counterparts. Analysis of the 2C source count, that had a log N – log S slope near –3 for all but the strongest sources, as well an independent statistical analysis derived by Ryle's student Peter

Scheuer,[17] appeared to indicate an excess of faint, presumably distant, radio sources compared to a universe that was filled with a homogeneously distributed population of sources. Undaunted by not knowing the nature of most of the radio sources in the 2C Catalogue, at his 1955 Halley Lecture at the Royal Astronomical Society in Oxford, Martin Ryle made the startling announcement that the 2C survey showed that "there seems to be no way in which the observations can be explained in terms of a Steady-State theory"[18] (Figure 5.3). Subsequently, Ryle and Scheuer presented their detailed analysis and reiterated that "Attempts to explain the observations in terms of a steady-state theory have little hope of success."[19]

However, Ryle faced immediate challenges from two fronts. From the theoretical side, his Cambridge colleagues Fred Hoyle and Thomas Gold, as well as his former Cambridge colleague Hermann Bondi, questioned Ryle's interpretation of the data. The Cambridge radio astronomers considered that theoreticians were not qualified to evaluate the complex analysis of radio source surveys and dismissed their arguments as uninformed.[20] Since Bondi, Gold, and Hoyle had all played major roles in wartime radar research, the Cambridge radio astronomer's attitude was somewhat disingenuous. A more serious challenge came from the Australian radio astronomers who questioned the reliability of the Cambridge data. Contemporaneous observations from Bernard Mills and Bruce Slee in Sydney disagreed with Ryle's dramatic conclusions that apparently doomed the steady-state universe. The Sydney observations were made at nearly the same frequency as the 2C survey, but used the new Mills Cross array,[21] and had a log N – log S slope of only –1.7, that they argued, in view of experimental errors, did not differ significantly from –1.5.[22] Moreover, a one-to-one comparison of the overlapping survey regions "was almost completely discordant." Mills and Slee concluded that the "discrepancies . . . reflect errors in the Cambridge Catalogue, and accordingly deductions of cosmological interest derived from its analysis are without foundation."[23]

The two sides faced each other at a 1955 international conference on radio astronomy held at the Jodrell Bank Observatory. Based on the steep 2C source count, Peter Scheuer's statistical analysis, the observed isotropy of the count, the dearth of optical identifications, and the claimed lack of significant instrumental effects on these results, Ryle forcefully argued that "distant regions differ from those in the neighborhood of the Galaxy; such a result is incompatible with the predictions of the steady-state theories of cosmology proposed by Bondi and Gold and Hoyle but might be interpreted in terms of evolutionary theories."[24] The Australian radio astronomer Joe Pawsey challenged Ryle's conclusions, noting that the Sydney source count, based on about 1,000 sources, was "substantially free of confusion" and did not show any significant departure from a –1.5 dependence.[25] Ryle defended the Cambridge results by pointing out that the steep slope does not occur near the confusion limit. He was challenged by two theoreticians, Thomas

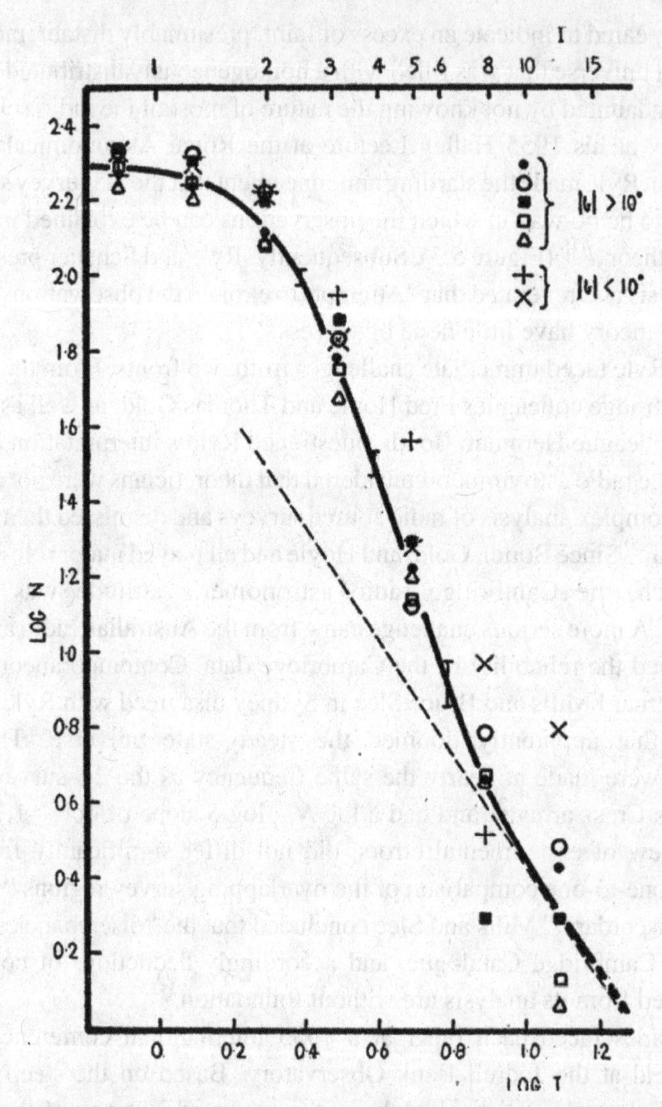

Figure 5.3 Radio source count presented by Martin Ryle at his 1955 Halley Lecture. The dashed line indicates the nominal 3/2 power law for a static Euclidean Universe. Credit: Ryle (1955).

Gold and William McCrea, who were resolved to not let data get in the way of their elegant theory, and maintained that the combined effects of noise and confusion, as well clustering and the resolution of extended sources, were too great to draw any conclusions about the validity of the steady-state cosmology. Similarly, Ryle would not accept the criticism of theoreticians and the debate turned vicious and personal.[26]

To bring his claims to a broader audience, Ryle wrote an article in *Scientific American*, somewhat dogmatically explaining not only how the Cambridge radio source counts excluded the steady-state theory, but also dismissing the claims by Mills that his Mills Cross observations did not require an evolutionary cosmology.[27] Mills was offended that Ryle had ignored his criticisms, and responded with a strong letter to *Scientific American* pointing out the poor agreement between the Sydney and Cambridge source positions and the discrimination against extended sources by the Cambridge interferometer, stating that "the Australian observations are quite consistent with a steady state universe and, if correct, the Ryle cosmology is invalid."[28] Not to be out-maneuvered by Mills, Ryle retorted in an even longer letter that Mills did not understand the complex Cambridge interferometer system, that it was the source intensities, not positions, that were relevant, and that, in any event, Scheuer's statistical analysis gave an independent confirmation of the excess of weak sources.[29]

Mills later argued that they did understand Scheuer's method, but believed that the steep source count claimed by Ryle was the result of clustering and the resolution of the strong extended sources by the Cambridge interferometer.[30] Unfortunately, the publication of Scheuer's paper was delayed by his three-year stint in military service, and ultimately appeared in a prestigious mathematics journal unfamiliar to radio astronomers, which may have contributed to the lack of acceptance of his statistical analysis.[31] Criticism of the Cambridge work was not confined to Australia. Based on a comparison of their own interferometer and total power observations, Jodrell Bank radio astronomers concluded that "the Cambridge 2C survey is severely resolution limited and that the apparent increase in the density of sources with distance deduced from this survey must therefore be considered extremely doubtful," and "the conclusions drawn from the [Sydney] survey would therefore appear to be more reliable than those drawn from the Cambridge 2C survey."[32]

Faced with the increasing criticism from Sydney, as well as from their colleagues at Jodrell Bank, the Cambridge group doubled their frequency to 159 MHz (1.9 m) to make a new more reliable 3C survey with twice the angular resolution of the 2C survey.[33] The 3C survey only contained 242 sources away from the Galactic Plane indicating that three-fourths of the Cambridge 2C sources were not real, but were blends of weaker sources.[34] The sources in the 3C Catalogue had a $\log N - \log S$ slope of only –2.0, much less than claimed for the 2C survey, but still greater than –1.5.[35] The 3C survey demonstrated, very clearly, that the 2C survey was useless for cosmology, thus confirming what the Australians had been arguing, even though the observed slope was steeper than –1.5. Nevertheless, based on the 3C survey, in his 1958 Bakerian Lecture, Ryle maintained that the radio source count was inconsistent with steady-state cosmology although, interestingly, he continued to refer to the discrete sources as "radio stars."[36] However, by this time, the Sydney group had completed their survey, that contained 1,658 sources in the southern sky and reported a log N – log S slope of –1.8 ± 0.1.[37] After correcting for the effects of noise and confusion, Mills continued to argue that the Sydney log N – log S relation did not differ significantly from –1.5 and concluded that "the source counts indicate no divergence from uniformity and no obvious cosmological effects."[38] Ryle went on to build a new more powerful instrument operating at 178 Mc/s (1.7 m), that his student, Andrew Bennett, used to produce the Revised 3C Catalogue (3CR) whose log N – log S slope was now reduced to –1.9.[39]

The two sides squared off again at an IAU Symposium on radio astronomy that was held in Paris in the summer of 1958 (Figure 5.4). Mills reviewed the results of their survey, which had an observed slope of –1.8.[40] After correction for the effects of noise and confusion, he reported a best fit slope of –1.65, which, he argued, in view of the uncertainties, did not differ significantly from –1.5. Moreover, he claimed, "the apparent small excess of faint sources could equally well be a statistical deficiency of close and strong sources." Aware that the source count slope was now converging toward –1.5, Ryle now appealed to the effect of the redshift and noted that, for the weaker sources "in the Australian survey, the steady-state model would indicate a deficit of 3.4:1, whereas again a slight excess was found," and that "the discrepancies between observation and the predications of the steady-state model are considerably greater than could be established hitherto."[41] Ryle went on to promise "more extensive observations" with "a new radio telescope [having] greater resolution and sensitivity than any previous instrument."

Ryle's new instrument was the first true synthesis radio telescope, and produced the 4C survey containing about 5,000 radio sources above a limiting flux density nearly five times fainter than the weakest sources in the 3CR Catalogue.[42] The 4C survey, as well as a 408 MHz survey made with Bolton's new Parkes 210 foot radio telescope in Australia,[43] each indicated a slope for the stronger sources of –1.8 ± 0.1 consistent with the value reported by Mills, but still different from –1.5 (Figure 5.5). Ryle and his student Rupert Clarke, along with their colleague Antony Hewish, continued to maintain that the effects of source clustering and resolution are negligible and that "the results provide conclusive evidence against the Steady-State model."[44] But Hoyle and his student, Jayant Narlikar, claimed that the source count could equally well be interpreted as a local deficiency of only a few dozen strong sources rather than a cosmic excess of many weak sources.[45]

Ryle and Mills met again at an IAU Symposium that was held in Santa Barbara, California in August 1961. Ryle again pressed his case against the steady-state cosmology now based on the new 4C survey, but he was challenged by both Henry Palmer from Jodrell Bank and by Mills.[46] Palmer noted that an analysis by his Jodrell Bank colleague Robert Hanbury Brown showed that Ryle's new radio

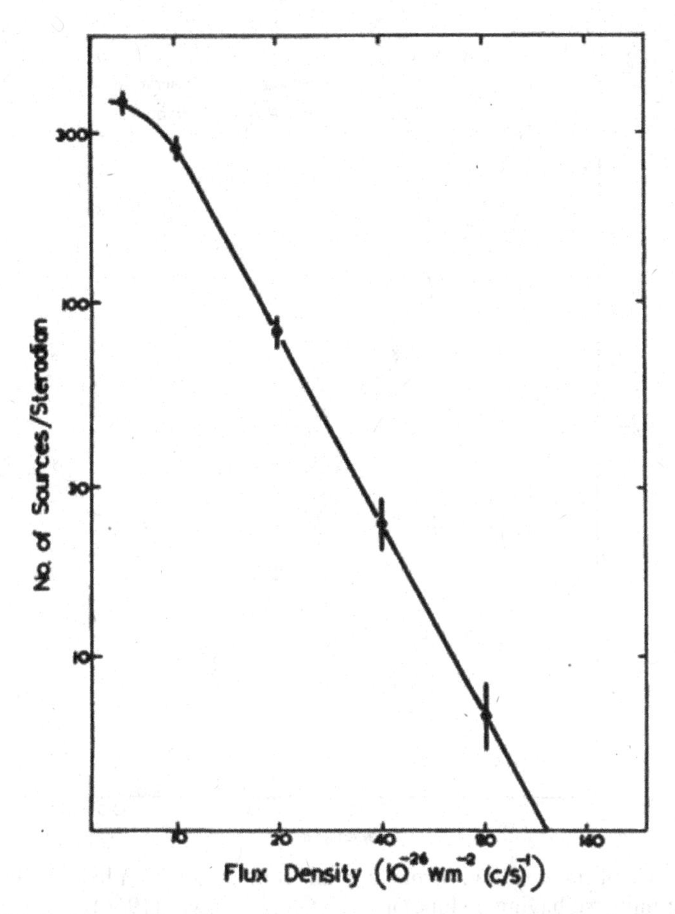

Figure 5.4 Sydney radio source count presented at the 1958 Paris Symposium. Credit: Mills (1959).

source counts could easily "be interpreted in terms of a local deficiency of radio sources." Ryle fought back by arguing that even the strong sources are so distant that they cannot be considered local and that their observed isotropy argued against a local interpretation. Mills then reminded Ryle that his initial claims were based on the 2C slope of –3, whereas the Sydney survey gave –1.8, so it seemed at the time that "any hope of reconciliation was impossible." Now there were only small differences in their source counts, and he reminded Ryle that, considering the effects of clustering, the Cambridge interferometer bias against extended sources, and the experimental uncertainties of both surveys, both the Cambridge and Sydney surveys were consistent with steady-state cosmologies. But Ryle insisted that their conclusions were independent of the number of strong sources and that there was no evidence of clustering in their data. In the following year, in a Jodrell Bank summer school on radio astronomy, attended by one of the present authors (KIK),

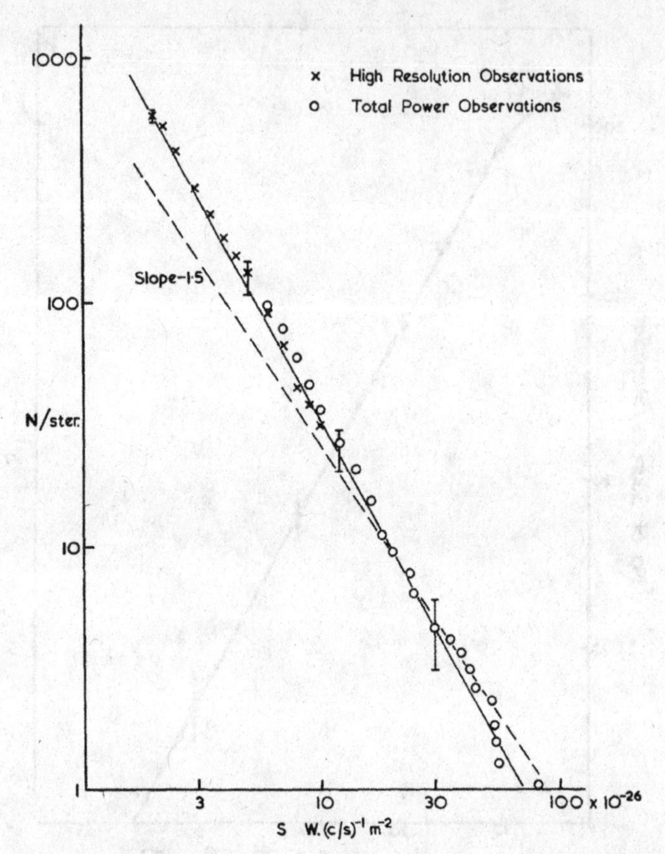

Figure 5.5 Cambridge 4C radio source count compared with a hypothetical static Euclidean universe having a slope of –1.5. Credit: Gower (1966).

Ryle stated as a known fact that "we must recognize that we live in an evolving universe in which the number of sources or their luminosity depends on Epoch."[47]

Both the Sydney and, especially, the Cambridge data contained serious instrumental errors as well as errors of interpretation. The effects of noise and confusion make the observed source count appear steeper than the true value. Because the source count is so steep, there are many more weak radio sources than strong sources, at any specified flux density level. Thus, even random errors due to noise or confusion cause more weak sources to appear stronger than cause strong sources to appear weaker. This was originally noted by Sir Arthur Eddington back in 1913 when evaluating the optical counts of stars brighter than a given magnitude, and was long known to astronomers as the Eddington Effect,[48] but, until the mid-1960s, was apparently not appreciated by the radio astronomers.

David Jauncey, a recent Australian graduate in cosmic ray physics, also pointed out that both Cambridge and Sydney were using cumulative counts, where each point was the sum of all the stronger sources. So, the data points were not independent and the estimated errors were unrealistically small. Moreover, in a cumulative source count, features at any flux density level propagate to lower flux densities and appear to steepen the count.[49] Also, the Sydney count was artificially flattened due to the inclusion of their extended sources, at least some of which were blends of weaker sources in their nearly one degree beam.[50] Finally, as already noted, in a real expanding universe, whether steady-state or big-bang, the effect of the redshift is to decrease the number and energy of incoming photons, so sources appear weaker than would be the case if their flux density fell off as an inverse square of the distance. The corresponding slope of the log N – log S relation in a uniformly filled non-evolving universe is then smaller than –1.5, and so even an observed slope of –1.5 would be evidence against the steady-state theory. Both the Cambridge and Sydney radio astronomers understood the effect of the redshift to lower the source count slope, but Mills argued that this would not be important if the sources "were not too far away," and that, since the radio luminosity function was unknown, it was unclear if this criterion was relevant.[51]

Figure 5.6 shows a modern 1.4 GHz (21 cm) radio source count presented in a differential form and normalized to a static Euclidean universe. Figure 5.6 was derived from a variety of different surveys and goes about a million times fainter than the early Cambridge or Sydney surveys. The strong source count is based on nearly all sky surveys but, for the weaker sources, the source density is derived from long integrations made in very small, presumably representative, regions of the sky. We can identify five regions of the source count:[52]

(1) At the highest flux densities, corresponding to source densities of only a few tens of sources per steradian, the source count is close to the Euclidean value.
(2) Between source densities of about 20 to 50 sources per steradian, there is a sharp increase in the number density of sources. It is this abrupt increase that is reflected in the apparent steep slope reported for the cumulative Cambridge counts and which led to the claims of a large excess of weak radio sources.
(3) Over a range of flux density of about 100 to 1, the count remains Euclidean, implying that the effects of evolution just balance the effects of the redshift and spatial curvature.[53]
(4) As first indicated by the Cambridge North Polar Survey,[54] and then dramatically demonstrated by the Cambridge One-Mile Radio Telescope 5C survey,[55] above source densities of about 1,000 sources per steradian, the source count rapidly converges, reflecting the effect of the redshift in the diminution of flux density faster than (distance)$^{-2}$.

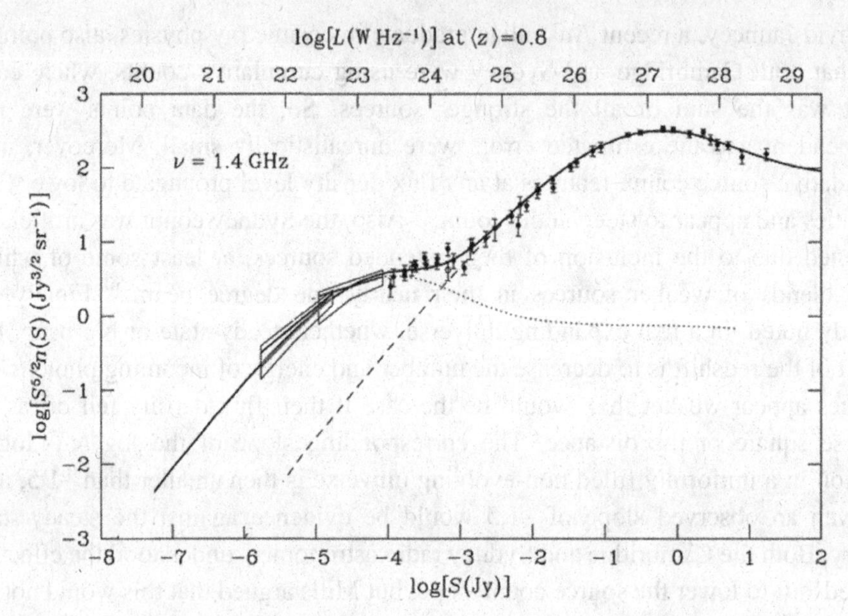

Figure 5.6 1.4 GHz radio source count normalized to a hypothetical static Euclidean universe. The two open areas indicate the constraints on the count from a statistical analysis of the image noise. Dotted and dashed lines indicate the count due to star forming galaxies and powerful AGN driven galaxies and quasars respectively. Credit: Adapted from Condon et al. (2012).

(5) With the much-improved sensitivity available from NRAO's Very Large Array (VLA), new very deep surveys indicated that above source densities near 10,000 per steradian, essentially all the radio galaxies and radio loud quasars have been counted. But when extended to even fainter sources, NRAO radio astronomer Jim Condon and his student Kenneth Mitchell found that the count again steepens due to the emergence of a new population of low luminosity radio sources related to the formation of new stars and the subsequent supernovae that produce cosmic ray electrons and synchrotron radiation.[56] Later, using an extension of Scheuer's P(D) analysis, Condon et al. showed that VLA surveys made with improved sensitivity meant that the source count corresponding to this star forming population converges for the weakest sources. This has been interpreted to mean that the first massive levels of star formation appeared at around the same epoch as the radio galaxies and AGN and has evolved in much the same way as their more powerful counterparts.[57]

Exploiting the improvement in centimeter wave technology in the 1960s, John Shimmins, John Bolton, and Jasper Wall used the Parkes radio telescope at 2,700 MHz (11 cm) to survey a few regions of the sky down to 0.06 Jy and claimed a slope of -1.4 ± 0.1 for the source count. Since the 11 cm survey had a slope much flatter than had been reported for any of the earlier low frequency surveys, the authors argued that if the source count depended on frequency "their value as evidence against the steady state cosmology may be questioned."[58] However, since the 11 cm survey covered only a small region of the sky, it did not contain many of the strong sources that accounted for the apparent steep slope in the earlier full sky surveys. Guy Pooley from Cambridge immediately pointed out that the Parkes 2,700 MHz survey covered a source density described by region 3 above where the count appears Euclidean, and as later shown when proper account is taken of the spectral index distribution, the radio source counts at different frequencies are consistent.[59]

It is clear now that the steep slope reported by the Cambridge observers for the strongest 100 or so sources did not have any relevance for cosmology, as the apparent steepening of the cumulative count appears to be due to the sharp jump between 20 and 50 sources per steradian, that is almost surely the effect of local clustering (Figure 5.7).[60] Moreover, the claimed isotropy of radio sources, that Ryle used as an argument against a local interpretation of the source count, applies only to the weaker sources where there are a sufficient number of sources to establish good statistics. For the strong sources the distribution is far from isotropic,[61] so, as argued by Hoyle and others, the interpretation of the strong radio source count in terms of a deficiency of about 50 sources rather than a cosmic excess of weak sources was not unrealistic.[62] Also, it is important to appreciate that there is only one Universe, and that the strong source count was based on observations covering essentially the entire sky, so there is no opportunity to improve the statistics by more observations.

Another reason that the radio source counts have been so controversial is the unique nature of the radio luminosity function. For a relatively flat luminosity function, where there are numerous intrinsically very luminous radio sources, such as Cygnus A, there will be a "normal" flux density–redshift relation in the sense that the sources with the lower flux densities will, on average, be more distant. If the luminosity function is very steep, that is, there are an excessive number of intrinsically weak radio sources, the sources with lower flux density will, on average, be the closer and intrinsically less luminous sources. As it happens, the radio source luminosity function is close to the critical intermediate value so that, contrary to intuition, over a wide range of flux density there is no relation between average flux density and distance. The average redshift of 1 milli-Jansky sources is about the same as for 10 Jansky sources. And at lower flux density, the sources become systematically intrinsically weaker and closer.

It is perhaps an accident of history that for many years the radio source counts were interpreted in terms of a large excess of weak sources rather than a small deficiency of strong sources. This was because the first surveys only contained about 100 sources,

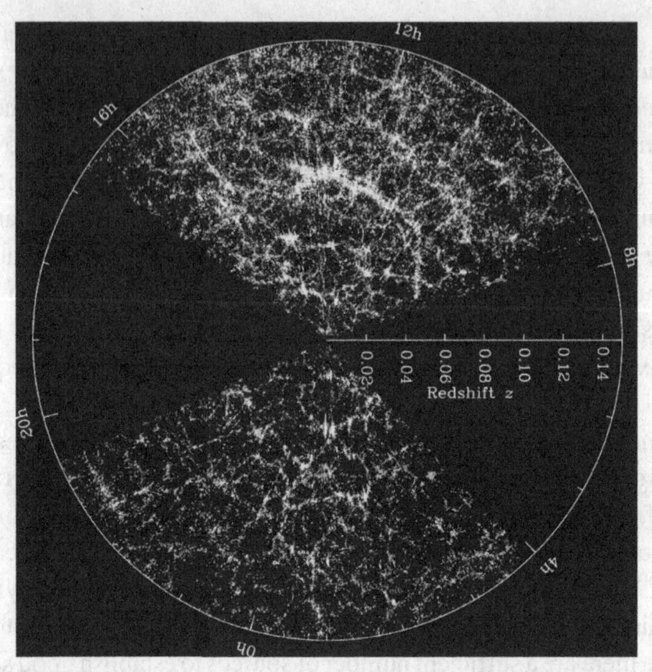

Figure 5.7 Distribution of galaxies with redshift within a narrow slice showing the large scale clustering of galaxies. Credit: Geller and Huchra (1989), courtesy of SDSS.

so later surveys that reached the more numerous weaker sources appeared to show an excess over what was expected from the few observed strong sources. If, on the other hand, the early surveys had reached the weaker population, the lack of a few dozen strong sources may not have had such an impact.[63]

For nearly a decade, the long controversy over the radio source counts was the face of radio astronomy, and within the broader astronomical community was a black mark on radio astronomers that was not erased until the exciting discoveries of the 1960s. Among radio astronomers, the 2C fiasco gave interferometers a bad name, which took years to overcome. The 3C and 3CR Catalogues were, however, a more positive lasting legacy to radio astronomy. Although the IAU formally adopted the ra-dec format (hrmin±deg) for radio source nomenclature, the 328 radio sources listed in the 1962 3CR Catalogue are still known and commonly referred to by their 3C number, in much the same spirit as the 102 objects in Messier's 1781 catalog of optical nebulosities.

The source count controversy, which went on for more than a decade, was intense, personal, and bitter.[64] The debate between the steady-state and big-bang cosmologies was only put to rest in 1965, but not from radio source counts. Rather, as described in Chapter 6, the discovery by Arno Penzias and Robert Wilson of the cosmic microwave background was almost universally accepted as convincing evidence for an evolving or big-bang universe.[65] In the end, Martin Ryle and his Cambridge colleagues were right; we appear to be living in an evolutionary universe. However, his data changed with time and, along with the data, the arguments, that were wrong and were based on unreliable data and inappropriate analysis, also changed. The claim that Scheuer's statistical analysis supported the steep 2C source count was also wrong. Mills had much better data, consistent with contemporary source counts. But he, as well as Bondi, Gold, and Hoyle, and later Geoffrey Burbidge,[66] were wrong in arguing that the radio source counts were consistent with the steady-state or any non-evolutionary cosmology, as the Cambridge source counts were the first astronomical evidence for the existence of cosmic evolution. But it would take a more thorough analysis by Cambridge student, Malcolm Longair, to show that the evolution was confined to only the most powerful radio sources and occurred only at relatively modest redshifts.[67]

The Size of Distant Radio Sources

In 1959, Fred Hoyle called attention to the relationship between the angular size and redshift of distant radio sources and its dependence on the geometry of the Universe described by different cosmological models.[68] Hoyle showed that depending on the value of the deceleration constant, q_0, which in turn depends on the average density of matter in the Universe, the angular size may continue to monotonically decrease with increasing redshift, or as in the steady-state model, the angular size asymptotically approaches a limiting value, that for a source like Cygnus A is about 4 arcseconds. In a closed Universe with $q_0 = 1$, there is actually a minimum in the observed angular size near redshift $z = 1.25$ beyond which the angular size actually increases with increasing redshift. Although seemingly non-intuitive, this is equivalent to terrestrial azimuthal–equidistant maps, where locations near the antipodes of the center subtend an increasingly large angle with increasing distance, so a small island located halfway around the world is seen equally in all directions and appears to have infinite angular size.

Efforts to test the angular size–redshift dependence from optical observations of high redshift galaxies were thwarted for two reasons.[69] First, the more distant galaxies were only about a second of arc in diameter, comparable to the atmospheric seeing limit of even the largest optical telescopes at good sites. Second, because the surface brightness of galaxies blends into the sky background, the measured size, or what Allan Sandage called an "isophotal diameter," depends on the redshift, so galaxies are not good "standard rods." However, as Hoyle pointed out, interferometers used by radio astronomers have sufficient angular resolution to resolve radio galaxies at any redshift, and their double component structure provides a well-defined metric standard rod.

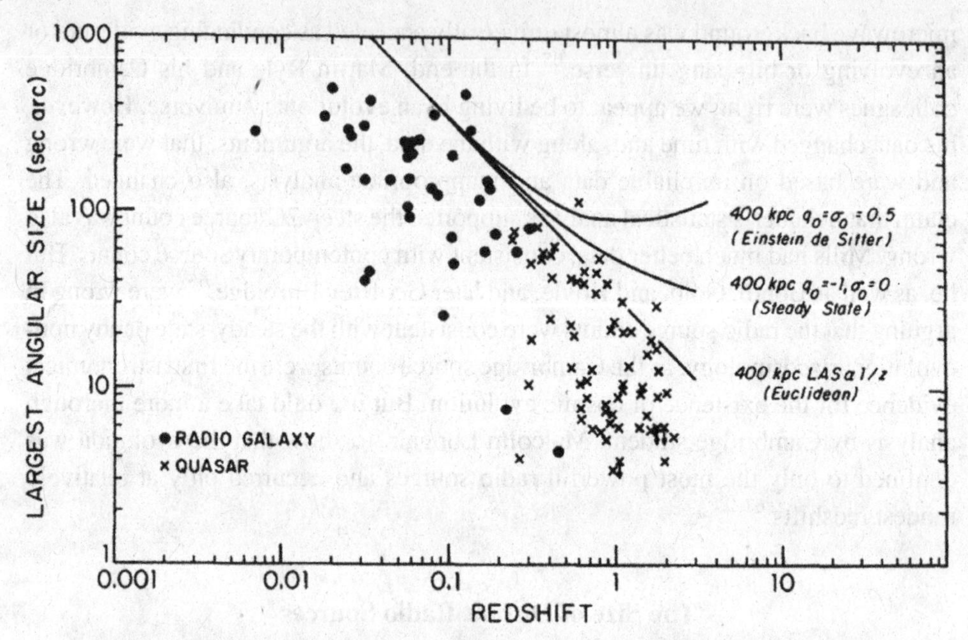

Figure 5.8 Plot of radio source angular size versus redshift for extended radio galaxies. Credit: Kellermann (1972).

George Miley received his PhD from the University of Manchester, based on his interferometric studies of quasars at Jodrell Bank, and then spent two years at the US National Radio Astronomy Observatory, where he examined the relation of the angular separation between radio galaxy and quasar lobes and their redshift. As shown in Figure 5.8, Miley found that the apparent lobe separation of radio galaxies and quasars surprisingly decreased linearly with increasing redshift, consistent with a simple static Euclidian universe and apparently inconsistent with any of the standard Friedmann cosmologies.[70] Miley also noted that the angular size–redshift plot indicated a continuity between radio galaxies and quasars in support of a cosmological interpretation of quasar redshifts.

Within the framework of standard cosmologies, Miley's analysis meant that the linear size of radio galaxies and quasars had to decrease with increasing redshift by just the right amount to offset the effects of spatial curvature, but appeared to support an evolutionary cosmology.[71] As noted by Miley and others, the decrease of linear size with redshift (distance) has a natural explanation in terms of the increased inverse Compton losses due to the higher background radiation density and the higher density of the intergalactic medium, both of which may constrain the growth of the radio lobes. The narrow range of radio lobe separation is in contrast to the wide range of radio source luminosity that limited the interpretation of the radio source counts

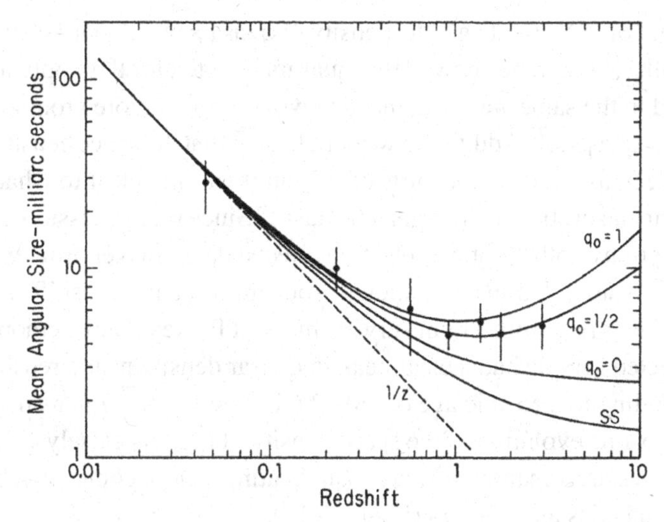

Figure 5.9 Plot of angular size versus redshift for compact quasars and radio galaxies. Credit: Kellermann (1993).

but, as with the radio source counts, the effects of cosmic evolution made the angular size–redshift test useless in distinguishing among world models.[72]

In an attempt to avoid the pitfalls of evolution, one of the present authors (KIK) examined the angular size–redshift relation using only compact radio sources whose angular structure had been determined by very long baseline interferometry.[73] The motivation for this was that these compact sources are less than 100 parsecs in extent and are located deep within the central region of active galaxies and so are not affected by the intergalactic medium or the strength of the cosmic background radiation. In contrast to the steep decline in angular size with redshift shown for the extended radio lobes, the compact AGN showed little change in angular size over the redshift range $0.5 < z < 3$ with evidence for a broad minimum near $z \sim 1$, as expected from cosmological models with $0 < q_0 < 1$ and without the need to introduce evolutionary effects (Figure 5.9). However, uncertainties in defining the size of these core-jet AGN morphologies made it difficult to reach firm conclusions about the true geometry of the Universe.

Quasars and Cosmic Evolution

In his classic paper written five years after he determined the redshift of 3C 273, Maarten Schmidt used a sample of 40 quasars to derive their luminosity function and to show that their space density dramatically evolves with cosmic time much in the same manner as do powerful radio galaxies and that, when the Universe was only about one tenth of its present age, the density of quasars was about 100 times greater than it is locally.[74] Later, he showed that quasars selected for their optical properties alone evolved in the same way, and that the evolution was more pronounced for the most luminous quasars.[75] Additional work indicated that the space density of quasars appeared to decrease beyond redshifts of 2.7 but it was not clear to what extent this might be due to absorption by intergalactic dust.[76] Since radio emission is unaffected by dust absorption or other optical selection effects, Peter Shaver, from the European Southern Observatory, led an international group that obtained redshifts for 442 radio loud quasars identified from 11 cm surveys made at Parkes. They demonstrated that the radio selected sample had a clear peak in quasar density near a redshift of about 2.5, corresponding to a cosmic age of only 2.6 billion years.[77] Interestingly, as they showed, the cosmic evolution of the space density of quasars closely coincided with the evolution in star formation in galaxies, indicating a close connection between the formation of galaxies and quasars (Figure 5.10).

Non-cosmological Redshifts

The interpretation of quasar redshifts as indicators of distance was not universally accepted. In view of their apparent extraordinary properties, a number of well-respected and some not well-respected astronomers argued that the large observed quasar redshifts must be intrinsic and are not due to Doppler shifts reflecting the

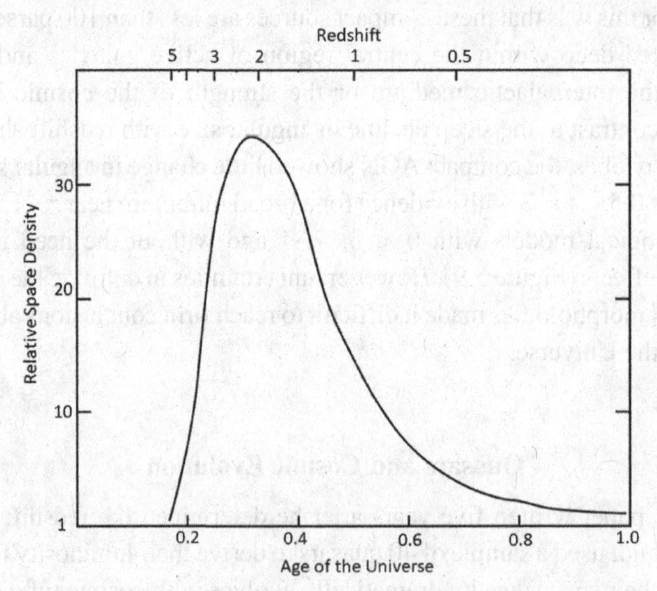

Figure 5.10 Schematic diagram showing possible evolution for strong radio galaxies and quasars based on radio source counts and quasar redshifts.

expansion of the Universe. They supported their arguments by a variety of intriguing observations challenging the conventional cosmological interpretation of the large quasar redshifts. These included:[78]

(a) *The absence of any redshift–magnitude (Hubble) relation for either the radio or optical data*: This is now understood in terms of the wide range of apparent quasar radio and optical luminosity.

(b) *QSO clustering near galaxies*: For many years, Halton (Chip) Arp, Fred Hoyle, Geoffrey Burbidge, and others maintained that the density of quasars in the vicinity of galaxies significantly exceeds that found in random fields. Thus, they argued that quasars must be ejected from relatively nearby galaxies with velocities close to the speed of light. But their quasar fields were clearly not chosen at random, and it was difficult to understand the absence of blue shifts (approaching quasars) in such a model. One not very convincing explanation was that light is emitted only in the opposite direction from the motion in the manner of an exhaust, hence only redshifts are observed. In a variation on this interpretation, James Terrell at Los Alamos suggested that quasars are ejected from the center of our Galaxy and have all passed the Earth, hence we only see redshifts from the receding objects.[79]

(c) *Distribution of observed redshifts*: Analysis of the distribution of observed quasar redshifts suggested that there were preferred values with peaks near 1.955 and at multiples of 0.061, indicating that the observed redshifts were not cosmological but were due to some intrinsic quantized phenomena.[80] This is now understood as observational selection effects.

(d) *Radio variability and superluminal motion*: The discovery of radio variability and especially the rapid inter-day variability observed in some quasars implied correspondingly very small linear dimensions. If quasars are at cosmological redshifts, Hoyle and others argued that they would self-destruct by inverse Compton interactions between the relativistic electrons and the radiation photons. This issue was addressed with the discovery of apparent superluminal motion, which is most straightforwardly understood as the effect of relativistic beaming by material moving at close to the speed of light so that the radiation is focused along the direction of motion, which can increase the apparent luminosity by orders of magnitude. But Arp, Burbidge, Hoyle, and others argued that the apparently faster than light motion deduced from radio interferometer observations of quasars is more easily understood if quasars are closer than indicated by their redshifts, so that observed large angular velocity would mean only relatively modest linear speeds, not even close to the speed of light. Indeed, as they argued, the relativistic beaming interpretation still requires velocities unrealistically close to the speed of light and it is difficult to accelerate and maintain such high speeds within the dense environment of galactic nuclei.

The arguments for non-cosmological redshifts lasted for more than three decades; many conferences were held and books written to debate the issues.[81] Most astronomers agreed that the apparent anomalies were due to the result of *a-posteriori* statistics and, in the case of redshift distributions, selection effects due to the limited number of strong quasar emission lines that could be observed considering the narrow range of the observable optical window and the blocking of certain spectral regions by night-sky lines. The arguments for non-cosmological redshifts only died when the proponents died or at least retired but, from time to time, they still surface.[82] But, like the specious arguments about the slope of the source count, the controversy over cosmological redshifts was intense and personal and has had a lasting impact on the sociology of astronomy and astronomers.

6

The Cosmic Microwave Background

> We compute the remaining unaccounted-for antenna temperature to be 3.5 ± 1.0 K at 4,080 MHz.[1]

Arno Penzias and Robert Wilson discovered the cosmic microwave background (CMB) at Bell Telephone Laboratories (BTL) in 1965 using an antenna and sensitive maser[2] receiver that was designed and built for testing the AT&T intercontinental telephone communications network. Penzias and Wilson were not trying to check any theory, nor were they trying to investigate cosmic evolution or cosmology. The existence of a CMB had been predicted much earlier, but the prediction was unknown to Penzias and Wilson. Other investigators, including both radio and optical astronomers, as well as other Bell Labs engineers, had seen evidence of the CMB radiation before Penzias and Wilson, but they did not recognize the significance of their observations. One group at nearby Princeton University, led by Robert Dicke, was developing a similar experiment designed to detect the CMB, although they were also unaware of the earlier predictions and apparent detections. Another group, in the Soviet Union, did make the connection between the theory and early Bell Labs published measurements but, apparently due to a language problem, they misunderstood the published data as reporting a negative result.[3] An even earlier measurement of optical absorption lines by interstellar CN (cyanogen) gave the first clues to the thermal background radiation, but its meaning was not recognized until after the experimental discovery of the CMB at Bell Labs in 1965. David Wilkinson and James (Jim) Peebles have given a personal account of the complex sequence events surrounding the CMB and their recollections of the activities at Princeton.[4]

Figure 6.1 Arno Penzias. Credit: AIP Emilio Segrè Visual Archives, Physics Today Collection.

Radio Astronomy Returns to the Telephone Company

Arno Allan Penzias was born in 1933 in Munich, Germany. After narrowly avoiding deportation to a Polish concentration camp, he emigrated to the United States with his family in 1940. Following his graduation from City College of New York, Penzias spent two years in the US Army Signal Corp working on radar research and then, in 1961, he received his PhD in physics from Columbia University working with Charles Townes to build one of the first maser amplifiers used for radio astronomy (Figure 6.1).

Robert (Bob) Woodrow Wilson grew up in southern Texas and received his Bachelor's degree in physics from Rice University in 1957. In graduate school, he worked with John Bolton at the Owens Valley Radio Observatory (OVRO) and obtained his PhD from Caltech in 1963 based on a survey of galactic radio sources made with one of the new OVRO 90 foot radio telescopes (Figure 6.2).

As youths, both Penzias and Wilson earned "spending money" by fixing radios and television sets for their friends and neighbors, learning skills that would serve them well in the future. Penzias joined the Bell Telephone Laboratories Radio Research Department in 1961 and Wilson two years later.[5]

The horn reflector antenna was invented at Bell Laboratories by Karl Janksy's former colleague, Al Beck, and Jansky's boss, Harald Friis. Smaller versions of horn reflectors are still seen on telephone circuit relay towers. Unlike conventional parabolic

Figure 6.2 Robert Wilson. Credit: AIP Emilio Segrè Visual Archives, W. F. Meggers Gallery of Nobel Laureates Collection.

dish antennas used by radio astronomers, the horn reflector has extremely low spurious responses so that two horn reflectors can be mounted back-to-back on a relay tower with no interference between the transmitting and receiving antennas. Also, unlike parabolic dish antennas, the gain of a horn reflector can be accurately calculated from its dimensions. The 20 foot horn reflector used by Penzias and Wilson for their remarkable discovery was built by Arthur Crawford and associates at the BTL Crawford Hill facility in Holmdel, New Jersey, not far from the site of Karl Jansky's pioneering discovery of cosmic radio emission from the Galaxy. It was originally intended to be used as a ground station to test satellite relay links.[6] The other key instrumental feature was the travelling wave maser amplifier that had been developed at Bell Labs by Derek Scoville to give unprecedented receiver sensitivity at 4,080 MHz (7.35 cm wavelength). It was this combination of sensitivity and low sidelobe response that enabled Penzias and Wilson to ultimately detect an excess antenna noise that came from the sky and not from the ground or the Earth's atmosphere.

In several publications, Bob Wilson and Arno Penzias have given their personal accounts of the sequence of events that led to their discovery of the CMB.[7] The 20 foot horn reflector was first used with the passive Echo 100 foot balloon reflector which had limited success and was superseded by the active Telstar relay satellite. In 1962, the US

Figure 6.3 Robert Wilson (left) and Arno Penzias near their Holmdel horn antenna. Credit: Nokia Corporation and AT&T Archives.

Congress passed the Communications Satellite Act which created the COMSAT Corporation to hold exclusive rights for communications using orbiting artificial satellites. AT&T was out of the satellite business, and the 20 foot horn reflector at Crawford Hill became available for Penzias and Wilson to use for radio astronomy. Wilson's interest in the radio sky background radiation began with his doctoral work at Caltech, where he observed the smooth increase in continuum radio emission as his antenna scanned across the plane of the Milky Way. But at Caltech he was unable to determine the level of isotropic radiation in regions away from the Milky Way. This required a measurement of the absolute sky background, uncontaminated by emission from the atmosphere or the ground (Figure 6.3).

As we have seen in Chapter 1, the nonthermal galactic radiation studied by Karl Jansky, Grote Reber, and others decreases rapidly with increasing frequency, so Penzias and Wilson did not expect to see any significant background radiation at their relatively high frequency of 4.1 GHz. A few years earlier, Edward Ohm had used the same 20 foot horn reflector and maser amplifier at 2.3 GHz to test a voice link between California and New Jersey using the Echo I satellite.[8] Since Echo I was only a passive balloon reflector, Ohm needed the best possible sensitivity at the receiving end. After carefully testing his system and measuring the contribution of the atmosphere, Ohm found a total system noise temperature of 22.2 ± 2.2 degrees K. This was an extraordinarily low system temperature, but it was more than the 18.9 K that Ohm had calculated based on the known or measured properties of the maser amplifier, the horn, the atmosphere, and other components of their radiometer.

In order to reconcile his measured total system temperature of 22.2 K with the expected value, Ohm concluded that the combined contribution from the ground and from their maser amplifier was 3.3 degrees hotter than had been measured or calculated. The scientists who had designed and built the maser amplifier were not pleased to be told that their carefully engineered device was not operating as claimed. Art Crawford and his colleagues, who developed the horn reflector,[9] were equally chagrined at the possibility that their horn was picking up more ground radiation than they had calculated. Ohm published his technical report in the *Bell System Technical Journal*, where it would remain obscure until discovered, but not understood, a few years later by two Russian cosmologists.[10]

Penzias and Wilson were puzzled by Ohm's results and thought that they could do a better job. Penzias built a reference load cooled by liquid helium to 4.2 K,[11] and Wilson built a switch to rapidly alternate their receiver between their horn antenna and the comparison load. Their approach was based on a technique known as a "Dicke radiometer," pioneered by Robert Dicke in 1946 at the MIT Radiation Laboratory, and used in essentially every single aperture (non-interferometer) radio telescope since that time (Figure 6.4).[12] But their observations in May 1964 were disappointing. At the zenith, the signal from the antenna was warmer than the signal from the cooled load, whereas it should have been colder (2.3 degrees from the atmosphere). Although Penzias was proud of his carefully designed cold load, he knew it could not be colder than the 4.2 K temperature of the surrounding liquid helium bath. The earlier measurement of excess temperature by Ohm could not distinguish between an origin in the maser amplifier or the antenna. Penzias and Wilson clearly showed that the excess noise came from the antenna and not the maser. But what could be wrong with their antenna temperature measurement? They considered many possibilities.

Perhaps their measurement of the temperature of the atmosphere was incorrect. However, their measured temperatures fit the expected dependence on the elevation angle, another technique that had been first used by Dicke back in 1946,[13] and gave 2.3 ± 0.3 K in good agreement with the earlier value determined by Ohm, thus eliminating any uncertainties about the contribution from the Earth's atmosphere.

As Wilson recalled, Crawford Hill is only 20 miles from downtown New York City, so they suspected that their excess antenna noise might be interference from the heavily populated New York area. They scanned their antenna along the

Figure 6.4 Robert Dicke. Credit: AIP Emilio Segrè Visual Archives, Physics Today Collection.

horizon and above the horizon but found no evidence of any excessive signal from the direction of New York City.

As we have seen, extrapolation of the low frequency nonthermal radio emission from the Milky Way or from discrete radio sources in the Galaxy should produce a negligible signal at 4,080 MHz so that seemed like an unlikely explanation.

The throat of their horn antenna where most of the loss (and corresponding thermal radiation) occurs was made from electroformed copper and it was not a perfect conductor – but it was close. Penzias and Wilson measured the loss of a sample of waveguide of similar material in the laboratory and estimated that the contribution from the antenna should be only 0.9 K – not nearly enough to account for the apparent excess. But there was a catch. A couple of pigeons had made their home in the antenna and, over the course of time, had deposited their droppings on the inner surface of the horn. Not much was known about the reflectivity and absorptivity of pigeon droppings, but Penzias and Wilson "evicted" the pigeons and cleaned out the droppings. Still there was no significant improvement. They placed aluminum tape over the joints in the horn to eliminate any possible radiation from the joints, but that too produced no change.

In 1962, the US Atomic Energy Commission and the Defense Atomic Support Agency launched Project Starfish Prime that exploded a high altitude thermonuclear explosion over Johnston Atoll in the Pacific Ocean. The resultant electromagnetic pulse (EMP) caused widespread electronic damage in Hawaii, nearly 1,000 miles northeast of the explosion, which released charged electrons that became trapped in the Earth's magnetic field, forming an artificial radiation belt.[14] Subsequent nuclear tests by the United States, as well as by the USSR, had further enhanced the level of

Figure 6.5 James Peebles. Credit: AIP Emilio Segrè Visual Archives, Physics Today Collection and Tenn Collection.

radiation, and Penzias and Wilson speculated that these artificial radiation belts might be contributing to their observed excess antenna temperature. However, after a year had passed there was no change in their measured antenna temperature, although the artificial radiation belts had decayed significantly. Puzzled about the source of their excess antenna temperature, they were annoyed that, with their combined expertise in electronics, they could not find the source of the excess noise.

Meanwhile, some 30 miles away, at Princeton University, Dicke's students, Peter Roll and David Wilkinson, were building a 3 cm radiometer to attempt to measure the residual cosmic microwave background radiation that they had speculated might result from the hot primeval fireball during the highly compressed phase of an oscillating or bouncing Universe. Dicke estimated that the initial temperature would be as high as 10 billion degrees. Due to the expansion of the Universe, after the approximately 14 billion years since the big-bang, Princeton postdoc Jim Peebles (Figure 6.5) estimated that the 10 billion degrees K would have cooled to at least 10 degrees K, which Dicke suggested could be detected by careful measurement. The Princeton group chose to work at 3 cm wavelength where, they argued, the thermal radiation from the atmosphere and the nonthermal radiation from the Galaxy and extragalactic radio sources would be at a minimum. On 19 February 1965, at a colloquium he gave at Johns Hopkins University in Maryland, Peebles discussed the Princeton plan to try to detect the leftover radiation of the big-bang. Kenneth Turner, who had received his PhD from Princeton under Bob Dicke and was now working in

Figure 6.6 Bernard Burke. Credit: G. Runion NRAO/AUI/NSF.

the radio astronomy department of the Carnegie Institution Department of Terrestrial Magnetism (DTM), heard Peebles talk and mentioned it to his DTM colleague Bernard Burke (Figure 6.6). That afternoon, Burke received a phone call from Penzias on unrelated business. He told Penzias what he had heard from Turner and suggested that Penzias should get in contact with the Princeton group.[15]

According to Penzias, he then contacted Dicke and told him about their excess antenna temperature.[16] Wilkinson later related that Dicke turned to his group and explained, "Well boys, we've been scooped!"[17] Dicke, along with Roll and Wilkinson, visited Bell Labs and quickly recognized that the excess noise found by Penzias and Wilson was not due to any fault in their experiment or analysis, but was in fact the cosmic background radiation that the Princeton scientists were hoping to detect. The Princeton group was disappointed to have been scooped, although Peebles was excited that his prediction was confirmed and that there was actually something there to study. Penzias and Wilson were pleased that they now had an explanation for their antenna excess noise. Following a reverse visit by Penzias and Wilson to Princeton, the two groups agreed to publish their work in back-to-back papers in the *Astrophysical Journal*. The short publication by Penzias and Wilson only reported the experimental details of their measurement saying, "we compute the remaining unaccounted-for antenna temperature to be 3.5 ± 1.0 K at 4,080 MHz."[18] Penzias and Wilson made no mention of any astrophysical interpretation. Instead they only wrote near the start of their paper, "A possible explanation for the observed excess noise temperature is the one given by Dicke, Peebles, Roll, and Wilkinson (1965) in a companion letter in this issue."[19] Nevertheless, the paper by Penzias and Wilson became one of the most highly cited papers in astronomy and led to their 1978 Nobel Prize in Physics. The companion paper by the Princeton group explained the Bell Labs result in terms of the cooled radiation left over from the primeval big-bang fireball, and discussed the 3 cm investigation still underway by the Princeton group.[20] The Princeton paper referred to the 1946 experiment by Dicke et al. to measure the contribution from the atmosphere,[21] but apparently Dicke had forgotten that in the same paper they had also placed an upper limit of 20 degrees on any isotropic background radiation in the wavelength range 1 to 1.5 cm and made no mention of this in their 1965 paper. Interestingly, Wilkinson and Peebles later suggested that Dicke most likely could have detected the CMB at the time of his 1946 experiments, which were intended to determine the feasibility of short centimeter wavelength radar.[22]

Roll and Wilkinson went on to complete their Princeton experiment and reported the CMB temperature to be 3.0 ± 0.5 K at 3.2 cm wavelength, in good agreement with the Bell Labs result and supporting evidence of a thermal spectrum of the CMB.[23] Penzias and Wilson repeated their experiment at 1.42 GHz (21.1 cm wavelength) and obtained a value of 3.2 ± 1.0 K for the temperature of the CMB.[24] In the United Kingdom, at Cambridge University, John Shakeshaft and his student, T. K. Howell, measured 2.8 ± 0.6 K at 1.45 GHz (20.7 cm) and, after allowing for the nonthermal contribution from the Galaxy, estimated the CMB temperature at 408 and 610 MHz (73.5 and 49.1 cm) to be 3.7 ± 1.2 K.[25] Subsequent measurements made over a wide range of wavelengths from the ground, from balloons, and from rockets, culminating with the extraordinarily precise observations from the Cosmic Background Explorer (COBE) Far Infrared Absolute Spectrophotometer (FIRAS) program and the Cosmic Background Radiation (COBRA) rocket experiment, showed that the CMB spectrum was remarkably close to that of a perfect blackbody, at a temperature of 2.725 ± 0.002 K (Figure 6.7).[26]

As happened 30 years earlier with Jansky, Penzias and Wilson had made a major astronomical discovery while working at Bell Laboratories with equipment designed to enhance long distance telephone communication. Jansky was working at 15 m wavelength using the reflection from the ionosphere to communicate beyond the terrestrial horizon. Penzias and Wilson were working at microwave wavelengths using equipment designed to communicate via orbiting satellite links. Like Jansky, Penzias and Wilson depended heavily on the preceding work of others at Bell Labs, including the development of the horn reflector antenna by Friis, Beck, and David Hogg, as well as the development of the traveling wave maser by Derek Scoville. Penzias and Wilson enjoyed the same burst of publicity surrounding their discovery of the CMB as happened three decades earlier with Jansky's discovery of the galactic nonthermal radio emission. The existing economic depression and the impending war, along with the lack of any immediate impact to mainstream astronomy, led Jansky to spend the rest of his career on defense- and communication-related research with very little opportunity to pursue any further work in radio astronomy. By

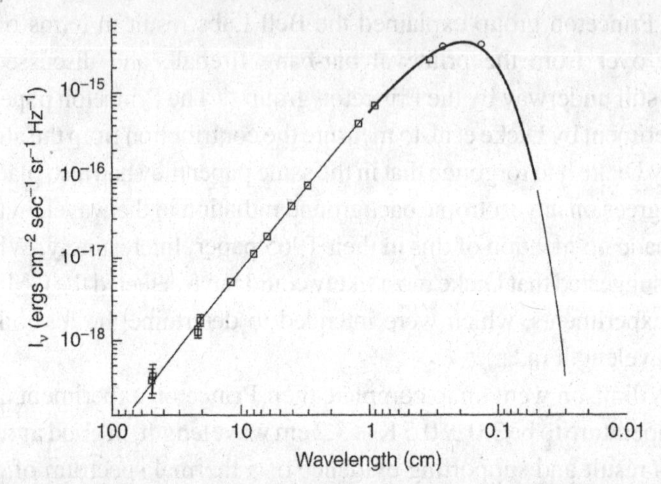

Figure 6.7 CMB spectrum showing intensity versus wavelength. Open circles and squares are from ground based measurements; heavy line is from the NASA COBE/FIRAS experiment; thin line is the theoretical spectrum for a 2.73 black-body. Courtesy of James Peebles.

contrast, the discovery of the cosmic microwave background by Penzias and Wilson had an immediate and major impact on astronomy, leading to their Nobel Prize a little more than a decade later. However, many of the same people who challenged the cosmological origin of quasar redshifts and argued against the evidence contradicting the steady-state Universe from the radio source counts also challenged the evidence from the CMB, suggesting that it could be the sum of many weak discrete sources, the thermal (free–free) emission from interstellar ionized hydrogen, or interstellar dust.[27]

After their 1967 measurement of the CMB at 21 cm, and a 1969 observation to determine the small-scale anisotropy of the microwave background at 3.5 mm,[28] Penzias and Wilson never returned to the CMB, which developed into a big "industry" with elaborate and correspondingly expensive experiments. To minimize the effects of atmospheric water vapor, CMB observations were made from remote mountain sites, from the South Pole, from high altitude aircraft, balloons, rockets, and ultimately from spacecraft. As discussed in Chapter 8, both Penzias and Wilson continued their radio astronomy research with their discovery of interstellar carbon monoxide (CO) and other molecules, initiating another new field of astronomical research, molecular spectroscopy. Nevertheless, like Karl Jansky three decades earlier, they were still expected to continue to contribute to the core research at Bell Laboratories to develop the technology used to support telephone communications. They worked together on measurements of atmospheric attenuation at infrared wavelengths for a relay system for the AT&T network and the effect of atmospheric attenuation on satellite-based relay systems.

Arno Penzias went on to become Director of the Bell Radio Research Laboratory, and in 1981 he became the Vice President for Research. Following the divestiture of AT&T mandated by the US Department of Justice, he was tasked with implementing the impact on Bell Labs of the loss of the AT&T operating revenue, that had traditionally supported research at the Laboratory. After he retired from Bell Labs, Penzias went on to provide technical support to a venture capital firm in California. Bob Wilson also rose to direct the BTL Radio Research Laboratory. Following his retirement from Bell Labs, Wilson joined the Harvard-Smithsonian Center for Astrophysics where he continued his research on millimeter spectroscopy and instrument development.

Inventing, Reinventing, and Rediscovering the CMB

Back in 1941, Mount Wilson Observatory astronomer Walter Adams measured the optical absorption line spectrum from a number of different interstellar molecules including excited CN.[29] As part of his PhD dissertation, Andrew McKellar noted that the splitting of the two absorption lines of interstellar CN at $\lambda = 387.400$ and $\lambda = 387.461$ nanometers, observed by Adams, corresponded to a rotational transition of the molecule which occurs at an energy corresponding to a frequency of 114 GHz (2.64 mm). From the ratio of the absorption depth at the two lines, McKellar was able to estimate the relative population of the two states and, from this, he deduced that the excitation temperature of CN at 114 GHz was 2.3 K.[30]

Neither McKellar nor anyone else appreciated the significance of this result, since apparently everyone assumed that the temperature he had derived was not a real radiation temperature, but that the CN molecules were excited by collisions with interstellar particles. McKellar's calculation was unknown to either the Bell Labs or Princeton groups and was not noticed until after Penzias and Wilson published their paper, when a number of people independently made the connection between McKellar's work and the CMB, and made new confirming observations indicating that the CN excitation temperature was close to the temperature of the CMB measured by radio astronomers at microwave frequencies.

Even before the 1965 direct detection of the microwave background at Bell Labs, George Field, then a young faculty member at the University of California at Berkeley, realized that collisions could not account for the excitation of CN and concluded "that there must be a previously unrecognized source of radiation" to excite interstellar CN.[31] However, he was advised not to publish his conclusion as being too speculative, since he could not show how the CN molecule would be excited by the radiation field. The solution came from an unexpected coincidence. In 1965, Field was completing an unrelated article for the *Annual Reviews of Astronomy and Astrophysics*, and the editor had sent him the proofs of a paper that discussed CN in comets merely to use as a style guide to aid him in correcting his own proofs. As he later recollected, after hearing about the Bell Labs result, Field immediately recognized the connection with the CN temperature paradox and retrieved the comet paper from where he had discarded it in his wastebasket. From the data presented in this paper, Field was able to show that the interstellar CN lines are indeed excited by the microwave background radiation and not by collisions.

Coincidently, that same day, University of California student John Hitchock, whose office was next door to Field's and who knew of Field's interest in CN, came into Field's office to tell him about the work he was doing using spectra of interstellar CN absorption obtained by George Herbig. Using Herbig's spectra of the stars ξ Ophiuchi and ξ Persei, Field, Herbig, and Hitchock concluded that the excitation temperature of interstellar CN was 3.22 ± 0.15 K.[32] In the same issue of the *Physical Review Letters* as the Field et al. paper, Patrick Thaddeus and John Clauser reported on their determination of the CMB temperature as 3.75 ± 0.50 K based on new CN spectra of ξ Ophiuchi from the McMath solar telescope at the Kitt Peak National Observatory in Arizona.[33] In the USSR, Iosef Shklovsky remembered McKellar's paper and also recognized the relevance to the cosmic background radiation.[34] He referred to the CMB as the "relict radiation" which became the standard term used by Russian cosmologists for CMB.

Ironically, soon after he had arrived at Bell Labs, Penzias had thought about trying to detect the 18 cm lines of interstellar hydroxyl (OH). In order to calculate the excitation for OH, he needed to know the radiation temperature of interstellar space and became aware from a discussion with George Field that interstellar CN was excited to a few degrees Kelvin. But he had forgotten about CN and never made the connection with their Bell Labs detection of the CMB until reminded of it in 1966 by Field. Similarly, Neville Woolf recalled that, in 1964, in response to a question from Bob Dicke about the temperature of the background, he had reminded Dicke about the excitation of interstellar CN. Dicke showed no interest and Woolf, who was then a young postdoc at Princeton, thinking that he had said something stupid, did not pursue it further.[35]

About the same time that McKellar was working on his PhD dissertation, but quite independently, George Gamow and his associates at George Washington University were thinking about the conditions of the big-bang fireball that would be required to form the chemical elements (Figure 6.8).[36] Robert Herman and Gamow's student, Ralph Alpher, extended Gamow's work with a paper that they submitted to *Nature* at the suggestion of Gamow.[37] In this paper, Alpher and Herman claimed to have corrected some errors in Gamow's *Nature* paper and were the first to explicitly discuss a residual background temperature. Based on a postulated big-bang temperature of 10 billion degrees, the then understood value of the Hubble constant, and other cosmological parameters, they estimated that, at the

Figure 6.8 George Gamow. Credit: AIP Emilio Segrè Visual Archives, Physics Today Collection.

present epoch, the residual temperature after cooling would be about 5 K, surprisingly close to what was measured more than a decade later. Subsequent publications by Alpher and Herman refined these calculations, based in part on revised estimates of the Hubble constant and other cosmological parameters,[38] and in 1950 Gamow quoted, without explanation, a value of 3 K for the "current temperature."[39] While much of Gamow's initial work was based on inappropriate assumptions, he was remarkably close to current theory and provided the initial motivation for Alpher and Herman to later make their more detailed calculations.[40]

Ironically, Alpher later recalled that he had nearly completed his original thesis research on the growth of galaxies in a relativistic expanding universe when Gamow showed him a Soviet publication by Lev Landau's student Evgeny Lifshitz on the same topic. As he later described it, Alpher had been scooped, and "so it was back to the drawing board," to pursue a thesis topic suggested by Gamow on the synthesis of the chemical elements in the early universe.[41] Lifshitz went on to write, together with Landau, their well-known 10 volume *Course of Theoretical Physics*, that sold more than a million copies in Russian, English, French, German, and other languages, and won the 1962 Lenin Prize – apparently the only time this prize was awarded for writing textbooks. Probably nearly every individual in the world who has earned a

PhD in physics has made use of one or more of the Landau and Lifshitz series of books on theoretical physics.

Independently, in the USSR, Andrei Doroshkevich and Igor Novikov were considering the conditions in the early Universe, and calculated that there would remain at present an observable microwave relict radiation of a few degrees Kelvin. Moreover, they understood that, because of its thermal spectrum, the relict radiation could be distinguished from other background radiations. They apparently had recognized that the Bell Labs 20 foot horn reflector was the best instrument in the world to detect the relict radiation from the cosmic big-bang. Unlike Western cosmologists, they read the literature and were familiar with the paper by Edward Ohm in the *Bell System Technical Journal* that reported an apparent excess system temperature over that expected from the instrument, the sky, and the ground.[42] Using the same technique pioneered 15 years earlier by Bob Dicke, Ohm had measured the contribution from the atmosphere to be 2.3 $\pm$ 0.2 degrees at the zenith, which he referred to as the "sky temperature." Unfortunately, possibly due to a language problem, Doroshkevich and Novikov misunderstood Ohm's "sky temperature" to mean the atmosphere plus the cosmic background. They commented that Ohm's measurement "coincides with theoretical computed atmospheric noise," and so they erroneously concluded that Ohm's experiment was apparently inconsistent with any measurable cosmic background temperature.[43]

Ten years earlier, as part of his PhD research, T. Shmaonov, at the Moscow Institute of General Physics, used a conventional horn antenna to measure a background temperature of 4 $\pm$ 3 K at 3.2 cm (9.3 GHz), but no one in the USSR or elsewhere, including Shmaonov himself, remembered this work until 1983.[44] Other earlier detections of the CMB came from the Paris Observatory in 1955, when Emile La Roux measured an isotropic background of 3 $\pm$ 2 K at 33 cm (910 MHz) and in 1962 at the US Naval Research Laboratory where William Rose estimated an uncertain background temperature of about 3 K in an unpublished attempt to measure a cosmic background radiation.[45] After Ohm's work at Bell Labs, William Jakes, using the same horn reflector at 7.2 cm (4.2 GHz) wavelength to test a Telstar satellite television link, measured an "over-all" system noise temperature "somewhat less than 17°K pointing at the zenith." This was more than 2.5 degrees greater than they could account for from waveguide losses (4.5 K), sky noise (2.5 K), side lobes (2.5 K), and the receiver (5 K).[46]

None of this earlier work was apparently known to Penzias and Wilson, nor to Dicke and his colleagues and students at Princeton.[47] Gamow was a colorful individual who traveled widely, gave many lectures, and wrote widely read popular science books, so he was better known than Alpher or Herman, who wrote many of the later papers and apparently, although incorrectly, felt that much of their work was ignored or inappropriately credited to Gamow.[48] At the time of the discovery of the CMB in 1965 and the broad subsequent interest by observers and theoreticians, Alpher and Herman were working for General Electric and General Motors, respectively, and were not in any academic position, which may have contributed to the initial lack of recognition of their work 20 years earlier. However it was Ralph Alpher, in 2005, who received the National Medal of Science from President G. W. Bush "For his unprecedented work in the areas of nucleosynthesis, for the prediction that the universe expansion leaves behind background radiation, and for providing the model of the big-bang theory." Ironically, Gamow, Alpher, and Herman had invented the hot primeval big-bang universe to explain the creation of the heavy elements by nuclear fusion starting from hydrogen, while Dicke had postulated the same 10 billion degree temperature at the condensed phase of an oscillating universe needed to get rid of the heavy elements created in the stars during the previous expansion phase.[49]

Figure 6.9 shows schematically the complex history of the invention, discovery, and rediscovery of the cosmic microwave background that occurred over the 25 year period from 1941 to 1966.

Smoother than a Billiard Ball

In their 1965 paper, Penzias and Wilson wrote, "This excess temperature is, within the limits of our observations, isotropic, unpolarized, and free of seasonal variation." In fact their measurements showed that the CMB radiation was isotropic to within about 10 percent and, had they appreciated the importance, they probably could have improved the limit by another factor of 10.[50] Several investigators subsequently suggested that there should be some small scale fluctuations of the order of 0.1 percent in the background radiation that would lead to the creation of galaxies and clusters of galaxies or quasars.[51] It was clearly difficult to experimentally measure such small signals, but several teams of investigators rose to the challenge. However, when the initial investigations gave negative results, the theoretical predictions reacted with successively smaller predicted fluctuations.[52] In his Stanford University PhD dissertation, Ned Conklin was unable to detect any arcminute scale fluctuations in the CMB, but working from a 12,500 foot site on Mount Whitney in California, Conklin measured the small anisotropy due to the Earth's motion through the fixed reference frame of the CMB, confirming that the cosmic background radiation detected by Penzias and Wilson was indeed a property of the Universe and not a galactic halo.[53] Later George Smoot and colleagues, flying in a U-2 aircraft at heights 20 miles above the Earth and above 95 percent of the earth's atmosphere, improved on Conklin's measurement of the Earth's motion.[54] Surprisingly, the inferred direction of the Earth's motion was in

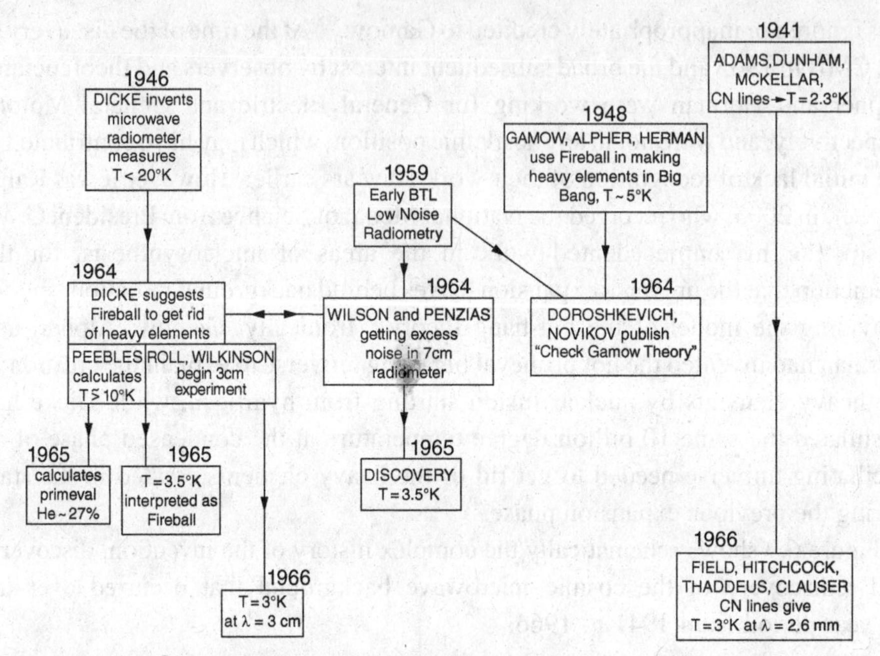

Figure 6.9 Schematic history of the CMB. Columns left to right: physicists, radio astronomers, theoretical physicists, and classical astronomers. Adapted from Wilkinson and Peebles (1984).

nearly the opposite direction and with nearly twice the velocity as the motion of the Earth around the center of the Milky Way Galaxy. This meant that the Milky Way along with other galaxies of the local group were moving toward a distant "Great Attractor" about 250 million light years away, reflecting the large scale local structure deduced from the source counts discussed in Chapter 5. Independent evidence of this local anisotropy came from the Great Wall of galaxies inferred from the analysis of the motions of galaxies obtained from extensive redshift surveys.[55]

The detection of the intrinsic much weaker small angular scale variations in the CMB corresponding to the seeds of galaxy and cluster formation proved to be much more elusive. Measurements from the ground, from high altitude aircraft, and from balloons were unable to detect the ever changing smaller theoretical predictions that managed to stay ahead of the increasingly precise measurements.[56] It would take the COBE satellite to detect and measure the small scale CMB temperature variations of only 30 millionths of a degree or one part in 100,000 of the CMB temperature.[57] As shown in Figure 6.10, later measurements by the NASA Wilkinson Microwave Anisotropy Probe (WMAP) and the European Planck satellites mapped out the variations on angular scales ranging from a few arcminutes to a few tens of degrees.

Figure 6.10 Planck image showing the fluctuations in the CMB. Credit: NASA.

These temperature variations of the CMB were smaller than the surface irregularities of a billiard ball and required extraordinary techniques to measure their structure.

An Unexpected Development

As we have seen, as a result of numerous measurements over a wide range of frequencies, in particular the COBE FIRAS results, it was widely assumed that the CMB had a perfect blackbody spectrum. But that changed in July 2006, when a group of scientists set out to measure the background temperature at multiple frequencies using their Absolute Radiometer for Cosmology, Astrophysics, and Diffuse Emission (ARCADE 2) instrument. ARCADE 2 was suspended from a balloon that flew 37 km above their Palestine, Texas launch site. To everyone's surprise, during the two-hour flight time, ARCADE 2 found an excess signal at 3.3 GHz (9 cm) that was 0.054 ± 0.006 degrees greater than the CMB temperature. Combining their 3.3 GHz measurement with the ARCADE 2 measurements near 8 GHz, as well as previously published background measurements at lower frequencies, suggested that, in addition to the 2.725 thermal cosmic background, there is also a nonthermal cosmic background radiation.[58]

Puzzled by this surprising result, in July 2017, many of the ARCADE 2 experimenters, together with other astrophysicists, gathered together in Richmond, Virginia, to address the challenges raised by this unexpected discovery.[59] The group agreed that an interpretation in terms of a galactic halo seemed unlikely, as this would make the Milky Way unique among spiral galaxies. Unless there is a new kind of previously unknown source, it is also difficult to explain the observed nonthermal background as the sum of a very large number of very weak radio sources, as the density of such sources would need to be more than an order of magnitude greater than that of the faintest (magnitude 29) galaxies found in the

Hubble Ultra Deep Field. More exotic explanations that were discussed by the gathered astronomers included dark matter annihilation and supernovae events in super massive stars, but they too have their problems. While the evidence for the nonthermal background radiation was convincing, the group discussed plans for new confirming experiments that would erase any uncertainties. Meanwhile, understanding the nature of the ARCADE 2 nonthermal cosmic background remains one of the outstanding problems of contemporary astrophysics.

7
Interplanetary Scintillations, Pulsars, Neutron Stars, and Fast Radio Bursts

It is worth emphasizing that if the aerial output had been digitized and fed directly into a computer, these sources might well not have been discovered, because the computer would not have been programmed to search for unexpected objects.[1]

Pulsars are rapidly rotating highly magnetized neutron stars whose radio and sometimes optical, X-ray, and gamma-ray emission is observed as short periodic pulses, much in the same manner as a rotating lighthouse beacon. The first pulsars were discovered in 1967 by a Cambridge University PhD student while looking for Interplanetary Scintillations (IPS), which were in turn discovered by a previous Cambridge PhD student while searching for new quasars by measuring the positions of small unidentified radio sources. Meanwhile, 4,000 miles away at an Alaskan Ballistics Missile Early Warning site, a US Air Force Sergeant also accidently detected the radio emission of about a dozen pulsars while looking for incoming enemy missiles but, due to Cold War security restrictions, the Alaskan discovery remained undisclosed for four decades. Subsequent observations at radio observatories throughout the world have disclosed thousands of radio pulsars that spin at rates of up to 716 times a second, as well as a variety of possibly related galactic and extragalactic transient sources, including giant isolated radio bursts, Rotating Radio Transients (RRATs), magnetars, and Fast Radio Bursts (FRBs). Pulsars, especially the millisecond pulsars with periods between one and ten milliseconds, are extremely stable clocks that have opened up multiple new fields of astronomy and physics research.

Interplanetary Scintillations

Following the exciting discovery of quasars discussed in Chapter 4, there was a great rush at radio observatories around the world to find more quasars. The radio astronomers at the Cambridge Cavendish Laboratory were especially keen to extend their earlier very productive program in extragalactic radio astronomy described in Chapters 3, 4, and 5. To join the chase for new quasars, Cambridge student Margaret Clarke used the existing 178 MHz interferometer system to measure the accurate position and flux density of 88 radio sources taken from the Revised 3C Catalogue as part of her PhD dissertation research.[2]

In the course of her observations, Clarke noticed that two, or perhaps three radio sources, known to be of very small angular size, less than 2 arcseconds in diameter, exhibited rapid fluctuations in intensity, and that no sources larger than 3 arcseconds showed any fluctuations. Clarke speculated that "some mechanism similar to that which causes [ionospheric] scintillations is operating," and correctly recognized that the absence of fluctuations in sources larger than about 5 arcseconds "eliminates atmospheric and ionospheric disturbances as a possible cause."[3] Recognizing that the fluctuations were most pronounced when the sources were close to the Sun she reasoned, "It is not inconceivable that the phenomenon is associated with solar coronal effects." Possibly as a cautionary cover suggested by her advisors, Clarke then stated, "It must be concluded that the cause of these observations remains an interesting but unsolved mystery."[4]

Clarke's two-part PhD dissertation was largely devoted to her observations of the intensity and positions of the 88 sources in the 3CR Catalogue and to "an investigation of ionospheric irregularities using a radio signal from an artificial satellite." The perplexing fluctuations of small diameter radio sources when close to the Sun appeared only in a six page appendix to Clarke's PhD dissertation.

Antony (Tony) Hewish, a senior member of Martin Ryle's radio astronomy research group, had a long-standing interest in the solar corona (Figure 7.1). Using observations of the angular scattering of radio sources, Hewish and his student, John Wyndham, had traced the solar atmosphere halfway out to the Earth showing some of the earliest evidence for a rapid solar wind.[5] When he learned of Margaret Clarke's discovery of the fast daytime variability of some small diameter radio sources, Hewish recalled that, as early as 1954, he had speculated that "if radio sources were of small enough angular size," turbulence in the solar atmosphere would lead to "a very rapid fluctuation of intensity."[6] At the time it was thought that radio sources were much too large to show this effect, which is analogous to the visual twinkling of stars caused by turbulence in the terrestrial atmosphere, so he did not take the idea seriously until Clarke's discovery eight years later. Observations of several other small diameter quasars by Hewish's students, Paul Scott and Derek Wills, confirmed that the amplitude of what they called interplanetary scintillations (IPS) decreased with decreasing angular distance from the Sun.[7]

With characteristic Cavendish secrecy, Hewish and his colleagues and students kept their discovery of IPS to themselves, with the intention of announcing it at the

Figure 7.1 Antony Hewish. Credit: AIP Emilio Segrè Visual Archives, Weber Collection, W. F. Meggers Gallery of Nobel Laureates Collection.

forthcoming General Assembly of the International Astronomical Union (IAU) to be held in Hamburg, Germany, in August 1964. Several weeks before the IAU meeting, one of the present authors (Kellermann) together with Alan Moffet from Caltech were visiting the Cambridge radio astronomy group. Moffet noticed that a shoebox was sitting on a shelf, with the letters IPS boldly written on the shoebox, and called it to Kellermann's attention. When Martin Ryle entered the room, he stood in front of the shoebox, blocking it from view, and Moffet later noticed that, after Ryle had moved, the shoebox had been discreetly repositioned with the IPS label turned to the wall. Realizing from Moffet's questions that the cat was out of the bag, the Cambridge radio astronomers explained about interplanetary scintillations, but only after Moffet and Kellermann were sworn to an oath of secrecy until after the IAU meeting.

The discovery of interplanetary scintillations was published in the 19 September 1964 issue of *Nature* by Hewish, Scott, and Wills, who confirmed Margaret Clarke's suggestion that IPS are caused by turbulence in the solar wind. Clarke, who got a thank you in the *Nature* paper, went on to a postdoctoral appointment in Australia, where she worked with John Bolton on the optical identification of Parkes survey sources.[8] She never returned to work on IPS.

Hewish now had a new way to study the solar wind as well as to determine the diameters of extragalactic radio sources. Anxious to exploit this opportunity, he

Figure 7.2 Cambridge 81 MHz array used by Jocelyn Bell to discover pulsars. Credit: Bell Burnell (1984).

applied for and received a grant of £17,286 to build a radio array in Cambridge specifically to study IPS, with the hope of finding more quasars through their small angular size.[9] This was a low budget project. Hewish's grant could pay for the nearly 100 miles of wires and cable, as well as the wooden poles and other material needed to build the array. However, the construction of the array used the time-honored tradition of low paid research students and even lower paid vacation students to plant the more than 1,000 wooden support posts and to construct and connect the 2,040 dipoles that comprised the 470 m by 45 m 81 MHz (λ3.7 m) radio array (Figure 7.2).

Little Green Men, Pulsars, and Neutron Stars[10]

In late 1965, Susan Jocelyn Bell joined Hewish's group as a new research student and participated with other students pounding the wooden support posts in the ground for a new antenna system (Figure 7.3). Later she was assigned the task of building baluns[11] and transformers, preparing the cables, and connecting up the dipoles that comprised the most sensitive radio telescope of the time.

Jocelyn Bell grew up in Northern Ireland and was an active Quaker. Before joining the Cambridge radio astronomy group, she received her Bachelor of Science degree from the University of Glasgow. Starting in mid-1967 she became responsible for operating the telescope and analyzing the data under supervision from her dissertation supervisor, Tony Hewish. The output from each of the four antenna beams looking at four different declinations in the sky was recorded on four separate chart recorders, producing 96 feet of chart paper each day. In this

Figure 7.3 Jocelyn Bell with paper chart recording. Credit: Bell Burnell (1984).

way, the transit radio telescope surveyed the entire sky between declinations –10 and +50 degrees every four days, so that each patch of sky was covered about 60 times a year.

Each recorder had three pen recordings. One track displayed the telescope output; a second filtered track showed only the fluctuating component; and the third track averaged the intensity of the fluctuating component. Because they were looking for rapid scintillations, the array was designed to use an unusually short time constant of 0.1 seconds, which would turn out to be critical to the discovery of pulsars. Although the Cambridge radio astronomy group had pioneered the use of electronic computers to analyze their radio source surveys, as Bell later recollected, they "decided initially not to computerize the output because until we were familiar with the behavior of our telescope and receivers we thought it better to inspect the data visually, and because a human can recognize signals of different character whereas it is difficult to program a computer to do so."[12]

This meant that, each day, Bell had to analyze the 96 feet of chart paper by hand, as she pointed out, "every day, seven days a week, and I operated it for six months, which meant that I was personally responsible for quite a few miles of chart recordings."[13] She quickly learned to distinguish between real scintillations and terrestrial interference, each of which had a different characteristic tracing on the chart records. Soon after she started to analyze her chart records, she noticed

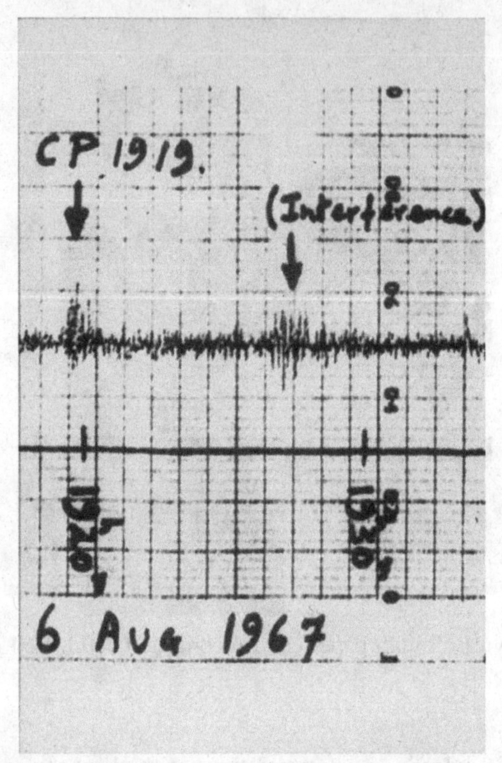

Figure 7.4 Chart recording from 6 August 1967 showing the pulsar CP 1919 and a burst of terrestrial interference. Credit: Bell Burnell (1984).

what she called a peculiar "scruff which didn't look like interference and didn't look exactly like scintillations," and that appeared to come from the same part of the sky. Curiously, it was happening only in the middle of the night. Since interplanetary scintillations are due to the solar wind, Bell was uncertain about the source of this "scruff" well away from the Sun. Adding to her confusion, she noticed that the "scruff" did not always appear when the telescope was pointed at this part of the sky (Figure 7.4).

After consulting with Hewish, they agreed to run the chart recorder faster in order to better determine the time dependence of the "scruff." Since the area of the sky that was showing the unusual "scruff" was transiting at night, it fell to Bell to go out each night to switch on the fast chart recording. She did this every night for most of the month of November but recorded only receiver noise. Convinced that the source was gone, she took a night off to attend a lecture, and when she went out the next morning to review the normal survey chart recording, there it was. Hewish let her know, with some annoyance, "you have gone and missed it." Fortunately, for Bell, on 28 November 1967, it reappeared. As she later recalled, "As the chart

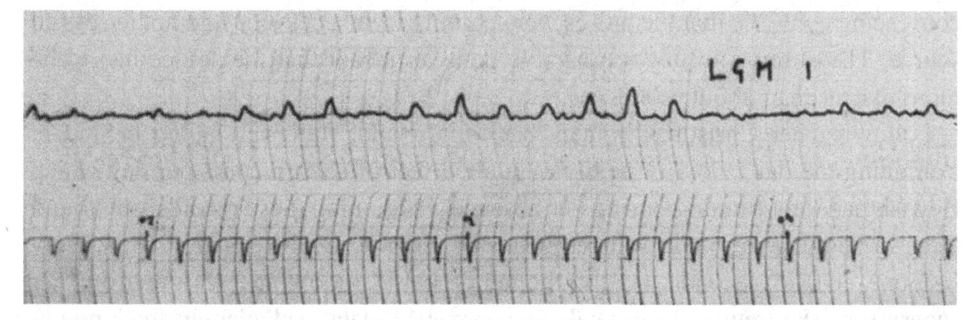

Figure 7.5 High time resolution recording of CP 1919, initially LGM 1. Lower tracing shows 1 sec time ticks. Credit: Bell Burnell (1984).

flowed under the pen I could see that the signal was a series of pulses ... 1 1/3 seconds apart."[14] She telephoned Hewish, who had just completed teaching an undergraduate laboratory class, to tell him the news. Hewish was an experienced radio astronomer and was familiar with terrestrial interference. Upon hearing from his inexperienced student, that the "scruff" was a string of pulses 1 1/3 seconds apart, he immediately declared it as typical of radar interference and explained, "Well, that settles it, it must be man-made" (Figure 7.5). But Bell noted that for many months the "scruff" occurred each day at the same sidereal time in their transit instrument (same right ascension) and not at the same solar time. So, as she explained, "It's not normal *earth-man-made* because normal *earth-man* works on a twenty-four hour schedule, not a twenty-three hour and fifty-six minute [sidereal] schedule – except of course for other astronomers." Following up, Hewish checked with other observatories in Britain, none of whom indicated that they were doing anything that might have generated the mysterious signal that Bell had detected.

To check whether it might possibly be an instrumental effect, colleagues Paul Scott and Robin Collins converted the old Cambridge 4C survey telescope to work at 81 MHz and confirmed that the source was real and was not an instrumental effect. It looked like a star, it went around the sky with the stars, and yet it looked man-made. What was it? The rapid pulsation rate was too fast for a star – at least a normal star. Around Christmas time, Bell walked into what she later called "a high level conference" in Hewish's office where they were discussing the possibility that the signals might be coming from an extraterrestrial civilization, whimsically dubbed Little Green Men or LGM. Bell later explained that, concerned about making fools of themselves, we "played the cards fairly close to our chests," and surrounded the project with even more than the traditional Cambridge secrecy. Several months of observations showed no evidence for Doppler shifts that might result from the orbital motion of the putative planet around its star but did show the Doppler shift corresponding to the Earth's motion around the Sun, providing convincing evidence that the pulses were coming from a celestial and not terrestrial source. The absence of any measurable parallax meant that the source had to be more distant than about 1,000 AU.[15]

Following her Christmas holiday back in Ireland, Bell was falling behind in examining the daily flow of nearly a hundred feet of chart records per day which Hewish had kept running during her absence. Examining these records, she found three more pulsating radio sources with periods of 1.19, 1.27, and a surprisingly fast one at 0.25 seconds. At this point, probably following pressure from her supervisor, she returned to analyzing her IPS data, calculating the angular diameters and preparing her dissertation, "The Measurement of Radio Source Diameters Using a Diffraction Method." Jocelyn Bell's remarkable discovery of pulsating radio sources was relegated to the Appendix of her dissertation, just as was Margaret Clarke's discovery of IPS.

The discovery of the first pulsating radio source, PSR 1919+21, was reported in the 24 February 1968 issue of *Nature* by Hewish et al., with Jocelyn Bell listed as the second of five authors.[16] This was better treatment than Margaret Clarke received for her discovery and interpretation of interplanetary scintillations. By timing the pulse arrival times over a period of many months, Hewish et al. were able to determine the pulse period to be 1.3372795 seconds with an accuracy of one part in 300 million, and noted that the pulse intensity varied widely on time scales of minutes, hours, and days.[17] They also showed that the signal had a bandwidth of only about 80 MHz and drifted downward in frequency at a rate of about 4.9 MHz/second. Hewish et al. realized that the delay in the arrival of the lower frequency signal was due to the propagation delay in the interstellar medium, known as "dispersion," that depends on the electron density along the propagation path and increases with the square of the wavelength. From estimates of the electron density in the interstellar medium, they concluded that the source was a fairly local object in the Galaxy at a likely distance of a few hundred light years. Based on the brief time scale of the individual pulses, the authors reported that the dimensions of the source must be smaller than about 3,000 miles and speculated that the radiation may be "associated with the rapid pulsation of a white dwarf or neutron star." Details on the other three pulsars, first noted by Bell in her survey, were published a few months later.[18]

On 20 February, shortly before the publication of the *Nature* paper, Hewish gave a widely attended seminar in Cambridge announcing the discovery of a new type of radio source. Until then, the Cambridge radio astronomy group had characteristically kept their discovery to themselves. Malcolm Longair, then a research student in the radio group, explained that, although his office was next door to Hewish's office, "I knew nothing about what was going on until he gave the lecture."[19] Fred Hoyle, a long-time antagonist (Chapter 5) of the Cavendish radio group, apparently annoyed that he had been kept in the dark, sarcastically remarked, "This is the first of these stars that I have heard of."[20] He then went on to explain that he didn't think they were white dwarfs, but were supernova remnants. The moniker "pulsar" was apparently left on a blackboard by a journalist whose name has been lost to history, but the name "pulsar" was immediately adopted in the scientific as well as the popular literature, with none of the acrimony that was associated with the use of the word "quasar" (Chapter 4).

The announcement of the discovery of pulsars had an immediate and broad impact. Within a few months, there was a flurry of theoretical and observational papers reporting discoveries of more pulsars, their polarization, spectral dependence, optical, X-ray and gamma-ray counterparts, the details of pulse profiles including the discovery of narrow sub-pulses, and a plethora of theoretical speculations about the nature of pulsars.

Cornell University's Thomas Gold was probably the first to suggest that pulsars are not pulsating white dwarfs or pulsating neutron stars but highly magnetized rotating neutron stars whose beams of radiation reach the Earth much like the light from a rotating lighthouse beacon.[21] Interestingly, in 1967, even before the discovery of pulsars, the Italian astrophysicist Franco Pacini, then working at Cornell, suggested that, following a supernova event, the star should collapse under gravitational pressure, electrons would combine with protons to form a dense neutron star, and the original stellar magnetic field would be compressed to form a very strong magnetic field. Due to the conservation of angular momentum, Pacini pointed out that the rotation of the neutron star would greatly speed up from the original stellar rotation rate to about once per second. Pacini went on to speculate that the rapidly rotating highly magnetized neutron star would radiate electromagnetic waves, and that this could provide the source of energy needed to drive the observed expansion of supernova remnants such as the Crab Nebula.[22] However, Pacini's prescient speculations went unnoticed until after the discovery of pulsars and played no role in their discovery.

The existence of neutron stars, composed of closely packed neutrons, with typical diameters less than 10 miles and a mass slightly more than that of the Sun, were first postulated by Walter Baade and Fritz Zwicky as far back as 1934, only two years after the 1932 discovery of the neutron by James Chadwick.[23] Later, pulsars were found to radiate at optical, X-ray, and gamma-ray wavelengths as well as in the radio spectrum.

Jocelyn Bell Burnell went on to work in X-ray and gamma-ray astronomy and became a leader in the British educational system. She has received numerous honors and awards for her discovery of pulsars, including being appointed by Queen Elizabeth II as Dame Commander of the Order of the British Empire.

In the half a century since the discovery of pulsars in 1968, thousands of pulsars have been cataloged.

The first pulsars were found as a result of the visual inspection of survey data to identify individual pulses. As Jocelyn Bell remarked in her dissertation, "It is worth emphasizing that if the aerial output had been digitized and fed directly into a computer, these sources might well not have been discovered, because the computer would not have been programmed to search for unexpected objects."[24] Ironically, subsequent pulsar searches have depended on the computer analysis of long data streams that search over a wide range of pulse periods and dispersions to enhance the sensitivity by averaging the data over many pulse periods.

Pulsars at the DEW Line

The Distant Early Warning (DEW) Line or Ballistic Missile Early Warning System (BMEWS) was established by the United States during the Cold War to provide advanced warning of nuclear armed missiles launched from the USSR toward targets in the United States. DEW Line radar stations stretched for 3,000 miles across North America from the Aleutian Islands, Alaska, and Canada to Greenland and Iceland. Each station was staffed by a team of operators. The BMEWS site at the Clear Air Force Station near Anderson, Alaska, had three large reflectors, each approximately 200 by 500 feet (60 m by 145 m) across, and a powerful 420 MHz pulsed radar system that produced multiple beams in the sky (Figure 7.6).

One year before his planned retirement, Air Force Sergeant Charles Schisler was assigned to the Clear Air Force Station. In 2008, at a scientific conference in Montreal, Canada, Schisler reported that in August 1967, he sometimes observed pulses on his monitor even when the radar transmitter was not operating. How could this be? Schisler made careful notes of these anomalous signals and saw that "it peaked four minutes earlier each day."[25] Before being assigned to the BMEWS site, Schisler had been a navigator on B-47 bombers and was trained in celestial navigation, so he quickly recognized that this strange signal had a celestial, not terrestrial origin, and was not a threat to the United States – at least from the USSR. By timing the peak signal in each of the antenna beams he was able to determine the celestial coordinates of these strange pulses. Having no experience in radio astronomy or familiarity with the astronomical literature, on his day off, Schisler drove to the University of Alaska at Fairbanks. With the help of the librarian and a local solar researcher he determined that his pulsating source was associated with the Crab Nebula discussed in Chapter 2.

Over the next months, Schisler logged an additional 12 pulsing sources and noted that they seemed to be concentrated in the general direction of the center of the Galaxy, but he was unable to make any clear association with any other known radio sources or optical object. In early May 1968, he heard on shortwave radio the news of the discovery of the first pulsar in Cambridge and recognized the

Figure 7.6 Alaskan Ballistic Missile Early Warning site near Clear, Alaska. Credit: US Air Force.

importance of his study. He had discovered more pulsars, including the important Crab Nebula pulsar, and his discovery preceded the Cambridge work by several months. But the BMEWS sites were shrouded in military secrecy. Schilser's meticulous work and important discovery remained highly classified and it would be nearly 40 years before he was able to share it with the scientific community.[26] By then all the prizes and recognitions had been awarded.

The Crab Nebula Pulsar: The First Optical Pulsar

The first indications of something special going on at the center of the Crab Nebula came from the multifrequency observations of interplanetary scintillations that suggested a small diameter, steep spectrum radio source located near the center of the Crab Nebula with a brightness temperature that appeared to exceed the limits of synchrotron radiation.[27] The actual pulsar at the center of the Crab Nebula was discovered or, more precisely, rediscovered in Green Bank in 1968, but not recognized as a pulsar until periodic pulses were found with the Arecibo 1,000 foot radio telescope in Puerto Rico. However, the story of the Crab Nebula pulsar goes back much further. In Chapter 3 we discussed the identification of the powerful radio source, Taurus A, with the Crab Nebula, long known to astronomers as the remains of a 1054 AD supernova observed by Chinese astronomers. In 1953, Iosef Shklovsky suggested that both the radio and optical radiation from the Crab Nebula was due to synchrotron radiation from relativistic electrons moving in a weak magnetic field.[28] But there was a problem. Due to radiation losses, the lifetime of the relativistic electrons was understood to be much shorter than the 900 year old supernova remnant.[29] Somehow, energy was being continuously supplied to reaccelerate the electrons so that they could continue to radiate. What was the source of that energy?

In 1942 Rudolph Minkowski noted the presence of a peculiar star with no spectral features located near the center of the Crab Nebula which he suggested was the stellar remnant of the 1054 supernova.[30] Known for decades as "Minkowski's star," the nature of the central star in the Crab Nebula would remain a mystery until the Dutch astronomer Lodewijk (Lo) Woltjer, then a professor at Columbia University, suggested that radio astronomers should search for pulsed radiation from the Crab Nebula and other supernova remnants.[31]

Prompted by the Cambridge discovery of dispersed pulsating radio sources, but probably oblivious to Woltjer's speculation, in October 1968, MIT's David Staelin and Edward Reifenstein III initiated a search for dispersed periodic signals using the NRAO 300 Foot radio telescope. Using a sampling time of 0.6 seconds at each frequency channel and a sophisticated computer algorithm to search for dispersed signals, they found a dispersed pulsating radio source, coincident with the Crab Nebula, but no evidence for any periodic behavior of the kind found by the Cambridge workers. In reporting their results in the journal *Science*, Staelin and Reifenstein noted that any periodicity had to be significantly shorter than their 0.6 second sampling time.[32]

Cornell radio astronomers took up the challenge. Using the 1,000 foot Arecibo radio telescope to observe a region around the Crab Nebula with a sampling rate of 100 samples per second, they found a strong signal corresponding to pulses with a 33.09 millisecond period that they associated with Minkowski's central star.[33] The mystery was solved: pulsar NP 0532 was the engine driving the acceleration of relativistic electrons that keep the Crab Nebula radiating throughout the electromagnetic spectrum from radio through optical to gamma-rays. But there were still surprises to come from one of the most extraordinary scientific investigations ever reported.

Michael Disney and John Cocke were two young theoretical astrophysicists who had recently joined the University of Arizona and who met accidently at the motel swimming pool where they were both staying with their families while looking for places to live. As Cocke and Disney later recalled, they agreed to get in on the excitement generated by the discovery of pulsars. Ignoring the discouragement from prominent astrophysicists, as well as from colleagues who scorned their lack of observing experience, Cocke and Disney decided to look for optical pulses from the Crab Nebula. Although neither one had any experience observing with telescopes, they managed to get permission to use the 36 inch University of Arizona telescope on Kitt Peak, the smallest telescope on the mountain, for four days in January 1969. As Disney later confessed, "looking for pulses in the optical is a rather complicated experiment, requires all sorts of equipment which we'd [sic] no idea at the time."[34] Fortunately, they were able to team up with Donald Taylor, the local "electronic wizard."

The group spent most of the first of their four nights getting the equipment working and trying to locate Minkowski's central star. They set the output of the photodetector at the focus of the telescope to average light from the star over the Crab pulsar period and displayed the result on a cathode ray tube (CRT), whose sweep rate was also set at the 33 millisecond pulse repetition period. In this way, if there were optical pulses at the same rate as the radio pulses, the pulse profile would appear on the CRT as the data were accumulated.

The next night they were ready to seriously observe Minkowski's star but found no discernable pulses. Apparently discouraged and with poor weather predicted, the next day Taylor returned to the University and left the inexperienced Cocke and Disney alone with the night assistant. Their final two nights were so foggy on the mountain that they got lost finding their way from the telescope to their dormitory, and they were unable to resume their observations. However, the next observer on the 36 inch called to inform them that his wife was ill and that they could have a few more nights of his time. Meanwhile, Cocke realized that he had made an error of 2π in calculating the correction for the Doppler shift of the pulse period due to the motion of the Earth. On 16 January 1969, after adjusting the timing of the CRT display to the new value, they pointed the 36 inch telescope to Minkowski's star. As Disney later described, "We all crouched in front of the electronics looking at this little screen, watching these green dots coming up the middle," and forgetting that they had accidently left a tape recorder running that then recorded their historic discovery of the first optical signals from a pulsar.[35]

DISNEY: We've got a bleeding pulse here.

COCKE: Hey. Wow. You don't suppose that's really it. Do you? Can't be!

DISNEY: It's right bang in the middle of the period. Look! I mean right bang in the middle of the scale. It really looks like something.

COCKE: Hmmm.

DISNEY: It's growing too. It's growing up the side a bit too!

COCKE: God! It is! Isn't it?
DISNEY: Good God! You know that looks like a bleeding pulse.
NIGHT ASSISTANT: It is. Hey. You are right!

Curiously, as it later developed, the 1969 detection of optical pulses from the Crab Nebula at the University of Arizona apparently was not the first time that this phenomenon had been observed. At the 2015 meeting of the International Astronomical Union in Hawaii, Jocelyn Bell Burnell herself told of two earlier incidents when optical signals were seen from the Crab Nebula pulsar.[36] Back in 1957, long before any pulsars had been observed, an unnamed woman visitor at the University of Texas McDonald Observatory Open House looked at Minkowski's star through the 82 inch telescope and reportedly remarked, "That star is flashing." Ten years later, Susan Simkin, an astronomer at Michigan State University, was using the Kitt Peak 84 inch telescope to obtain a spectrum of Minkowski's star and noted that it was "flickering." In neither case did any other astronomer confirm the reports from two young women who apparently had the rare visual acuity to see the rapid 30 Hz variability of the Crab pulsar.

Millisecond Pulsars

The radio source known as 4C 21.53 first appeared as an innocuous source in the Cambridge 4C Catalogue.[37] In 1979, James R. Rickard and William (Bill) Cronyn called attention to 4C 21.53 as an extreme example of what they called "scintars," a proposed new class of radio sources that lay in the plane of the Milky Way, yet exhibited interplanetary scintillations (IPS).[38] Scintars were hard to explain, since a radio source lying in the plane of the Milky Way should show an angular broadening due to scattering from turbulence in the interstellar medium by an amount that increased as the square of the wavelength. Rather, 4C 21.53 surprisingly appeared unaffected by interstellar scattering and displayed IPS characteristics of radio sources smaller than 1 arcsecond in diameter. Also, the radio frequency spectrum was unusually steep, similar to that of pulsars, and not characteristic of radio galaxies or quasars. Aware of the 1939 speculation by Fritz Zwicky that neutron stars are formed as a result of supernovae explosions,[39] University of California professor Donald Backer's attention was drawn to 4C 21.53 by the nearby extended radio source that Backer considered might be a supernova remnant similar to the Crab Nebula, surrounding a pulsar that appeared as the scintillating radio source. If the pulsar was relatively close to us, it would explain the lack of interstellar broadening. However, his hypothesized association of the steep spectrum scintillating source and the nearby extended radio source meant that the position of 4C 21.53 in the 4C Catalogue had to be in error. Moreover, no pulses had been observed from 4C 21.53. Undeterred,

Backer speculated that perhaps the pulsar period was so brief that the pulses were smoothed out by scattering in the interstellar medium, and he submitted a hastily written paper to the European journal, *Astronomy and Astrophysics*, with his speculative conjecture. Following a critical review by the journal referee and the editor's admonishment that the paper was speculative, Backer let the paper with his prophetic suggestion die unpublished.[40] However, still convinced of his idea, Backer asked colleagues at the Owens Valley Radio Observatory, Arecibo, and the Very Large Array in New Mexico to look for pulses from 4C 21.53, but they were all unsuccessful. Important encouragement came from his colleague, Miller Goss, who responded to Backer's request to observe with the Westerbork radio telescope in the Netherlands. Goss confirmed the small diameter of the IPS source in 4C 21.53 and more importantly reported a high degree of polarization, characteristic of pulsars.

Encouraged by Goss's report, Backer asked his Berkeley graduate student Srinivas Kulkarni to look again for pulses using the Arecibo 1,000 foot radio telescope, this time using a faster sampling time needed to disclose a fast pulsar. Kulkarni indeed found evidence for pulsations with a remarkably short apparent repetition period of about 1.6 milliseconds (0.0016 seconds), but the signal appeared only intermittently, and it was not clear if they were really seeing a pulsar or some spurious signal. Two months later, in November 1982, Backer and his colleagues again used the Arecibo antenna to confirm the 1.558 millisecond period corresponding to a remarkably fast pulsar spin rate of 642 rotations per second.

Following the established convention of naming radio sources based on their coordinates, they named the pulsar PSR 1937+214. The high spin rate was perilously close to a neutron star's maximum spin rate of around 2,000 times per second, above which the star would catastrophically destruct due to the extreme centrifugal force at the surface, that would be moving at half the speed of light. Backer and his group were also able to detect the effect of interstellar scintillations that explained the apparent intermittent signals they had observed earlier.[41] From the observed frequency dispersion of the pulse arrival times, assuming a galactic electron density, Backer et al. estimated that PSR 1937+214 lay at a distance of about 8,000 light years, and George Djorgovski at the Palomar Observatory was able to optically identify PSR 1937+214 with a magnitude 20 red object.[42] The nearby extended radio source that Backer suspected of being a supernova remnant associated with the pulsar, motivating his dogged investigation, turned out to be an unrelated H II region. Were it not for Backer's incorrect speculation about the association of the compact steep spectrum source with the extended radio source, he would not have pressed on to search for pulses from 4C 23.51 and ended up discovering an important new class of pulsars. As David Helfand wrote in *Nature*, "The story of the discovery [of millisecond pulsars] is one of perseverance combined with a dash of good luck."[43]

Pulsars with very short periods, less than about ten milliseconds (0.01 seconds) are referred to as millisecond pulsars and appear to form a separate population from other pulsars, which have typical periods between 20 milliseconds and several tens of seconds. Millisecond pulsars are frequently found in globular clusters where the high stellar density allows neutron star interactions with other stars to form binary systems. Unlike normal pulsars that slow down as they emit particles and radiation, millisecond pulsars have been spun up, probably the result of accretion by the pulsar from a binary system stellar companion. Millisecond pulsars are extremely stable clocks with error rates of less than one part in 10^{13} or 1 second in a million years.[44] This extraordinary stability has enabled new precision tests of General Relativity, the first detection of gravitational waves, and uncovered the first known planets outside of our Solar System (Chapter 10).

FRBs, Perytons, Magnetars, and Microwave Ovens

From time to time, a few pulsars, including the Crab Nebula pulsar, emit a single giant pulse thousands of times stronger than the periodic pulses.[45] In a search for other examples of giant pulses, a group led by the University of West Virginia professor Maura McLaughlin found a new type of radio source characterized by dispersed bursts and suggested that they were from rotating neutron stars with magnetic fields of up to 10^{13} times the Earth's magnetic field.[46] Since each pulse from a given source had the same unique dispersion measure, they were able to exclude the possibility that they were seeing terrestrial radio interference. Unlike pulsars that have regular periodic pulses, these sources had irregular intervals between bursts ranging from 4 minutes to 3 hours, which the group named "Rotating RAdio Transients (RRATs)."

Later, West Virginia University Professor Duncan Lorimer and his student, David Narkevic, were searching for more RRATs by re-examining five-year-old data from the Parkes 210 foot radio telescope taken in the general direction of the Small Magellanic Cloud (SMC).[47] Because they were looking for RRATs possibly located in the SMC, their search included an unusually large range of dispersion measures, well beyond those that might be associated with sources located within the Galaxy. They found a single strong pulse in data taken on 24 August 2001 that became known as the "Lorimer Burst." The Lorimer Burst was very narrow in time, lasting less than 5 milliseconds, and had the characteristic dispersion of a celestial source which was greater than that of any known pulsar located away from the Galactic Plane (Figure 7.7).

The data had been taken with the Parkes multibeam feed system that placed 13 separate beams on the sky. From comparison of the intensity in each of the three beams where it was seen, the team was able to locate the position of the burst in the sky about 3 degrees away from the SMC with an accuracy of about 7

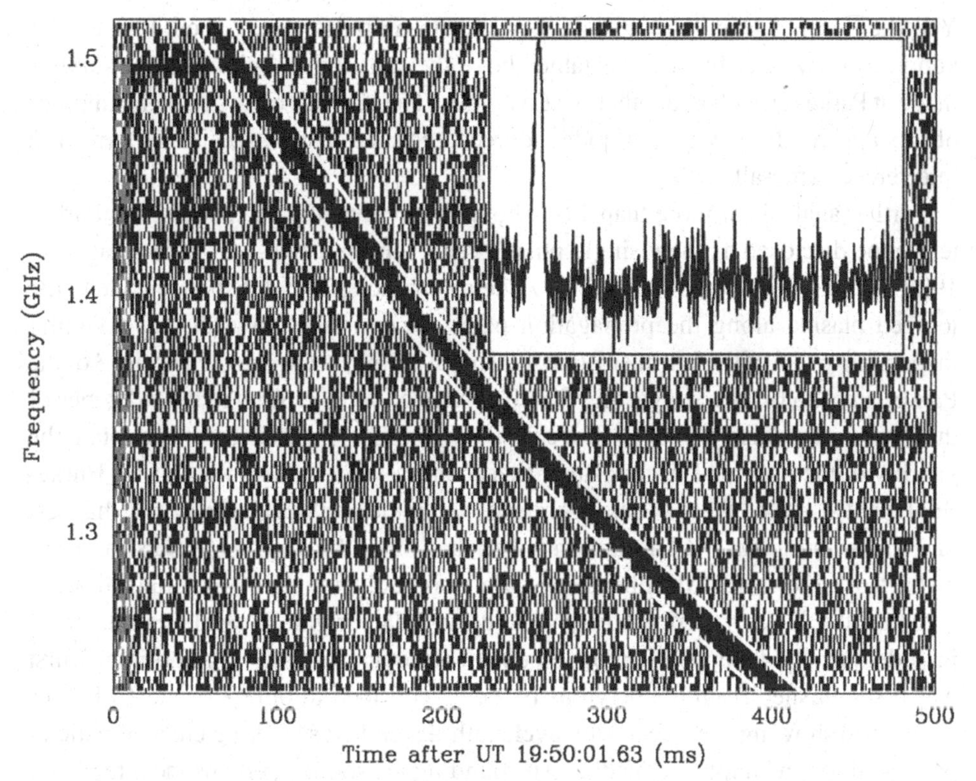

Figure 7.7 First FRB detected by Duncan Lorimer and David Narkevic from five-year-old Parkes data. Main plot shows the drifting burst frequency as a function of time. Inset shows the burst signal averaged over the whole band and corrected for the frequency dispersion. Courtesy of Duncan Lorimer.

arcminutes. Based on the fact that they had analyzed data taken over about 500 hours and covered only about 0.01 percent of the sky, Lorimer et al. estimated that, over the whole sky, there might be as many as 100 observable bursts per day. There was no obvious optical object or pulsar anywhere near the radio burst position. On the basis of the unusually large dispersion, as well as the increase in apparent pulse width with wavelength which was consistent with that expected from a signal propagating through the interstellar medium, Lorimer and colleagues concluded that the burst originated in a distant extragalactic object, and that they had discovered a new type of radio source.[48]

But there was a problem. The Lorimer Burst was very strong; more than 100 times greater than their detection limit. In fact, the signal was so strong that it saturated the data acquisition system of one of their antenna beams. For essentially all celestial objects, stars, galaxies, quasars, pulsars, etc., there are more faint objects than strong ones. The extragalactic radio source count that we discussed in Chapter 5 is typical.

Yet, the Lorimer Burst was a single isolated strong pulse. There should have been hundreds of weaker bursts detectable above the sensitivity limit of pulsar surveys made at Parkes as well as at other observatories. The absence of any fainter examples of this apparently new type of radio source suggested that maybe it was terrestrial interference after all.

Further analysis of more than 1,000 hours of archival data taken with the Parkes telescope disclosed 16 new single pulse events during a five-year time span from 1998 to 2003. Each pulse showed the λ^2 delay characteristic of dispersion due to the ionized plasma along the propagation path, yet curiously they all had a similar dispersion to that of the Lorimer Burst.[49] Moreover, they were detected in all 13 of the Parkes antenna beams, suggesting that they were coming in through the sidelobes of the antenna and were terrestrial, not celestial in origin. This cast further doubt on the validity of the original Lorimer Burst as a cosmic event. In Australia, Sarah Burke-Spolaor et al. called these events "perytons," after the mythical "winged elk that cast the shadow of a man."[50] The authors noted that, since the perytons were seen in the protected 1,400–1,427 MHz band, they were probably not intentional transmissions and that the lack of periodicity precluded a radar origin. Possible explanations included locally generated RFI or some meteorological activity. The Lorimer Burst was distinguished from the perytons by being confined to only 3 of the 13 Parkes beams and showing the effects of wavelength-dependent scattering characteristic of passage through an ionized medium surrounding the source, perhaps the interstellar medium of the host galaxy. Yet the concerns about the cosmic origin of the Lorimer Burst remained.[51]

In an apparent breakthrough in 2013, an Australian team discovered four more highly dispersed pulses. Now, instead of a single event, there was a population of these mysterious sources named Fast Radio Bursts (FRBs) which had measured dispersions characteristic of being very remote, with redshifts as great as one.[52] But all four of these new FRBs were found by the reanalysis of archival data obtained at Parkes. If they were real, why weren't FRBs seen by other radio telescopes? The conundrum was finally solved in 2014, when the Arecibo Observatory reported the detection of FRB 121102 in a sidelobe of one of their seven antenna beams and, in 2015, the NRAO Green Bank Telescope found the second non-Parkes FRB, 110523.[53]

The following year saw a bizarre solution to the mystery of the perytons. Since perytons were observed only at Parkes, only during normal office hours, and rarely on weekends, it was natural to suspect a local origin. However, it was difficult to explain their λ^2 dispersion law characteristic of a signal propagating through the intergalactic medium, although all of the perytons suspiciously appeared to have the same dispersion. It seemed unlikely that any artificial signal or anything associated with meteorological activity could reproduce these properties, but that is precisely what happened. As the detection of perytons was clustered around local noon, this cast suspicion on the microwave ovens used to prepare lunch at the Observatory. Further tests showed that, when a heating cycle was interrupted by opening the oven door before the end of the cycle, this prematurely shut down the magnetron that powers microwave ovens, and somehow generated an artificial pulse with the same characteristic frequency dispersion as FRBs. Because the shielded microwave oven door was open, it did not provide the RFI protection that would normally prevent any radiation from escaping.[54] Although the problem of the perytons could now be set aside, many questions still remained.

Since FRBs were not known to repeat, many researchers suspected that they might be the result of some cataclysmic event, such as a supernova, that should destroy the source. But follow-up observations of the Arecibo FRB disclosed 10 additional bursts, each having the same dispersion.[55] This meant that whatever was causing these bursts was not a catastrophic event like a supernova or exploding black hole. FRBs had only been observed with large filled aperture radio telescopes having limited angular resolution. Thus, the locations of FRBs were known only with arcminute accuracy, insufficient to identify optical counterparts. Over a six month period in 2015 and 2016, Shami Chatterjee and colleagues pointed the Very Large Array (VLA) in New Mexico at the repeating Arecibo FRB 121102, hoping to catch a burst and measure its position. Over the 83 hours that they observed, they detected nine bursts and determined their position in the sky with a precision of 0.1 arcseconds. They also imaged a faint continuum radio source, co-located with the bursts. Observations with the NRAO Very Long Baseline Array and the European VLBI Network indicated that the continuum source was less than 2 milliarcseconds in diameter with a corresponding brightness temperature greater than about 10 million degrees K. The relation, if any, between the continuum source and FRB, however, was uncertain. Optical observations then disclosed a very faint (magnitude 25, $z = 0.19$) otherwise unremarkable dwarf galaxy coincident with the FRB and the faint continuum sources.[56]

Subsequently, another team used the Australian SKA Pathfinder (ASKAP) interferometric array of 24 antenna elements to detect and localize a single dispersed burst from FRB 180924. In contrast to the Arecibo burst, hundreds of hours of observations of FRB 180924 with both ASKAP and the Parkes 210 foot antennas, both before and after the single detected burst, showed no evidence for any repeated bursts. The precisely measured interferometric radio position of the burst enabled the team to find the faint ($z = 0.32$) optical counterpart. But, unlike the Arecibo repeating burst, that was coincident with the center of the host galaxy and the continuum radio source, the ASKAP burst had no continuum counterpart and was located 0.8 ± 0.1 arcseconds away from the center of the host galaxy. Moreover, the optical spectra of the host galaxy indicated that it has an old stellar population, whereas the Arecibo host galaxy showed signs of active star formation. On the basis of these observed differences of the only two FRBs where the host galaxy was identified, the Australian ASKAP team suggested that "there could be two different populations of burst progenitors."[57]

Many other transient radio sources have also been discovered, some of which are periodic, while others are "one-off." Powerful transient radio, gamma-ray, and X-ray emission has also been detected from the region close to the Galactic Center, as well as from brown dwarfs that may or may not be related to pulsars.[58] Of particular interest are the so-called magnetars, that were first discovered by both Soviet and American interplanetary spacecraft as a result of their intense gamma-ray emission. Magnetars are named for their intense magnetic fields that can reach up to 10^{12}–10^{14} times that of the Earth's magnetic field. Radio pulses have been observed in only a fraction of the approximately 30 known magnetars that are thought to be young neutron stars, suggesting that magnetars may be the engine that drives FRBs.[59]

Strong radio bursts from black hole explosions and supernovae were predicted decades before they were serendipitously discovered by observers looking for something else, and played no role in the actual discovery of FRBs, RRATs, or magnetars.[60] Hundreds of FRBs have now been observed, and it is estimated that thousands of FRBs occur each day and could be detected with current radio telescopes if they were pointed to the right place at the right time. A new generation of very sensitive wide field radio telescopes such as ASKAP and CHIME, the Canadian Hydrogen Intensity Mapping Experiment (Chapter 10), will be able to detect and study many new FRBs. CHIME has already discovered more than one thousand new FRBs.[61] Most, but not all, are probably extragalactic. Only about 10 percent of FRBs are known to repeat. Are these yet a different class of objects, or will all FRBs repeat if observed long enough and with sufficient sensitivity? The basic physics responsible for these energetic bursts remains unclear. Indeed, for the first decade after the initial discovery by Lorimer, the number of proposed models, ranging from extragalactic magnetars to exhausts from alien beams generated to propel interstellar spacecraft, exceeded the number of known FRBs. It is not even clear if all FRBs are produced by the same process or if pulsars, RRATs, magnetars, and FRBs are all manifestations of the same or similar processes and differ only by the propensity of astronomers to assign new names when detecting variations in observed phenomena.

8

Interstellar Atoms, Molecules, and Cosmic Masers

> And by the way, radio astronomy can really become very important if there were at least one line in the radio spectrum.[1]

The prediction of the 21 cm hydrogen line by the Dutch astronomer Henk van de Hulst and the subsequent discovery by Harvard University PhD student Harold (Doc) Ewen is well known. However, the ultimately successful quest was marred by poorly understood theory, a somewhat embellished funding proposal, a distracted and unenthusiastic student investigator, and an innovative, but almost fatal, instrumental design. Radio recombination lines were later observed, first in the Soviet Union, but only after a long standing theoretical misunderstanding of the effect of line broadening. The highly competitive discovery of hundreds of molecular lines in the radio spectrum led to the very fruitful new field of astrochemistry and the recognition of Giant Molecular Clouds as the largest objects in the Galaxy. The unexpected discovery of surprisingly strong cosmic hydroxyl (OH) and water (H_2O) masers could have happened decades earlier, had anyone thought to look. Subsequently, high resolution observations of powerful megamasers resulted in the first clear dynamical evidence for the existence of supermassive black holes in the nuclei of galaxies and led to a new, more accurate, and independent determination of the cosmic distance scale.

The 21 cm Atomic Hydrogen Line[2]

Previously, we have discussed the broad band or continuum radio emission from various celestial bodies that arises from both thermal and a variety of nonthermal physical mechanisms. Until the detection of the 1.4 GHz (21 cm) hydrogen line in 1951, there was no radio equivalent of the rich spectral lines that appear in the visual spectra that astrophysicists use to determine the constituents and properties of the Sun, the stars, and the interstellar medium. These spectral lines are caused by the

Figure 8.1 Dutch astronomer Henk van de Hulst. Credit: AIP Emilio Segrè Visual Archives, Physics Today Collection.

transition from one atomic or molecular energy level to another, with the emission or absorption of a photon at a frequency $f = \Delta E/h$, where ΔE is the change in energy between the two states and h is the Planck constant, equal to 6.63×10^{-27} erg sec.

In 1944, Henk van de Hulst was a student in German-occupied Netherlands. At the suggestion of Jan Oort, he discussed the possibility of observing features in the radio spectrum at a small meeting of the Dutch *Astronomenclub*. Only 20 people were present to hear van de Hulst's talk. After the Second World War he published the results of his study in the Dutch language journal *Nederlandsch Tijdschrift voor Natuurkunde*. Following some pessimistic remarks about the unimportance of radio observations to astrophysics, van de Hulst considered three cases of possible radio emission from interstellar hydrogen (Figure 8.1):[3]

(a) Free–free or thermal bremsstrahlung resulting from the interaction between electrons in ionized hydrogen clouds, that he showed could not explain the results reported by Jansky.
(b) Radiation resulting from the recombination of free electrons with atoms (mostly hydrogen) in the ionized interstellar medium and the subsequent cascade transitions of electrons from one high quantum level to a lower level. Typically, the transitions that give rise to radio emission occur at quantum energy levels of the order of 100. These so-called radio recombination lines (RRL) are the equivalent of the familiar Balmer (n to $n = 2$) series that are seen in optical spectra resulting from transitions to the low $n = 2$ quantum energy levels of hydrogen.[4] Due, in part, to an algebraic error, van de Hulst incorrectly

concluded that the recombination lines would be so broad that they would blend together and be unobservable as discrete lines in the radio spectrum.[5]

(c) In two short paragraphs near the end of his paper, van de Hulst, almost parenthetically, speculated on the detectability of the 21 cm hyperfine structure hydrogen line corresponding to the spin flip of the electron in the hydrogen ground state ($n = 1$). The difference in the energy levels between the two states where the electron and proton spins are parallel or anti-parallel corresponds to a radio frequency of 1,420.405 MHz (21 cm).

Based on his calculation of the low transition probability that an electron would flip from the higher hyperfine structure state to the lower one and thus lead to an observable radio line, van de Hulst concluded that if the sensitivity of receivers could be improved by a factor of 100, then "this possibility does not appear hopeless," and that "the existence of this line remains speculative." Interestingly, van de Hulst did not state the precise frequency that would be needed by observers to search for the radio emission resulting from the hyperfine transition, and only mentioned that it would result in a "quantum of wavelength 21.2 cm." Apparently the physical constants were not sufficiently well known, at least to van de Hulst, as the correct value is 21.1 cm.[6]

Shortly after the end of the War, van de Hulst visited the Yerkes Observatory in Williams Bay, Wisconsin, and met with Grote Reber to discuss the possibility of detecting the 21 cm hydrogen line. Inspired by his discussions with van de Hulst, Reber began to build the necessary equipment for a 1,420 MHz receiver but, as described in Chapter 1, in mid-1947, Reber sold his antenna and all of his equipment to the National Bureau of Standards and never returned to pursue the hydrogen line. Curiously, in their 1947 review paper, Reber and Jesse Greenstein cited the work of van de Hulst, but inexplicably quoted an incorrect frequency of 1,410 MHz, that is closer to 21.3 than 21.2 cm.[7] Somewhat later, Reber also alerted the Australian radio astronomers to the 21 cm hydrogen line, but apparently they did not follow-up until after they learned of the successful detection at Harvard.[8]

Iosef Shklovsky in the USSR did not have access to van de Hulst's Dutch paper, but was aware of the possibility of detecting the 21 cm hydrogen line from the brief reference to van de Hulst's work in a copy of the Reber and Greenstein paper that had reached the USSR. Excited by the prospect of observing the 21 cm hydrogen line, Shklovsky carried out his own calculations. Contrary to van de Hulst, Shklovsky concluded from his analysis that the 21 cm hyperfine transition should be observable with then existing equipment in the USSR, and he recognized the potential to determine the distribution and temperature of interstellar hydrogen clouds from the Doppler motions and broadening of the observed 21 cm lines.[9] Moreover, recognizing the opportunity for Russia to get world recognition, Shklovsky concluded his paper with,

Figure 8.2 Russian radio astronomer Viktor Vitkevitch. Credit: Pushchino Radio Astronomy Observatory, P.N. Lebedev Physical Institute, Russian Academy of Sciences.

"Soviet radiophysicists and astronomers should try to solve this intriguing and important problem."

Not having any experimental experience himself, Shklovsky encouraged his colleague Viktor Vitkevitch to attempt to detect the 21 cm hydrogen line (Figure 8.2). Vitkevitch, who was then the leading radio astronomer in the Soviet Union, enthusiastically responded to Shklovsky's suggestion and soon began to design the specialized instrumentation needed to detect the hydrogen line.[10] According to Shklovsky, Vitkevitch's design incorporated a novel scheme, later independently adopted by Ewen and Ed Purcell at Harvard, to eliminate the effect of changing gain of the receiver by rapidly switching between two frequencies, one at the expected line frequency and one well removed. The difference between the signal at the two frequencies represented the hydrogen line emission above the level of any continuum radiation or receiver noise.

However, suddenly and without explanation, Vitkevitch apparently lost interest and told Shklovsky, "It is easy for you theoreticians to argue, but what happens to us experimenters? I have enough ideas of my own and no time for your projects." Shklovsky claimed that for the next year and a half he was "breaking his head" to get Vitkevitch to complete the instrumentation and to try to detect the hydrogen 21 cm radio line.

In the meantime, Harold "Doc" Ewen, a talented and ambitious Harvard physics graduate student, was looking for a thesis topic. However, he also needed to earn money to support his growing family and took a full time job at the Harvard Nuclear Research Laboratory with the task of producing an external beam of protons from the Harvard cyclotron. Ewen was not your ordinary graduate student. After receiving his undergraduate degree in mathematics and astronomy from Amherst College in 1943, Ewen taught astronomy for a year at Amherst before serving for three years as a Second Lieutenant in the US Naval Air Corp, where he took courses in electronics at both Princeton and MIT. At Amherst, he taught navigation to Air Force pilots, including the Boston Red Sox baseball star Ted Williams, who went on to fly jet fighters during the Korean War. After completing his military service, Ewen retained his appointment in the Naval Reserve and enrolled in Harvard to obtain a PhD in physics under Harvard professor Donald Menzel, who was a Captain in the Naval Reserves and his commanding officer. Menzel sent Ewen to the National Bureau of Standards, where he got his first exposure to radio astronomy working under Grote Reber during a two week training assignment.

Seeking advice for a thesis topic, Ewen turned to physics professor Edward Purcell, with an expressed interest in working on something that would utilize his experience in microwave electronics. By then, the frequency of the 21 cm hyperfine structure transition had been accurately measured in the laboratory, and Purcell suggested that Ewen might try to detect the line in the interstellar medium.[11] Apparently neither Purcell nor Ewen were aware of the van de Hulst prediction until Ewen heard about it at the December 1948 meeting of the American Astronomical Society at Yale University. Through his naval connections, Ewen obtained a translation of the van de Hulst paper but was discouraged to learn that van de Hulst apparently considered it unlikely that the 21 cm line could be observed in the interstellar medium. Nevertheless, Purcell and Ewen agreed that they could put a significant upper limit on the strength of the galactic hydrogen line, and Ewen inferred, incorrectly as it turned out, that at least he would not have to compete with the Dutch, who would be unlikely to spend time on what appeared to be a fruitless search. To finance the project, Purcell sent a short application to the American Academy of Arts and Sciences (AAA&S) asking for $500 to construct a horn antenna ($150), purchase a war surplus transmitter to use as the local oscillator ($100), and construct the 21 cm receiver.[12] As a physics professor, Purcell partly justified the proposal by arguing that, by detecting galactic hydrogen, they could more accurately determine the line frequency. This was perhaps an exaggeration, or at least hopefully optimistic, since the line would be both broadened by thermal motions in the hydrogen gas as well as Doppler shifted by the relative motion of the Earth and the hydrogen cloud but, at the time, Purcell had no background in astronomy.[13] Fortunately, however, Harlow Shapley, the director of the Harvard College Observatory, was chair of the committee that seemed to ignore this aspect of the proposal. Purcell was informed six weeks later, in a short two sentence letter from the AAA&S, that his proposal had been approved.

While still working full time at the cyclotron, Ewen spent nights and weekends at the Lyman Physics Laboratory, just down the street, using the AAA&S funds to build

Figure 8.3 Doc Ewen, with the horn antenna used to detect the first 21 cm radio emission from interstellar hydrogen. Credit: NRAO/AUI Archives, Papers of H. Ewen.

a large horn antenna and to construct a 1.4 GHz receiver. Ewen's pyramidal horn antenna, which had a 4 by 5 foot aperture, was constructed from plywood, which he lined with copper foil for electrical conductivity. He pointed it out the fourth floor window of the Lyman Laboratory building at an angle where the Milky Way would pass through the beam each night, but it also became an inviting snowball target for passing students as well as funneling rain into Ewen's laboratory (Figure 8.3).

During the War, Purcell was Bob Dicke's supervisor at the MIT Radiation Laboratory. So, he was aware of the switching technique that Dicke had used to stabilize the output of his radiometer in his 1945 experiments that detected thermal radio emission from the Sun and the Moon.[14] Purcell made the important suggestion that Ewen use a novel adaptation of the Dicke system: instead of switching between the sky and a reference source, Ewen should switch between two frequencies, one

Figure 8.4 Doc Ewen, in his Harvard University Laboratory. Credit: NRAO/AUI Archives, Papers of H. Ewen.

tuned to the line frequency and the other to a reference frequency well removed from the line frequency. Ewen used a military surplus radar jamming transmitter as the 1,400 MHz local oscillator and a commercial National Radio HRO shortwave radio, borrowed from the Cyclotron Laboratory on weekends, as the final stage of his receiver. In his first stage mixer, Ewen used experimental crystals given to him by Harald Friis, Jansky's old supervisor at Bell Labs, that Friis told him were "two of the best 1N21 crystals ever tested" (Figure 8.4).[15]

Ewen's initial observations were negative and showed no evidence for any hydrogen line emission. Resigned to a negative thesis, Ewen was less than enthusiastic about his thesis research topic and more interested in producing a proton beam before his rivals in Enrico Fermi's group at the University of Chicago. Only after learning about Shklovsky's more optimistic calculation did Ewen become more excited about his 21 cm thesis project, but now he was also concerned about being scooped by Russian investigators he assumed, incorrectly as it turned out, would have been inspired by Shklovsky's paper. As Ewen later wrote to Purcell, "I was near

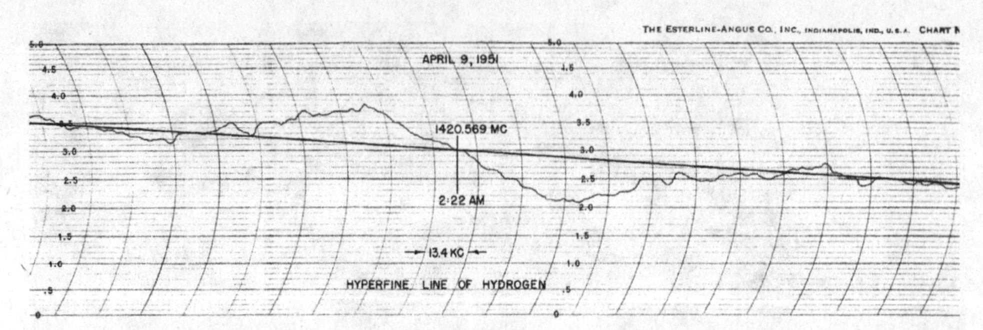

Figure 8.5 Chart recorder tracing of Ewen's first observation of 21 cm interstellar hydrogen showing the double response as the receiver was scanned first through the main channel and then through the reference channel. Courtesy of Paul Horowitz.

paranoia on occasion thinking the Russians would beat us out. The Dutch never entered my mind."[16]

Ewen was switching his receiver between two frequencies only 10 kHz apart. He understood that if the hydrogen line were broader than 10 kHz, the signal from two channels would cancel, but his switching range was limited by the HRO receiver which he was not able to modify since it belonged to the Cyclotron Lab.[17] After learning about Ewen's problem, Purcell considered it for a day, and then opened his wallet to give Ewen $300 in cash to allow him to buy a new shortwave radio that he could modify to enable frequency switching over a 75 kHz range. The two passbands were slowly tuned across a 400 kHz frequency range by means of a motor that drove the tuning capacitor.

Ewen was able to complete the receiver modifications during the long Thanksgiving break, uninterrupted by needing to return test equipment to the Cyclotron Lab. He returned the following week to the Cyclotron Lab to work on producing the external proton beam and planned to return to his hydrogen experiment during the long Christmas holidays. However, it was not until 16 February 1951 that the cyclotron successfully produced an external beam of protons and Ewen could go back to his hydrogen thesis project.

The long Easter weekend presented a special opportunity to use the borrowed Cyclotron Lab instruments without interruption for his 21 cm experiment. At around 2:30 a.m. on 25 March (Easter Sunday), the Galaxy passed through Ewen's antenna beams and, as Ewen swept his receiver across a range of frequencies, he thought he saw the characteristic S-shaped response as the signal passed first through one frequency band and then the other (Figure 8.5). But the signal frequency was 150 kHz above the laboratory value of the hydrogen hyperfine transition frequency. Suspecting that this might be due to the Doppler shift resulting from the relative motion of the Earth and the galactic hydrogen, Ewen telephoned the Harvard College Observatory to ask what the Doppler shift was at the position where he found the signal. Whoever answered the phone at the Observatory apparently did not like being bothered at this early morning hour and hung up. Calling on his astronomy background, Ewen then did the calculation himself. Within a week, Ewen could see that the peak signal strength came four minutes later each day, providing firm evidence that the signal was celestial, and not instrumental or from an external terrestrial source. Finally, after more time, Ewen noted that the observed signal frequency changed, corresponding to the changing Doppler shift as the direction of the Earth's motion changed as it moved in its orbit around the Sun, providing further proof that he had succeeded in detecting radio emission from galactic hydrogen.

At the time both van de Hulst from Leiden and Frank Kerr from Australia were visiting Harvard. Van de Hulst was giving a course on radio astronomy, while Kerr, whose background was in radiophysics, was there to learn some astronomy. Both van de Hulst and Kerr keenly followed the Harvard investigation and kept their colleagues in Leiden and Sydney informed, including the critical need for frequency switching. When Purcell called a meeting to tell Kerr and van de Hulst of Ewen's success, he suggested that they ask their Sydney and Dutch colleagues to confirm the Harvard detection.

According to Ewen, van de Hulst did not offer any congratulations and seemed "shocked" and not particularly "delighted to learn that his 'long shot speculation' proved to be correct."[18] Ewen claimed that van de Hulst asked a lot of questions, but did not disclose that "at the time they were working on the line, and had been for several years, but that was obvious from the tone of the discussion." But the Leiden group did not have the same strong technical background as the Harvard Physics Department and were slow to develop the needed instrumentation. Moreover, they had suffered a setback when a fire destroyed their laboratory. As Ewen later remarked, he was surprised to learn that the Dutch were actively trying to detect the hydrogen line themselves, and added, "Had we known, we would not have tried."[19] Van de Hulst finally explained to Ewen that Jan Oort felt so strongly about the importance of having a feature in the radio spectrum that would facilitate measurements of motions and distances within the Milky Way that he continued to pursue the 21 cm line investigation, in spite of van de Hulst's pessimistic expectations.

Ewen and Purcell submitted their paper to *Nature* on 14 June, with the suggestion that publication be delayed until confirmation was received from Leiden and Sydney. However, Ewen established their priority with papers presented on 16 June 1951 at a meeting of the American Physical Society in Schenectady, New York, and a few days later at the American Astronomical Society in Washington.[20] After learning from van de Hulst about Ewen's success and the experimental technique of frequency switching, as well as the correct frequency of the hydrogen transition, the Dutch were able to confirm Ewen's result on 11 May using a 7.5 m Würzburg-Riese antenna, and submitted their paper to *Nature* on 20 June. In Australia, where they had not previously initiated a program to detect the hydrogen line, Jim Hindman and Chris Christiansen, after being alerted by Frank Kerr, quickly assembled the needed apparatus and soon confirmed the Harvard result. The Harvard and Leiden papers appeared together in the 1 September issue of *Nature*, along with a copy of a brief telegram from Joe Pawsey also reporting the successful Australian detection of the 21 cm hydrogen line.[21] Later, the Australians followed up with a full survey of the southern Milky Way, while the Dutch group expanded their analysis of galactic motions.[22] Then the Dutch and Australian groups combined efforts to make a complete 21 cm map of the northern and southern skies that clearly showed the spiral structure of the Milky Way, unobscured by the intervening dust that hides the spiral nature from visual observers.[23]

Ewen later claimed that he wrote his 47-page thesis in three days but had to spend two weeks studying German before he was able to finagle his way through the Harvard language translation requirement by piecing together information from the illustrations and his own knowledge of the subject material.[24] He defended his thesis in May, just two months after first detecting the 21 cm hydrogen signal. His thesis concluded with recommendations for further observations with a larger steerable antenna and a frequency displacement of 150 kHz instead of 75 kHz.

Twenty years after Ewen and Purcell announced the successful observation of the 21 cm hydrogen line, Shklovsky and Viktor Vitkevitch found themselves as the co-examiners of a successful PhD candidate in Moscow. At the customary post-exam celebration, Vitkevitch invited Shklovsky to come sit next to him. No doubt encouraged by multiple vodka toasts to the new PhD, Vitkevitch asked, "Do you remember the affair with the 21 cm line? Do you want to know the truth why I stopped the work?" He then explained that he had enthusiastically related the planned experiment to detect the hydrogen line to his father-in-law, the famous Russian physicist, Academician Lev Landau. Landau immediately dismissed the idea, asserting that any student could do the calculation using the equations that were in his book, and that Shklovsky's calculations were clearly pathological. Vitkevitch, of course, could not ignore his wife's father, the great Lev Landau, but he did not have the courage at the time to explain the reason for his withdrawal to Shklovsky.[25]

Another missed opportunity occurred in Australia in 1949, when Paul Wild independently wrote two internal memoranda discussing the 21 cm hyperfine structure line, the recombination lines, as well as possible fine-structure lines of hydrogen.[26] Although Wild was motivated primarily by the possibility of detecting these spectral lines in the Sun, he discussed applications to the interstellar medium as well, and concluded that probably only the 21 cm hyperfine structure line would be observable. Like van de Hulst, he apparently assumed that the radio recombination lines would be so broadened that they would be blended together and so would be unobservable. In 1952, Wild finally published his paper in the *Astrophysical Journal*, but only after the 21 cm line had been successfully detected by the three groups from Harvard, Leiden, and Sydney.[27] There is no record that these reports were ever distributed outside of the Radiophysics Laboratory, although they were certainly known among the Radiophysics staff. Also, according to John Bolton, his Radiophysics colleague Kevin Westfold had translated Shklovsky's paper and suggested that someone at Radiophysics should build the equipment needed to detect the 21 cm line.[28] Moreover, Bolton and Frank Kerr, who had visited both Harvard and Leiden, were aware of the planned work and separately encouraged their colleagues to search for the 21 cm line at Radiophysics. However, apparently everyone at Radiophysics was already busy with their own projects, and no one considered doing a search for the 21 cm line until after they heard from Kerr about Ewen's success.

Even earlier, in India, Meghnad Saha independently discussed the possibility of radio spectral lines, including the 21 cm hydrogen line, also in the context of detecting the radio line in the Sun or even from other stars.[29] But his paper apparently went unnoticed or was ignored by the radio astronomers. In Sweden, Olof Rydbeck had read Shklovsky's paper, and in 1949 he began a program to detect the 21 cm hydrogen line using one of five captured German Würzburg-Riese antennas that he had brought to Sweden by barge from Norway. But his application for funding was turned down by the Swedish Science Foundation because one of the referees apparently stated that there was no astronomical interest in detecting the 21 cm line. Rydbeck proceeded anyway to construct a 21 cm receiver, but with the limited resources available did not complete the instrumentation until after the announcement by Ewen and Purcell that they had detected the line at Harvard.[30]

Although the theoretical framework had been established in the Netherlands, the USSR, Australia, and India, it seems that these efforts played little or no role in Ewen's important discovery, which was motivated mainly by Ewen's desire to exploit his interest in microwave electronics and by Purcell's interest and understanding of microwave spectroscopy. The Harvard *Nature* paper was primarily concerned with the physical conditions of the interstellar medium, while the companion Dutch paper was quick to utilize the opportunity to explore the structure and motions of the Milky Way Galaxy. Both the Sydney groups under Joe Pawsey and the Leiden group under the leadership of Oort went on to develop important 21 cm research programs aimed at understanding galactic structure and kinematics.

The 21 cm research program at Harvard was, however, slower to develop. Instead of expanding his discovery with more observations and publications, Ewen applied to and was accepted as a student in the Harvard Business School to pursue an MBA degree. Meanwhile, 7,000 miles away, the Korean War was raging, and Ewen was called back to active military duty, although he served for only a month. After returning to Cambridge, Ewen started a small business together with his friend Geoff Knight, that initially built radio astronomy instrumentation for Harvard as well as for other radio observatories.

Under pressure from several students who expressed an interest in radio astronomy, Harvard astronomy professor Bart Bok initiated a radio astronomy program. After learning that all of Ewen's equipment had been given to Merle Tuve at the Carnegie Department of Terrestrial Magnetism (DTM), Harvard purchased a new receiver from Ewen-Knight. In spite of a slow start, under Bok's leadership, Harvard built first a 24 foot and then a 60 foot radio telescope that were used primarily for 21 cm research, primarily by a group of talented students who later went on to become among the leaders in US radio astronomy.[31]

Ewen maintained a part time faculty position with the Harvard Astronomy Department and, together with Bok, taught a course in radio astronomy and became co-director of the radio astronomy project. Ewen later participated in several radio astronomy conferences and was part of the committee that chose the Green Bank site for the National Radio Astronomy Observatory, but he never returned to any active research in radio astronomy. Meanwhile Purcell, who received the 1952 Nobel Prize in Physics for his work on nuclear magnetic resonance, continued his career in atomic and nuclear spectroscopy.

Hydrogen line research has developed into a major field of radio astronomy for studying the structure and motions within the Milky Way Galaxy as well as the kinematics of external galaxies, where it played a major role in the discovery of dark matter (Chapter 10). One observation that created a lot of initial excitement, but which turned out to be more of an embarrassment than a discovery, was the report by two radio astronomers at the Naval Research Laboratory that they had detected a redshifted 21 cm absorption line in the radio galaxy Cygnus A at 1,341 MHz that corresponded to the optically measured redshift, $z = 0.057$.[32] For the first time, it appeared that radio astronomers were able to measure the redshift, and thus the distance of external galaxies, without needing to depend on optical observations. Unfortunately, the claimed absorption line detection of Cygnus A was never confirmed and was apparently spurious.[33] Moreover, due to a calculation error, their claimed line was 4.23 MHz below the frequency corresponding to the $z = 0.057$ optical redshift.[34] Ed Lilley and Ed McClain found what they were looking for and where they expected to find it – except they were looking in the wrong place and there was nothing really there!

Figure 8.6 Nikolai Kardashev. Credit: AIP Emilio Segrè Visual Archives, John Irwin Slide Collection.

Radio Recombination Lines

In his discussion of Radio Recombination Lines (RRLs), van de Hulst incorrectly argued that the Stark effect due to the perturbations caused by the "electric field of nearby electrons and ions" would broaden the energy transition, thus blending all the lines together, and so he concluded, "it is evident that all of the α [those with $\Delta n = 1$] are *unobservable* ... The same is true for lines arising between other transitions between the high energy levels of hydrogen."[35] That remark, which was based partly on the above-mentioned error in his calculations, apparently discouraged any observers from seriously trying to detect any recombination lines, although many observatories had sufficient capability. Likewise, in his 1949 paper, Shklovsky reached similar pessimistic conclusions, as did Paul Wild in 1952. It would be another decade before Shklovsky's brilliant, and characteristically optimistic student, Nikolai Kardashev, reconsidered the problem and showed that van de Hulst, Shklovsky, and Wild had all overestimated the amount of line broadening by the Stark effect, and that it might be possible to detect radio lines due to high order energy transitions, especially those with $\Delta n = 1$ (Figure 8.6).[36]

Encouraged by Kardashev's optimistic calculations, Russian observers eagerly sought to detect the radio recombination lines from interstellar hydrogen. Two independent detections of the H90α line at 8,872.5 MHz by Roman Sorochenko and Eduard Borodzich in Crimea and the 104α line at 5,763 MHz by Alexander Dravskikh at Pulkovo were first reported at a meeting of the Radio Astronomy Commission at the 1964 IAU General Assembly in Hamburg, Germany. Because none of the investigators were allowed to travel to Western countries, the papers were presented by Viktor Vitkevitch and Yuri Parijisky, respectively, the directors of the two observatories.[37] It was an unusually hot day in Hamburg with no air-conditioning in the conference room; the two RRL talks were nearly the last ones scheduled at the end of a long series of talks; the audience was hot and irritable and ready to go home; the visual projection material was poorly prepared; the speakers had limited English language proficiency, and no direct understanding of the observations that they were presenting to a somewhat hostile audience. Moreover, due to the restricted technical information allowed by the Soviet authorities, the two presentations appeared to lack credibility, and the reported RRL detections were widely discounted by Western radio astronomers, although both groups subsequently published their papers in the well-respected Soviet journal *Doklady Akademiia Nauk SSSR.*[38]

Peter Mezger, at Bonn University in Germany, was also impressed by Kardashev's paper, and as early as 1960 he considered looking for recombination lines using the University's 25 m radio telescope on Stockert Mountain. He began to build an 11 cm spectrometer but got sidetracked by other observing projects. Before he had time to look for radio recombination lines, he left Bonn for the Siemens Research Laboratories in Munich in order to build a low-noise parametric amplifier for the Stockert antenna. Meanwhile, in Australia, John Bolton and colleagues unsuccessfully tried to detect 6 cm recombination lines using the Parkes 210 foot radio telescope. Had they not made a careless approximation in calculating the line frequency and miscalculated the expected line frequencies, they surely would have had sufficient sensitivity to be successful.[39]

At Harvard, Ed Lilley also considered trying to observe RRLs and discussed his plans with Donald Menzel, who told him that line broadening would be a problem and discouraged Lilley from pursuing it further.[40]

In 1963, Mezger had left Germany to join the scientific staff at the National Radio Astronomy Observatory in Green Bank, West Virginia. One of his first projects was to join Bertil Höglund, a visiting scientist from Sweden, to build a 20 channel receiver that could be used to search for radio recombination lines. The following year, Höglund and Mezger put their receiver on the NRAO 85 foot Tatel radio telescope, but their results were inconclusive. In July 1965, three months before the dedication of the telescope, Höglund and Mezger arranged to use the almost completed NRAO 140 Foot radio telescope. On their first night of observing, they unambiguously detected the unexpectedly strong 105α line at 5,009 MHz in a number of different ionized hydrogen (H II) clouds including the Omega and Orion Nebulae. Anxious to

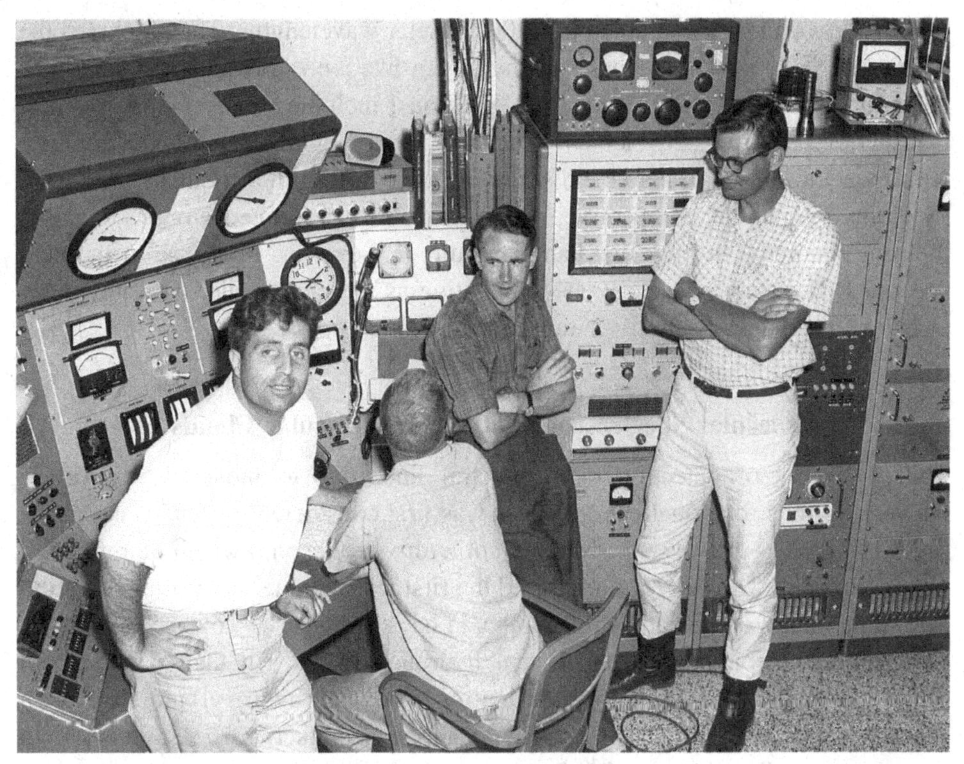

Figure 8.7 From left to right, Peter Mezger, Troy Henderson, Bertil Höglund, and Neil Albaugh at the controls of the Green Bank 140 Foot radio telescope during their detection of the 105α hydrogen recombination line. Credit: NRAO/AUI Archives, Records of NRAO.

stake out their discovery, just four weeks later, Höglund presented a talk at the 111th meeting of the American Astronomical Society in Ann Arbor, Michigan, and also submitted a paper to the British journal *Nature.* After a short but impatient wait, worried about their results being leaked to their British competitors, they abandoned *Nature* in favor of the American journal, *Science*, which quickly accepted their paper.[41] All the theorists had apparently missed the fact that, since adjacent energy levels see nearly the same electric field of nearby electrons and ions, they are perturbed by the same amount, and the energy transition between them remains sharply defined (Figure 8.7).

The exciting confirmation of the long-overlooked radio recombination lines by Höglund and Mezger led to a flurry of activity by other observers, primarily from Harvard and MIT, using the NRAO 140 Foot and Haystack 120 foot radio telescopes, but also from Parkes, the University of California at Berkeley and other observatories. These programs detected other hydrogen recombination line transitions with principal quantum numbers, n, ranging from about n = 40 at millimeter wavelengths to about $n = 300$ at meter wavelengths. They included the β lines ($\Delta n = 2$), as well as lines from hydrogen-like ionized helium,[42] and led to a better understanding of the physical conditions, including temperature, velocity distribution, density, and element abundance within ionized hydrogen regions. RRLs were also used to map the distribution of hydrogen in the Galaxy. The RRL programs were highly competitive, with many radio astronomers anxious to gather the long ignored, but now low hanging fruit. However, as discussed in the next section, RRLs were not nearly as competitive as the race to discover previously unknown molecules in the interstellar medium.

Molecular Spectroscopy and Giant Molecular Clouds[43]

The possibility of detecting radio spectral lines due to molecular transitions between various rotational energy states was first considered by Shklovsky in the same 1949 paper that discussed the 21 cm hydrogen line and which he refined in 1953.[44] In 1955, van de Hulst organized the first international symposium on radio astronomy that was held at Jodrell Bank and he asked Charlie Townes "to suggest what might be found after the 21 cm hydrogen line."[45] In their papers, both Shklovsky and Townes drew special attention to the OH (hydroxyl) molecule, with Townes suggesting that, "This molecule appears, in fact, to be susceptible to detection by radio telescopes."[46] But it would be another decade before Townes' prediction would be verified. In 1959, Townes and his colleagues measured the OH microwave spectrum in the laboratory and confirmed the theoretical prediction that there are actually four OH lines near 18 cm, the two main lines at 1,665 and 1,667 MHz and two satellite lines at 1,612 and 1,720 MHz.[47] As MIT's Alan Barrett remarked in 1983, "The frequency was then known, but there appeared to be no interest among the radio astronomers to take a look even though anybody with a reasonable-sized reflector would have found the line had they looked" (Figure 8.8).[48]

Barrett, who had obtained his PhD on microwave spectroscopy under Townes at Columbia University, had recently joined the MIT faculty. Earlier, he had tried to detect interstellar radio emission from the OH molecule, but was thwarted by not accurately knowing the OH frequency.[49] In 1963, Sander Weinreb had just received his PhD in Electrical Engineering from MIT based on his development of a novel digital autocorrelation spectrometer. After Townes and his group at Columbia were able to determine a better experimental determination of the OH transition frequencies, Weinreb teamed up with Barrett to look for interstellar OH. Using the Lincoln Laboratory 25 m Millstone Hill antenna and Weinreb's autocorrelation spectrometer, they found an absorption line from OH in front of the strong continuum radio source Cassiopeia A (Figure 8.9).[50] Coincidently, at the time of their discovery,

Figure 8.8 Charles Townes. Credit: Nokia Corporation and AT&T Archives.

Figure 8.9 Sander Weinreb. Credit: NRAO/AUI Archives, Records of NRAO.

Figure 8.10 From left to right, Sander (Sandy) Weinreb, Alan Barrett, M. Littleton Meeks, and J. Henry discussing their 1963 discovery of OH. Courtesy of Sandy Weinreb.

many radio astronomers were at the 1963 World Administrative Radio Conference (WARC) meeting in Geneva to discuss the protection of important radio astronomy frequencies from other users of the radio spectrum. Barrett sent George Swenson, the American representative at the meeting, a telegram to inform him about the OH detection, in the hope that the Conference would make an effort to reserve the 18 cm band for radio astronomy. Swenson informed the other radio astronomers in Geneva, and the word quickly circulated around the world. Within a month three separate groups were also able to detect the OH line (Figure 8.10).

When John Bolton in Australia was alerted by the news from Geneva, he was anxious to get in on the discovery. Bolton and his colleagues at the Parkes radio telescope quickly got to work that same day, 20 November, to modify a 21 cm receiver to work at 18 cm. With the aid of a hacksaw, Bolton "tuned" the 21 cm dipole feed to work at 18 cm. That night, they turned the 210 foot Parkes radio telescope toward the Galactic Center and scanned their makeshift receiver across a frequency range covering the expected OH frequencies. They quickly detected both the 1,665 and 1,667 MHz

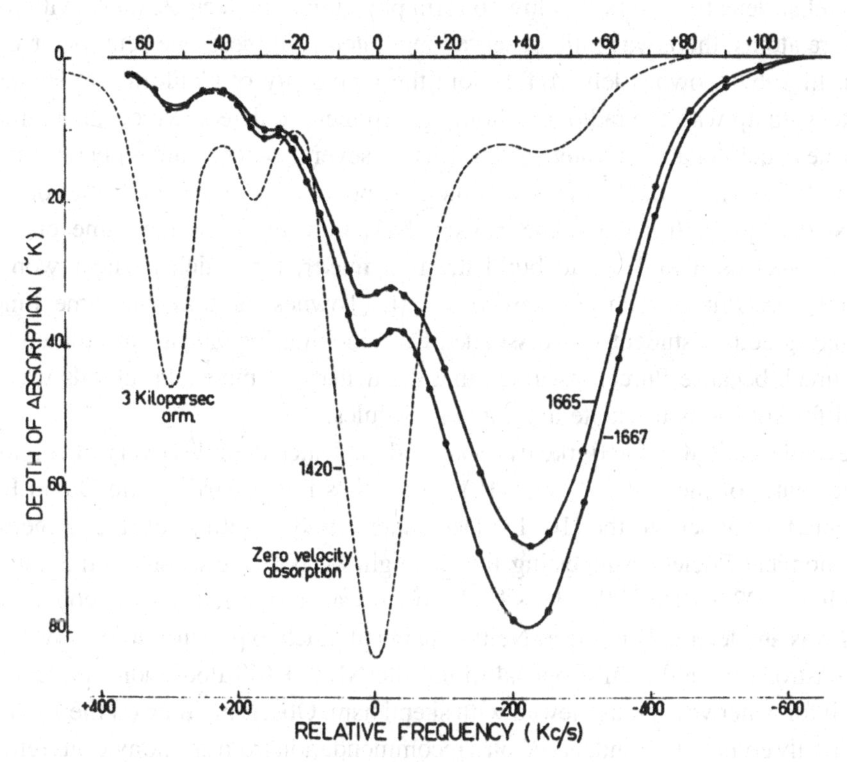

Figure 8.11 Solid lines: OH absorption spectrum toward the Galactic Center. Dashed line: 21 cm HI absorption spectrum. Credit: Barrett (1984).

absorption lines from Sagittarius A corresponding to the well-known 21 cm absorption lines of hydrogen at zero radial velocity lying in front of Sgr A. Although bothered by a steeply sloping baseline at both frequencies that they assumed was an instrumental effect of their hastily built receiver, they dashed off a short letter to *Nature* and subsequently reported the detection of the weaker 1,612 MHz and 1,720 MHz OH satellite lines as well.[51]

However, observations by a group from Harvard that covered a wider frequency range, as well as follow-up observations in Australia, showed that the sloping baseline was in fact the side of a much deeper absorption line that was outside the frequency range covered by the original Australian observations (Figure 8.11). It was instead due to an OH cloud flowing outward from the center of the Galaxy with a velocity of 40 km/sec.[52]

In his 1957 paper, Townes also discussed the 1.3 cm lines of ammonia (NH_3) at 23.6 GHz and water (H_2O) at 22.2 GHz. Since it was widely thought that interstellar densities are far too low to form polyatomic molecules, those with three or more atoms, there was little observational interest in these one centimeter wave lines. In 1967, Townes left MIT to join the University of California at Berkeley and teamed up with the radio astronomy group there to detect two of the ammonia lines near the Galactic Center, as well as in several other sources, using only a small 20 foot (6 m) dish.[53] This was the first polyatomic molecule to be found in interstellar space. Ironically, the 1.3 cm NH_3 transition was the same one that Townes had used in 1946 to build the first maser, for which he later won the 1964 Nobel Prize in Physics. According to Townes, around the same time a Harvard graduate student was dissuaded from looking for interstellar ammonia by Ed Purcell because Purcell assumed that the density of molecular clouds was too small for collisions to excite ammonia molecules.[54]

Several weeks after the Berkeley group had published their discovery of ammonia in the center of the Milky Way, NRAO scientists Lewis Snyder and David Buhl presented a paper at the 10–13 December 1968 meeting of the American Astronomical Society, suggesting that it might be possible to detect the galactic H_2O line at 22.3 GHz.[55] Snyder's background was in infrared spectroscopy, while Buhl was an electrical engineer. Neither one had much experience in observational radio astronomy and their proposal to use the NRAO 140 Foot radio telescope to search for water vapor was viewed with skepticism. Observing time on the 140 Foot was highly competitive and, based on a recommendation from an anonymous referee, their proposal was turned down just weeks before the Berkeley group announced their successful detection of microwave radiation from interstellar water using a much smaller antenna.

Following their discovery of interstellar ammonia, the Berkeley group did not have to deal with referees and the scheduling priorities associated with obtaining observing time at a national observatory. Although Townes had considered searching for other molecules such as CO, they had no equipment for the 2.6 mm CO wavelength, so instead turned their attention to the nearby 1.35 cm water line at 22.235 GHz. Just a few weeks after detecting the ammonia line, they found unexpectedly strong H_2O emission, also in the direction of the Galactic Center, as well as in several H II regions, including the Orion Nebula. The water vapor lines were, in fact, so strong that they could easily have been detected before the 21 cm hydrogen line, perhaps even by Bob Dicke's 1946 observations of 1.35 cm continuum emission from the Sun and Moon, as well as from the terrestrial atmosphere (Chapters 2 and 9). As Al Cheung exclaimed when he telephoned Berkeley during a Christmas week party to tell Townes about the strong water line he saw in the Orion Nebula, "It must be raining in Orion!"[56]

NRAO was so embarrassed by this incident that Snyder and Buhl were given a few days of observing time on the 140 Foot telescope to do whatever they wanted.[57] They teamed up with two former Harvard students, Pat Palmer and Ben Zuckerman, who had already been assigned observing time for another project and, on their first night of observing, they found strong interstellar absorption lines from formaldehyde (H_2CO). Formaldehyde was the first interstellar organic polyatomic molecule discovered, and it is so plentiful that it can be seen in almost all directions in the sky.[58] The discovery of the widespread existence of interstellar formaldehyde demonstrated that polyatomic molecules containing at least two atoms other than hydrogen can form in interstellar space, and that their formation does not require extremely unusual conditions. Ironically, formaldehyde could also have easily been detected much earlier with any of the many 25 m class radio telescopes found at a number of radio observatories. But again, no one thought to look until the discovery of interstellar hydroxyl, ammonia, and water changed the thinking of astronomers. Indeed, in 1969 radio astronomers detected formaldehyde in absorption against the cosmic microwave background. This rediscovery of the microwave background could have occurred before the Penzias and Wilson discovery only four years earlier, if anyone had looked.[59]

The big game changer occurred in early 1969 when a Bell Labs team, led by Arno Penzias, wrote to NRAO to ask for two months of observing time on the NRAO 36 Foot millimeter-wave radio telescope in Tucson, Arizona, to search for CN, CO (carbon monoxide), and HCN (hydrogen cyanide) in "decreasing order of our interest."[60] The Bell Labs group justified their search for CN based on the well-known existence of the optical absorption lines discussed in Chapter 6, but added, "although CO has a much smaller dipole moment than CN, it is probably worth looking for." Since it was anticipated that any molecular lines would be very weak and require long integration times, the Bell Labs team provided their own computer to average and display the results. So, they were surprised when they pointed the telescope at the Orion Nebula and found a strong CO signal at 115 GHz in real time on the chart recorder without any need to average a long integration. They also readily found strong CO emission in eight other galactic regions of ionized hydrogen, including the Galactic Center, and the floodgates were open.[61]

Following the Bell Labs detection of carbon monoxide, molecular spectroscopy and the search for new molecular lines became the hottest topic in radio astronomy, at least within the United States. CO is observed throughout the Galaxy and is also used to trace the distribution of mass and kinematics of external galaxies.

The competition to be the first to detect a new molecule or a new isotope became intense and, at times, devious and even unscrupulous. As the most powerful millimeter-wave radio telescope in the world, the demand on the NRAO 36 Foot

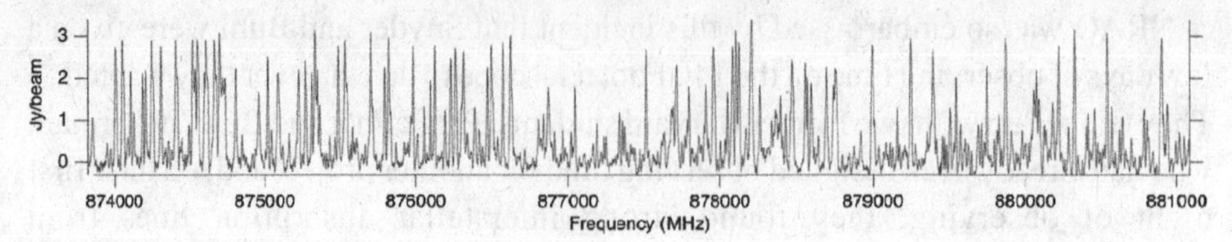

Figure 8.12 High resolution ALMA spectrum toward the massive star forming region NGC 6334I over a 7 GHz frequency range. About 100 different spectral lines of more than 10 different molecules are seen in this single spectrum. Credit: McGuire et al. (2018).

radio telescope was heavy and, for several years, only about 20 percent of the proposed observations were scheduled. Although all proposals were sent to multiple anonymous referees, the referees and observers came from the same relatively small common pool of millimeter-wave radio astronomers, and this led to some difficulties and charges of unfair treatment.

Competing observers established overlapping collaborations with theoreticians or laboratory spectroscopists in order to learn the correct frequency of their favorite molecule or isotope, and then kept the frequencies a carefully guarded secret or even leaked false information to their competitors. Although observers were permitted to look only for molecules described in their proposal, there was a lot of claim jumping. One observer purposely entered an incorrect frequency in the telescope logbook in order to mislead the next observer. Others, who were caught observing a source not in their proposal, would claim that they were using it as a calibrator. One very competitive radio astronomer claimed that he suspected that someone was going through his desk drawer at night, so he invented a bogus molecule and a fake observing proposal. Another observer reportedly spent a lot of observing time looking for this molecule.

Radio astronomers and molecular spectroscopy ultimately matured and revolutionized our understanding of the structure, dynamics, temperature, molecular abundance, and chemistry of regions where new stars are born and where old stars evolve. The new field of astrochemistry spawned by these observations has identified thousands of lines in the radio spectrum, with hundreds of molecules and isotopes identified, including many polyatomic organic molecules. Huge clouds of gas found in Giant Molecular Clouds can span tens of light years, and are the largest structures in the Galaxy, each containing up to 100,000 times the mass of the Sun, where new stars are being continuously born. Numerous molecules and isotopes found in the low density interstellar medium are unstable in terrestrial conditions and have no terrestrial or laboratory counterparts. Many other interstellar lines have not been identified with any molecule (Figure 8.12).

Cosmic Masers

Soon after the discovery of OH absorption, Harvard graduate student Ellen Gundermann was using the Harvard 60 foot radio telescope to study OH absorption in ionized hydrogen regions as part of her thesis research. In April 1964, she found unexpectedly strong and surprisingly unresolved narrow band OH emission in the direction of the strong radio source known as W49 and noted that the observed ratio of line strengths in the two 1,665 and 1,667 OH lines departed from the theoretical and laboratory value of 5:9. In order to pin down the unexpected narrow bandwidth of the emission features, publication was delayed until Harvard could build a narrow band filter spectrometer.[62] In the meantime, the Berkeley group and later the Australian group found similar strong unresolved narrow band OH emission.[63] Mysteriously, they found that the OH emission appeared to be highly polarized and the ratio of the strengths at the four frequencies departed significantly from the expected theoretical values. Moreover, the OH emission was variable, and the surprisingly narrow bandwidth implied little or no motion that would cause Doppler broadening of the lines.[64] Faced with this confusing paradigm, especially their inability to detect the expected 1,667 MHz counterpart to an observed 1,665 MHz line in several sources, the Berkeley group suspected that, instead of OH, the unidentified 1,665 MHz line might be from a new element which they named "mysterium." However, the MIT group was soon able to detect the 1,667 MHz line in all sources as well as the 1,720 MHz satellite line, and responded, "In our opinion, 'mysterium' is anomalously excited OH."[65]

Even before the recognition of the anomalously excited OH emission lines by Harvard and Berkeley, Jodrell Bank observers were looking for linear and circular polarization that might be expected from Zeeman splitting of the narrow OH line. They were surprised to find up to 100 percent circular polarization in some OH features, which they, perhaps prematurely, ascribed to Zeeman splitting.[66] Using the NRAO 140 Foot radio telescope, MIT Professor Alan Barrett and his student Alan Rogers wrote, "We have not found any OH emission which is not circularly polarized," and "many of the lines at all four OH frequencies, exhibit 100 per cent circular polarization."[67] Noting that the existence of a "single line with right-hand circular polarization . . . without a line of the opposite sense of circular polarization being present, . . . is completely contrary to Zeeman predictions." Based on the detection of anomalous polarization characteristics,[68] Barrett and Rogers made the bold suggestion that, "It might be possible that a maser-type amplification by the OH is occurring in the interstellar medium."

A series of interferometric observations made with increasing baselines (angular resolution) showed that the OH emission regions were very small, less than 0.01 arcseconds across. This meant that the corresponding brightness temperatures were up to 10^{12} K, thus adding support to the Barrett and Rogers interpretation of the anomalously excited OH emission lines as coherent radiation from cosmic masers.[69] Cosmic masers (acronym for **m**icrowave **a**mplification by **st**imulated **e**mission of **r**adiation) work in the same way as the laboratory maser invented in 1964 by Townes, by the stimulated emission from molecules in an inverted energy state. While laboratory masers are "pumped" by appropriate microwave radiation, in intergalactic space, collisions with other atoms and molecules or radiation from nearby stars provide the energy to excite OH and other cosmic masers to populate the upper energy levels. Cosmic masers are associated with stellar outflows either in newly-formed stars or in old dying stars.

Interestingly, the idea of stimulated or maser emission appears to go back to at least a 1937 paper by Donald Menzel, who considered but then rejected possible applications to interstellar gaseous nebulae.[70] Menzel wrote,

> the condition may conceivably arise when the value of the integral turns out to be negative. The physical significance of such a result is that energy is emitted rather than absorbed ... The process merely puts energy back into the original beam, as if the atmosphere had a negative opacity. This extreme will probably never occur in practice.

Following their discovery of strong H_2O emission, Townes contacted the radio astronomers at NRL, who had a larger antenna that worked well at 1.35 cm. Observations with the NRL 84 foot radio telescope and later with the NRAO 140 Foot dish, showed that, like OH, the H_2O emission was also highly polarized, came from a very small region, that the lines had very narrow bandwidths, and that the intensity was variable on time scales as short as a few days. The small dimensions inferred from the rapid variability suggested brightness temperatures up to 10^9 K, which was in sharp contrast to the upper limit of about 100 K kinetic temperature implied by the narrow line widths.[71] As was the case for OH, the inferred small dimensions were confirmed when Very Long Baseline Interferometer observations (Chapter 11) showed that some H_2O emission regions were as small as a few thousandths of an arcsecond (or about an Astronomical Unit) with corresponding brightness temperatures up to 10^{12} K, about an order of magnitude greater than had been inferred from the rapid time variability.[72] It was now becoming clear that the galactic H_2O, as well as the OH radiation, was probably due to maser emission, although interestingly, in their H_2O discovery paper, Cheung et al. only made the passing remark that, "there may even be maser action."[73] Apparently the NRAO referee of the Snyder and Buhl 140 Foot proposal did not consider the possibility of H_2O masers. But almost no one except Lew Snyder and David Buhl, not even Charlie Townes, thought about looking for cosmic water vapor until 1969. One can only speculate what the reaction would have been if Dicke had discovered the strong interstellar H_2O emission in 1946, before Townes' invention of the maser in 1953.

Both silicon monoxide (SiO) and methanol (CH_3OH) are also observed as strong interstellar masers.[74] Powerful hydroxyl and water vapor masers are found in other galaxies as well as in the Milky Way, and are called "megamasers." H_2O megamasers found in active galactic nuclei were particularly game-changing, as they led to a powerful new technique to measure the scale of the Universe and, at the time, were the best evidence for the existence of supermassive black holes (SMBH) of 10^7 to 10^9 solar masses in the nuclei of galaxies. As we saw in Chapter 4, it was widely suspected that powerful radio galaxies and quasars were powered by SMBHs, but it would take an elegant interferometric observation to directly show the presence of a SMBH in the galaxy NGC 4258.

The story begins in 1992 at the Nobeyama Radio Observatory in Japan where they had built a powerful new eight section spectrometer that could simultaneously cover 16,384 different frequency channels and spanned a wide range of frequency (velocity). Old water megamasers were surveyed and new ones sought. Although there was little reason to expect a maser velocity spread beyond the normal galactic rotation range of about 300 km/s, the Japanese astronomers observed NGC 4258 with their new spectrometer. They found a signal from a previously discovered maser in the active galaxy NGC 4258 at a frequency corresponding to the redshift of the galaxy.[75] Only many months later, when they examined the data from the remaining filter sections, they discovered, to their amazement, two groups of masers symmetrically located around the systemic one, and offset by ± 75 MHz (1,000 km/s). The Japanese observers suggested several possible interpretations, including that "the masers are emitted from a circumnuclear molecular torus rapidly rotating around a compact massive object, such as a black hole."[76]

Later observations with the continent wide Very Long Baseline Array (VLBA, Chapter 11) showed that the maser spots in NGC 4258 lie along a remarkably thin line about 0.015 arcseconds (0.3 light years) in extent. From the observed Doppler shift of the spots of maser emission, a team of Japanese and American astronomers found that, on one side, the group of maser spots was rotating toward the observers and, on the other side, they were moving away. From an analysis of the motions, they were able to show that the maser spots were part of a flat disk rotating around a central mass of 36 million solar masses that was contained within a few light months. Only a supermassive black hole could have such a high mass density of more than 100 million solar masses per cubic light-year. Moreover, by comparing the linear size of the rotating disk of 0.26 pc with the measured angular size of 0.0082 arcseconds, they determined the distance to NGC 4258 as 6.4 ± 0.9 Mpc (later revised to 7.2 ± 0.3 Mpc and finally 7.57 ± 0.15 Mpc) (Figure 8.13).[77]

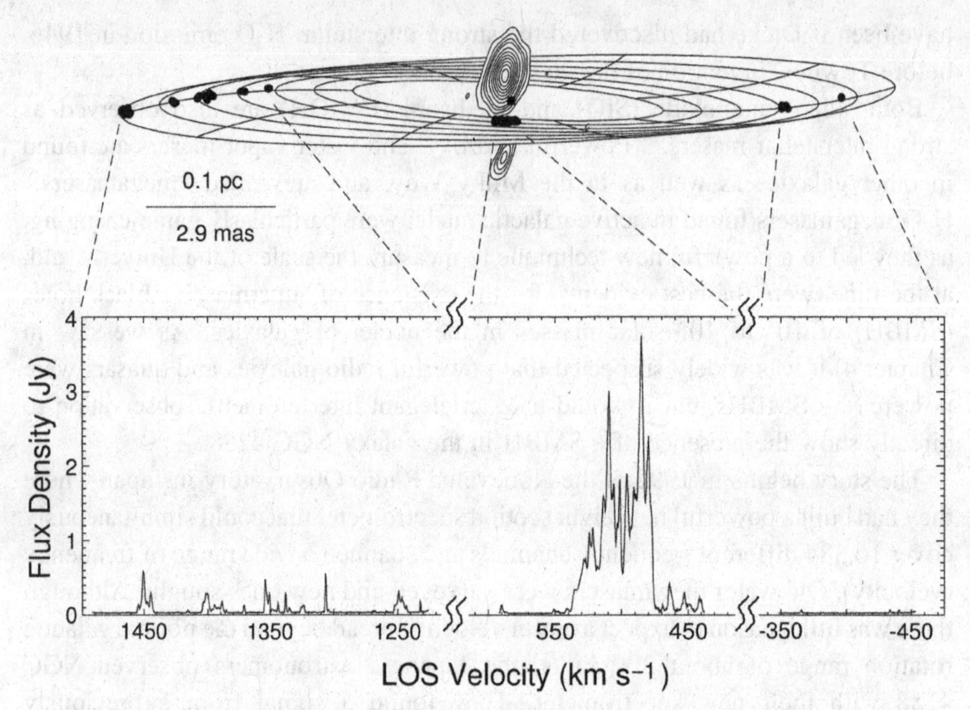

Figure 8.13 NGC 4258. Top: Schematic picture showing the central radio source, the warped disk, and the H_2O maser emission spots. Bottom: H_2O spectrum showing the strong emission lines near the 500 km/sec galaxy velocity and the much weaker satellite maser lines. Courtesy of James Moran.

These observations of the NGC 4258 H_2O masers established three important results:

(a) The first observational evidence for the existence of an accretion disk suggested earlier by theoreticians to surround quasars and AGN.
(b) The best observational evidence for the presence of the long suspected supermassive black hole in the nucleus of a galaxy thought to power radio galaxies and quasars.
(c) The first direct geometric measurement of the distance to an external galaxy independent of the conventional extragalactic distance ladder.

However, there was a problem. Traditionally, the distance to external galaxies and, thus, the value of Hubble constant was determined by a complex hierarchical series of measurements known as the "cosmic distance ladder," ranging from trigonometric distances to nearby star clusters to the apparent magnitude of supernovae in distant galaxies. A key link is the relation between the intrinsic luminosity and the period of variability of what are known as Cepheid variable stars. Astronomers use this

Cepheid period–luminosity relation to determine the distance to nearby galaxies by measuring the period of Cepheid star variability to calculate the intrinsic luminosity of the stars and comparing it to the apparent magnitude.

The maser distance to NGC 4258 of 7.2 ± 0.3 Mpc found from the radio interferometer measurements was inconsistent with the value of 8.5 ± 0.5 Mpc obtained by Hubble Space Telescope observations of Cepheid variable stars in NGC 4258. Prompted by this discrepancy, a subsequent reanalysis of the Cepheid data led to the recognition of a long standing systematic error in the assumed Cepheid period–luminosity law used to calibrate the distance scale and to better agreement with the radio measurement.[78]

But a new, potentially more serious problem, has emerged. An international team has extended the VLBA maser observations to include megamasers in more distant galaxies and has derived a new, more precise, value of the Hubble constant, $H_0 = 73.9 \pm 3.0$ km/sec/Mpc, independent of the traditional comic distance ladder.[79] Conventional observations such as those from the Hubble Key Project, with $H_0 = 73.5 \pm 1.4$ km/sec/Mpc, and other determinations of H_0 in the "local" Universe, now agree well with the megamaser value, but differ significantly from the value of H_0 determined from observations of the cosmic microwave background anisotropies and the large scale distribution of distant galaxy clusters that give $H_0 = 67.4 \pm 0.5$ km/sec/Mpc.[80] This "tension" between the maser/supernova value of H_0 and the predictions based on Planck satellite measurements of the anisotropies in the cosmic microwave background is one of the outstanding current problems in cosmology and may reflect the need for a fundamental change in the basic physics underlying our understanding of conditions in the early Universe.[81]

9

Radio Studies of the Moon and Planets

> The other side of the planet is in eternal darkness, and its temperature cannot be far from absolute zero, –273°C, even colder than the surface of the remote Pluto.[1]

We have seen in previous chapters how radio astronomers uncovered a previously unknown universe of nonthermal radiation, magnetic fields, and high energy electrons that dominates the radio emission from the Sun, the Milky Way, radio galaxies, quasars, supernova remnants, cosmic masers, and pulsars. No one suspected that radio emission from the relatively nearby planets would uncover any surprises but, as in the more exotic realm of cosmic radio astronomy, the observers led the way to unexpected discoveries. Radio emission from the planets in our Solar System was expected to follow the simple thermal radiation law from bodies having a temperature that is determined primarily by their distance from the Sun that provides their source of heating. Surprisingly, with the exception of Mars, all of the other planets showed various forms of unexpected behavior. Particular surprises were the discovery of meter wavelength radio noise from electrical storms on Jupiter and the later detection of decimeter radiation from the extensive radiation belts that surround Jupiter. In the meantime, investigations using active radar systems established a new value for the Astronomical Unit and revealed surprising new information on the rotation of Mercury and Venus. Radio astronomers also discovered the first extrasolar planets through their gravitational influence on the motion of their pulsar "sun" and the consequent effect on measured pulse arrival times.

Taking the Temperature of the Planets

All of the bodies in the Solar System are warmed by heating from the Sun at a rate that depends primarily on their distance from the Sun and the amount of incident energy that is reflected rather than absorbed by the planet. Assuming that the planets behave as black bodies, that is, they absorb and radiate all of the incident radiation from the Sun, their expected temperature can be calculated from the Stefan-Boltzman law, which says the temperature should be proportional to the fourth root of the incident radiation which itself is proportional to the inverse square of the distance from the Sun. The detailed distribution of temperature across the planet will also depend on the electrical properties near the surface, the rate of the planet's rotation, and on the nature of any atmosphere that can circulate heat as well as provide insulation to mitigate the rate of cooling from the nighttime side.

Knowing the luminosity of the Sun from measurements made on the Earth allows us to calculate the temperature at the subsolar point on any planet (where the Sun appears directly overhead) as $392/\sqrt{R}$ K, where R is the distance from the Sun in Astronomical Units (AU).[2] Averaged over the illuminated surface of the planet, the average incident radiation is reduced by a factor of $2^{1/4}$, and so the mean temperature on the daytime side becomes $343/\sqrt{R}$ K, while on the unlit nighttime side, in the absence of an atmosphere, the temperature is close to absolute zero. If the planet is rapidly rotating so that both sides are alternately heated, or if there is a circulating atmosphere that carries heat from the daytime to the nighttime side, then, since energy is absorbed only on the daytime side, but is radiated from both the sunlit as well as the dark nighttime side, the mean temperature over the whole surface is further reduced by a factor of $2^{1/4}$ and becomes $277/\sqrt{R}$ K.[3]

These calculations assume that all of the incident solar radiation is absorbed by the planet. If instead there is a fraction of incident radiation that is reflected, with an albedo, A, then the temperature is reduced by a factor of $(1-A)^{1/4}$. Typically, the albedo is less than 10 percent, meaning most of the incident radiation is absorbed by the planet. However, if there are oceans or dense cloud cover that reflect the sunlight, the albedo can be as much as a few tenths.

For a pure blackbody (e.g., one that absorbs and re-radiates all of the incident radiation) we expect that the flux density measured by radio telescopes will increase with the square of the frequency and is proportional to the mean temperature of the visible planetary disk reduced by a small factor depending on the emissivity or departure from a pure blackbody. While it might appear that this means that one should go to the highest frequencies (shortest wavelengths) to detect the thermal radio emission from the planets, unfortunately in practice, the sensitivity of radio telescopes deteriorates as the frequency is increased. In the 1950s and 1960s the "sweet spot" was at centimeter wavelengths.

Measurements of planetary temperatures made in the infrared usually refer to the temperature at the visible part of the planet's surface or atmosphere, while radio measurements refer to the temperature at a depth below the surface, typically 10 to 20 wavelengths, and depends on the electrical conductivity of the planet near the surface. So, radio measurements made at millimeter and short centimeter wavelengths probe the temperature near the surface, that closely tracks the solar illumination, while measurements made at longer wavelengths reveal the temperature well below the surface which, as on the Earth, is normally more constant.

The Moon As part of their 1946 experiment to measure the atmospheric water vapor discussed in Chapter 2, along with observing the Sun, Robert Dicke and Robert Beringer made the first observations of the thermal radio emission from the Moon from an MIT building rooftop in Cambridge, Massachusetts. Using just a small 18 inch diameter parabolic reflector at 1.25 cm (24 GHz) they determined an effective blackbody temperature of the nearly full Moon as 292 K, comparable with the average temperature on the Earth, which is at the same distance from the Sun.[4] Observing at the same frequency two years later, Jack Piddington and Harry Minnett, at the CSIR Radiophysics Division in Sydney, tracked the Moon around its orbit using a 44 inch parabolic reflector and found that the maximum temperature occurred about 3.5 days after the full moon and the minimum temperature lagged the new moon by a similar amount.[5] Piddington and Minnett contrasted their result with infrared observations that closely followed the phase of the incident radiation, since they are essentially a measure of the temperature at the lunar surface, while the microwave observations referred to the temperature about 40 cm below the surface. From their measured time lag of 3.5 days and the ratio of maximum to minimum temperature averaged over the lunar surface, Piddington and Minnett concluded that the ratio of radio to thermal absorption coefficients must somehow increase with depth below the lunar surface, and that a simple explanation would be that there is a thin layer of poorly-conducting dust that covers a solid surface. Subsequent radio observations of the Moon made over a range of frequencies further refined these pioneering studies and gave some assurance that the first missions to land instruments on the Moon would not be enveloped in a thick layer of dust.

Mercury: Old Observations and Theories Refuted Mercury orbits the Sun in a very elliptical 88-day orbit that varies between 0.39 AU (36 million miles) and 0.46 AU (42 million miles) from the Sun. Only under the most favorable conditions, when it is at its greatest elongation from the Sun, can Mercury be observed in the twilight just before sunrise or just after sunset and, even then, at a maximum elevation of only 28 degrees above the horizon. Although Mercury's existence was known for thousands of years, it is said that even Copernicus never saw Mercury.

Giovanni Schiaparelli was a nineteenth century Italian astronomer who used his visual observations to make drawings of the surface features of Mars and Mercury. He is probably best known for his description of observed linear features on Mars as "canali," meaning "channels" in Italian, that were famously but incorrectly translated into English as "canals," leading to speculations of intelligent life on Mars. But his observations of Mercury would be even more long lasting and misleading to future generations of astronomers and students.

Schiaparelli's observations of Mercury were of necessity made under difficult twilight conditions near the horizon or in full daylight when the planet was higher in the bright sunlit sky, but still near maximum elongation from the Sun. Based on his visual observations spanning some five years, Schiaparelli made 150 drawings of the surface of Mercury. Since he thought he was seeing the same features on the partially illuminated surface of Mercury whenever he was able to make an observation, Schiaparelli wrote that,[6]

> Observations in 1882–3, confirmed in 1886–7, showed that the planet revolved about the sun at least somewhat as the moon revolves around the earth, namely; in turning always the same face, or nearly the same face, to the primary body.

On 8 December 1889, Schiaparelli presented his remarkable finding to a special meeting of the *Accademia dei Lincei* of Rome that was attended by the King and Queen of Italy. Schiaparelli's announcement meant that nearly one half of Mercury would be in perpetual sunlight and be very hot while the other side would experience a continual very cold night.

Later astronomers, including Eugène Antoniadi, André-Louis Danjon, René Jarry-Desloges, Walter Haas, and Henry McEwen all reported that they were able to confirm Schiaparelli's findings.[7] As late as 1953, the French astronomer Audouin Dollfus compared his new observations from the excellent site at the 9,350 foot summit of the Pic du Midi with the pictures drawn by Schiaperelli, and wrote "*La durée de rotation moyenne de Mercure entre ces deux dates est égale à sa durée de revolution avec une precision supérieure à un dix-millième*,"[8] or, "the period of Mercury's rotation is equal to its period of revolution with a precision better than one part in ten-thousand."

Considering that the rotation of Mercury was thought to be locked to its revolution around the Sun and that Mercury was thought to have no atmosphere, for nearly a century every relevant scientific paper and astronomy textbook explained that, due to the locked rotation, the perpetual day side of Mercury would be fiercely hot, while the nighttime side would remain in cold darkness.

For example, in her well known 1954 popular book on astronomy, Cecilia Payne-Gaposchkin wrote,[9]

> The sun beats fiercely on the face of Mercury; on the average the planet receives seven times as much sunlight per unit area as the earth. But Mercury always turns the same face to the sun, (just as the moon does to the earth), and so, . . . one side of the planet is in continual sunlight; the other is in perpetual shadow. Under the sun's rays the surface of Mercury is kept at a temperature near 350 C or 660 F . . . But the other side of the planet is in eternal darkness, and its temperature cannot be far from absolute zero, –273°C, even colder than the surface of the remote Pluto.

The first hint that there might be something wrong with this picture came from observations made at 3.8 cm with the University of Michigan's 85 foot radio telescope.[10] The Michigan scientists observed Mercury on six occasions, mostly during the spring and summer of 1960. To minimize contamination from the Sun, they confined their observations to times when Mercury was between 19 and 28 degrees from the Sun and found a mean equivalent blackbody disk temperature of about 400 K. After correcting for the variation in solar illumination with the angle of the Sun in the Mercurian sky, and assuming that the temperature of the unilluminated surface was near zero, they calculated that the temperature at the subsolar point was a surprising 1,100 $\pm$ 300 K, far above the 600 to 700 K expected from solar heating or the 610 K determined from infrared measurements.[11]

One of the present authors (KIK) was perplexed by this curious result and used the Australian 210 foot radio telescope at Parkes to check the Michigan claim by measuring the apparent disk temperature of Mercury over nearly half a cycle in its orbit around the Sun. Observations made at 11 cm (2.65 GHz) over a six week period in May and June 1964 included the time when Mercury was near inferior conjunction, so that only about 10 percent of the visible disk was illuminated by the Sun. Nevertheless, the apparent temperature of the nearly fully dark side of Mercury appeared to be nearly the same as on the sunlit side. As reported in the *New York Times* and later in *Nature*, the temperature on the unlit nighttime side of Mercury, long thought to be about the coldest place in the Solar System, was about 300 K (80°F or 27°C), close to normal room temperature (Figure 9.1).[12]

How could this be? Somehow heat was being conveyed from the sunlit to the dark side of Mercury. It appeared clear from nearly a century of visual observations that one side of Mercury never saw sunlight, but evidence against an atmosphere was less certain. In fact, George Field at Princeton University had suggested that Mercury might contain a thin argon atmosphere that would circulate heat from the daytime to the nighttime side of Mercury, and Nikolai Kozyrev, in the USSR, claimed to have detected a thin hydrogen atmosphere that could heat the nighttime surface

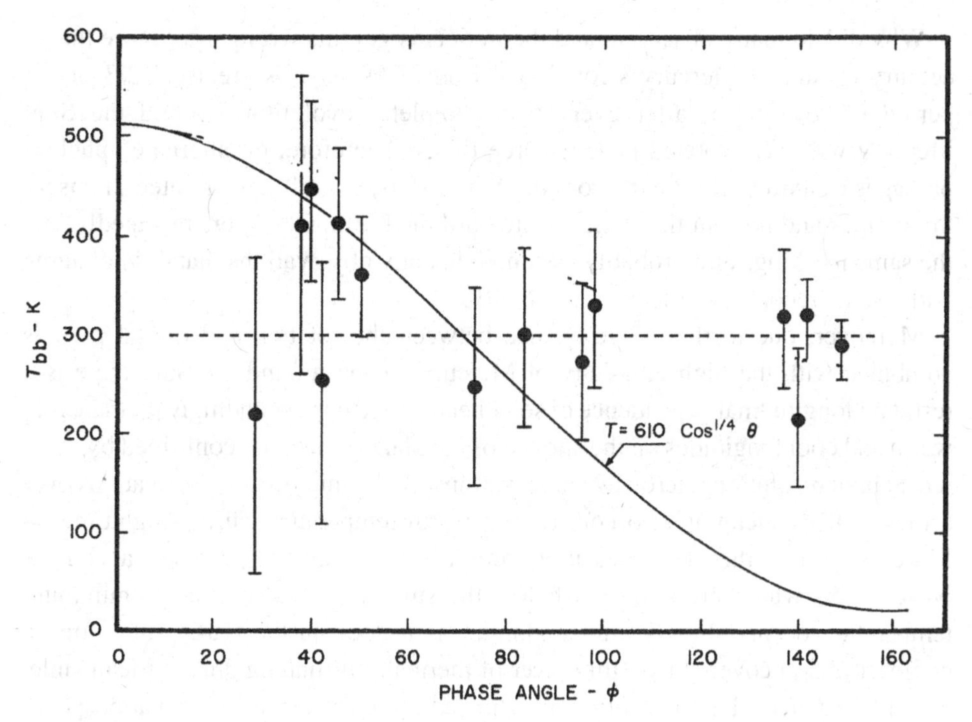

Figure 9.1 Measured 11 cm brightness temperature of Mercury versus the angle with respect to the line of sight. The solid curve represents the expected dependence for a 610 degree subsolar (noon) temperature for a non-rotating planet with no atmosphere to distribute the heat to the nighttime side. The dashed curve shows the expected response for a uniform temperature of 300 K. Credit: Kellermann (1966).

temperature to 300 K, just about the temperature determined from the Parkes radio observations.[13] But it was not that simple.

When Cornell scientists Gordon Pettengill and Rolf Dyce pointed the giant 1,000 foot Arecibo radar antenna with its 2 megawatt peak power transmitter toward Mercury in April 1965 they found that, in fact, Mercury rotated with a period of 59 ± 5 days, quite different from the 88 day period of revolution around the Sun.[14] Faced with this new observational result, the theoreticians were quick to rise to the occasion and explained that, because of strong tidal torque when Mercury is at perihelion (closest to the Sun), the rotation period should be exactly 2/3 of 88 or 59 days.[15] As Pettengill later remarked, "Any theoretician who had looked at this fact, even at this stupidly simple level, would have immediately realized that Mercury had to be rotating with a period of 59 days."[16] A reanalysis of the old drawings of visual observations showed that they were actually consistent with the radar result if the rotation period was 58.4 ± 0.4 days, close to the modern value of 58.258 days.[17]

Why did so many observers and theoreticians get this wrong after more than a century of study? Mercury's rotation period of 59 days is precisely 2/3 of the period of revolution; after every two complete revolutions around the Sun, Mercury will have rotated exactly three times. Therefore, on alternate aphelion passages (maximum distance from the Sun), Mercury will have rotated precisely three times and present the same face toward the Earth. Observers repeatedly saw the same markings and probably assumed that any observations that did not agree with their preconceived ideas were faulty.

Moreover, due to the 3/2 resonance between the rotation and orbital periods combined with the high ellipticity of Mercury's orbit around the Sun, there is a resultant longitudinal dependence of solar heating and correspondingly permanently warm and cool longitudes on the surface of the planet which are confirmed by more recent high resolution microwave observations.[18] The microwave observations over a range of wavelengths also confirm the warm temperature during nighttime on Mercury, show that the measured temperature slightly increases at longer wavelengths which probe deeper below the surface, and also that the minimum temperature occurs slightly after midnight, as it does on the Earth, suggesting a compact region covered by a thin layer of thermally insulating dust.[19] Meanwhile, ground- and space-based observations showed Mercury has little or no atmosphere that could contribute to the circulation of heat around the planet.

Venus: A Planet-Wide Greenhouse Probably the first claim of radio emission from Venus was in a series of three letters to *Nature*, all in July 1956, by John Kraus from Ohio State University. Kraus claimed that, from February to June 1956, he had detected strong impulsive 11 m (26.7 MHz) radio signals from Venus lasting up to a few minutes, with individual pulse durations of less than 1 second.[20] Although Kraus reported that the peak intensity was several times that of the strong Cygnus A radio source, there have been no other reports of Venusian radio bursts. Most likely Kraus's detections were the result of local electrical interference, thunderstorms, or some instrumental effect, perhaps inspired by the similarity with the then recently discovered meter wavelength radio bursts from Jupiter (see later in this chapter).

The first confirmed detection of radio emission from Venus came from the US Naval Research Laboratory (NRL) in Washington. The NRL 50 foot radio telescope had a very precise surface, so, unlike most other radio telescopes of the time, it could be used at short centimeter wavelengths where the thermal radiation from the surface of planets was the strongest (Figure 11.1). Following a long period of testing and calibration, one of the early astronomical targets for the dish was the planet Venus. Due to its heavy cloud cover, that perpetually obscures the planet, nothing was known about the surface of Venus prior to the first radio observations made in the spring of 1956. NRL scientists Cornell Mayer, Timothy McCullough, and Russell Sloanaker calculated that the thermal radio emission from Venus should be easily observable at 3.15 cm (9.52 GHz). Observations with the 50 foot antenna in May and June indicated a surprisingly high average temperature, probably at the surface, of about 580 K (310°C) near inferior conjunction, the closest approach when only the unlit side of Venus is seen from the Earth.[21] This was considerably higher than the 285 $\pm$ 9 K measured from infrared observations of carbon dioxide in the Venusian atmosphere or the 325 K expected from solar heating. Surprised at the high temperature, as a check that some of the observed 3.15 cm radio emission might be nonthermal, they also made some hasty measurements at 9.4 cm (3.2 GHz). Although, as expected from the factor of three difference in wavelength, the radio emission from Venus was about a factor of 10 weaker at 9.4 cm than at 3.15 cm, and the interference from the Sun was greater. The 9.4 cm observations indicated that the radio spectrum of Venus did not differ significantly from that of simple thermal radiation, and confirmed that the temperature near the surface of Venus was an astounding 600 K (or about 330°C).

The high temperature of the surface of Venus observed by radio astronomers was a dramatic confirmation of the greenhouse effect created by the thick carbon dioxide-laden atmosphere that was predicted more than a decade earlier by Princeton astronomer Rupert Wildt and vigorously promoted by Carl Sagan.[22] According to Mayer, "This was not an altogether popular result. In fact, it was very disappointing to many people, including a number of prominent astronomers, and other important people, who were reluctant to give up the idea of a sister planet and perhaps even the possibility of life."[23] One belligerent told Mayer that he knew someone who had just returned from Venus who reported that, "it is a very beautiful place with mountains and valleys and streams and rivers."

Many scientists, as well, were skeptical whether the microwave radio observations were really a measure of the temperature of Venus, or were instead due to free–free emission from a dense ionosphere, but this was shown by Sagan and his student Russell Walker to be untenable.[24] Later observations of Venus that reached around to the daytime side, made over a wide range of wavelengths at various radio observatories in the United States, as well as in Australia and the USSR, showed that Venus is the hottest place in the Solar System, with daytime temperatures averaging 800 K and peaking near 1,000 K.[25]

As was first noted by Frank Drake from his 10 cm Green Bank observations, radio measurements made from 1961 to 1963 over a range of positions in the orbit of Venus showed that the minimum surface temperature occurred after inferior conjunction. This was the first indication of the retrograde (counter-clockwise) rotation of Venus – opposite to the direction of the planet's revolution around the Sun.[26] The suspected retrograde rotation of Venus was soon confirmed by the direct radar observations discussed later in this chapter.

These ground-based radio observations that first disclosed the hot surface temperatures and indicated an atmospheric pressure on Venus nearly 100 times greater than on the Earth crucially paved the way for the later American and Russian missions to Venus, that could then be instrumented to deal with the extreme climatic conditions they would encounter in the Venus atmosphere and on the surface.

Mars: The Normal Planet The first detection of the thermal radio emission from Mars was by the NRL group using their 50 foot radio telescope.[27] In September 1956, near the time of opposition (closest approach) they measured the brightness temperature of Mars at 3.15 cm (9.5 GHz) as 218 ± 76 K. Subsequent radio observations of Mars over a wide range of wavelengths and with improved accuracy indicated a mean temperature of the Martian surface of 192 ± 10 K.[28] This was consistent with the infrared measurements and the value expected from solar heating at the distance of Mars from the Sun, assuming that thermal inertia near the Martian surface was longer than the 24.6 hour rotation period.[29] However, there was one alarming report that created a temporary panic within NASA and the planetary science research community.

At the 13th General Assembly of the International Astronomical Union, held in Hamburg, Germany in August 1964, Rodney (Rod) Davies from the Jodrell Bank Observatory reported that he had observed an unexpectedly high 21 cm temperature on Mars of 1,140 ± 50 K. Davies and others speculated this could be due to synchrotron radiation from an intense radiation belt around Mars similar to that known to exist around the Earth (Van Allen Belts) and Jupiter (see next section).[30] NASA was already planning the 1964 launches of Mariner 3 and 4 to Mars that would have been damaged by any Martian radiation belt, and so were considering expensive modifications to shield the spacecraft from any Martian radiation. Fortunately for NASA, subsequent observations made with the Parkes 210 foot radio telescope in Australia showed that the Jodrell Bank results were affected by confusion from nonthermal background sources and that the claimed high temperature was wrong. Because Mars, like all the planets, moves with respect to the fixed celestial background of stars and galaxies, the Parkes observations made at the same position in the sky after Mars had moved away were subtracted from the observation of Mars, to determine the net radio emission from Mars, which was consistent with that expected from solar heating.[31]

Spectroscopic observations made with the Very Large Array in 1990 provided the first ground-based observation of water in a planetary atmosphere and mapped the distribution of water vapor in the atmosphere of Mars.[32]

The Giant Planets Unlike the so-called terrestrial planets Mercury, Venus, Earth, and Mars, which all have solid surfaces, the giant planets Jupiter, Saturn, Uranus, and Neptune are composed primarily of very dense gas. The radio observations of the major planets probe to different depths in the planetary atmospheres depending on the observing frequency and the nature (content, pressure, and temperature) of the atmosphere, and constrain the composition of the giant planet atmospheres. This, in turn, helps constrain theories of planet formation. In the late 1950s and into the 1960s, the first radio observations of the major planets were made over a range of centimeter and decimeter wavelengths, primarily at NRL, Caltech, Green Bank, and the University of Michigan in the United States, and at Parkes, Australia, and Jodrell Bank in the United Kingdom.[33] Around the same time, Frank Low was using his newly-developed infrared bolometer to determine the temperature of the planets in the 17 to 25 micron atmospheric window.[34]

Due to the intense nonthermal radiation from Jupiter's radiation belts (see next section), the thermal emission from Jupiter itself is best measured at short wavelengths where the nonthermal contribution is insignificant, or by high resolution interferometric observations that separate the thermal emission from the small planetary disk from the nonthermal contribution that originates in extended radiation belts. Fresh off their successful detection of the thermal radio emission from Venus, NRL scientists turned their attention to Jupiter and measured an apparent disk temperature at 3.2 cm (9.4 GHz) of 145 $\pm$ 26 K, consistent with the temperature of Jupiter's clouds determined by infrared observations (128 $\pm$ 3 K) and with that expected from solar heating (120 K).[35] Later Caltech interferometer observations indicated a disk temperature of Jupiter of about 260 K at 10.4 cm and 21.3 cm, or about twice the value measured at 3.2 cm.[36] More recent high resolution observations at 1.3 mm (230 GHz) with the Atacama Large Millimeter/submillimeter Array (ALMA) revealed the various temperature bands associated with the well-known cloud structure seen visually (Figure 9.2).[37]

The first radio observations of Saturn, using a newly-developed traveling wave tube receiver at 8,000 MHz (3.75 cm) on Harvard's 28 foot radio telescope, barely detected the thermal radio emission from the cold distant planet.[38] Due to the uncertainty of whether or not there was any contribution from Saturn's rings, no attempt was made to determine a brightness temperature. Later measurements at the University of Michigan, made at the same frequency using a more sensitive radiometer, reported a disk temperature of 168 $\pm$ 11 K, while Frank Drake found a higher value of 196 $\pm$ 44 K at

Figure 9.2 230 GHz ALMA image of Jupiter showing the temperature bands coincident with the familiar bands of cloud cover. Credit: de Pater et al. (2019).

10 cm from his Green Bank 85 foot telescope observations.[39] Meanwhile, in Australia, Ken Kellermann reported temperatures of 179 ± 19, 196 ± 20, and 303 ± 50 degrees at 6, 11.3, and 21.3 cm, respectively. All of these radio measurements exceeded the 93 ± 3 K measured by Low in the infrared, that was close to the expected 90 K based on Saturn's distance from the Sun.[40]

Radio observations of Uranus at Green Bank, Caltech, and Michigan at 2, 3.1, and 3.75 cm indicated brightness temperatures of 220 ± 35, 158 ± 20, and 159 ± 16 K, respectively, all of which were well in excess of the 50 ± 3 K determined by Low in the infrared or the 63 K expected from the incident radiation from the Sun.[41]

One of the early targets for the newly-completed Green Bank 140 Foot radio telescope was the planet Neptune, that was found to have a disk temperature of 180 ± 40 at 2 cm wavelength, while later Caltech measured 115 ± 36 K at 3.12 cm, both considerably greater than the 50 K expected at Neptune's 30 AU distance from the Sun.[42]

In all of the giant planets, the temperature measured at centimeter and at decimeter wavelengths is greatest at the longer wavelengths, where it reaches a value about a factor of two to three times higher than measured in the infrared or the temperature expected from solar heating. At the longer wavelengths, the radio observations probe deeper into the planetary atmospheres and show that the temperature increases with increasing depth in the planet.[43] The Estonian astronomer Ernst Öpik has suggested that there may be a source of internal heating in the major planets, possibly due to radioactive decay, while Larry Trafton at Caltech pointed out that the dense atmosphere of the major planets could generate a greenhouse effect of the kind found on Venus.[44]

Elusive Pluto At a distance ranging from 30 to 49 AU, Pluto receives less than 0.1 percent of the energy from the Sun than the Earth, and it is among the coldest places in the Solar System. It is so far away that it takes nearly 250 years to complete one orbit around the Sun. The low surface temperature of Pluto, combined with its relatively small size and great distance from the Earth, means that the radio flux density is about 50 times weaker than from Neptune. So, for several decades Pluto eluded observations, even with the world's most sensitive radio telescopes.

For all of the other planets, the first radio observations were made at short centimeter wavelengths. But to compensate for the low surface temperature and small diameter of Pluto, it was necessary to go to much shorter wavelengths. However, it would be 20 years after the first radio observations of Neptune at 2 cm before radio astronomy had achieved the required sensitivity to detect Pluto at 1.2 mm, where the signal strength was nearly 300 times stronger than at 2 cm. In 1986, a team of radio astronomers from Germany and France finally succeeded in detecting the weak 250 GHz (1.2 mm) thermal radio emission from Pluto and its moon Charon. Using the 30 m radio telescope on Pico Veleta in Spain they found a temperature of 39 ± 4 K, which was close to the expected value.[45]

Meanwhile, the position of Pluto in its orbit around the Sun remained uncertain to about one second of arc, that was too large for NASA's planned New Horizons flyby of Pluto. In order to better locate the position of Pluto, ALMA turned its 50 antennas toward Pluto in 2014 and measured its position to an accuracy of 0.1 arcseconds, enabling New Horizons to arrive only 7,800 miles (12,550 km) from the surface of Pluto on 14 July 2015.[46]

Jupiter Presents Dual Surprises

Electrical Storms on Jupiter In 1954, Bernard Burke, then a young 26 year old scientist, who had just received his PhD from MIT doing research on microwave spectroscopy, joined the new radio astronomy group at the Carnegie Institution's Department of Terrestrial Magnetism (DTM) in Washington, DC. The radio astronomy program at DTM was under the leadership of Merle Tuve, who had earlier invented the radio-sounding technique for studying the ionosphere, and later led the important Second World War program that developed the proximity fuse. Together with Jesse Greenstein from Caltech, Tuve organized the first major conference on radio astronomy in January 1954 and, as Chair of the National Science Foundation Radio Astronomy Advisory Committee, he played a major role in the creation of the National Radio Astronomy Observatory.[47]

Tuve recognized the potential of radio astronomy to make new discoveries and, aware of the lack of experienced radio astronomers in the United States, he invited Francis Graham-Smith from Cambridge to advise on starting a radio astronomy program at DTM. Graham-Smith suggested that DTM build a Mills Cross (Chapter 5) transit radio telescope working at the relatively unexplored wavelength of 22 MHz (13.5 m). Since the first detection of cosmic radio waves by Karl Jansky some 20 years earlier, the trend had been to move to ever shorter wavelengths to exploit the tremendous progress resulting from the wartime radar advances in radio technology and to facilitate the higher angular resolution needed to image the radio sky. In a field about 20 miles (32 km) from Washington, near the small Maryland town of Seneca, the new DTM cross was built to make a multi-year survey of the radio sky. To achieve the desired angular resolution, the DTM cross consisted of two arms oriented north–south and east–west, each 2,047 feet (1,271 m) in length and each containing 66 half wave dipoles (Figure 9.3). The DTM cross had a slightly elliptical reception beam about 2 degrees in size that scanned the sky at a fixed declination as the Earth rotated and was able to locate the position of a radio source to a fraction of a degree as it drifted through the antenna beam.

Figure 9.3 DTM 22 MHz Mills Cross Array near Seneca, MD, used by Burke and Franklin to detect Jupiter radio bursts. Courtesy of Carnegie Institution for Science.

As Burke later recalled, they built a simple interferometer to check for interference at 22 MHz and, while observing the Crab Nebula in June 1954, they noticed a burst of noise that they originally thought came from the Sun. This piqued their interest, since they both knew about the reports of intense solar radio bursts discussed in Chapter 2. But, with accurate measurements made possible using an interferometer system, they realized that the bursts were not coming from the Sun or the Crab Nebula and, according to Burke, Graham-Smith shrugged it off as "interesting."[48]

After Graham-Smith returned to Cambridge in 1954, Burke was joined by Kenneth Franklin, who had just received his PhD in astronomy from the University of California (Figure 9.4). To prepare for a full sky 22 MHz multi-year survey, while calibrating the array with daily observations of the Crab Nebula, they noticed that an apparent burst of interference seemed to occur at the same time each day and lasted about the same length of time that it took a radio source to pass through their beam. Franklin was concerned and cautioned Burke, "Hey, you know we have to figure out what that thing is," to which Burke apparently responded only, "I suppose so."[49]

When the time came to start the sky survey in early 1955, Burke and Franklin went out into the field to readjust the array to a new declination. There were only two choices, and Burke asked, "Do we go north or south?" Not wanting to waste time

Figure 9.4 Ken Franklin. Credit: Image 2A7217, American Museum of Natural History Library.

standing in the winter cold, Franklin hastily responded, with no particular reason, "South." As they continued daily to move their beam south in small increments, the response to the Crab Nebula got weaker, but remained visible on the side of their beam. Several other weak radio sources were seen on their chart recordings, as well as the mysterious interference which appeared to be similar to automobile ignition interference. Since it seemed to come at the same time each evening, someone jokingly suggested that maybe it was a nearby farmhand returning each evening from a date.

Anticipating that they would have some results from their survey to present at the upcoming April meeting of the American Astronomical Society (AAS), Burke and Franklin submitted an abstract to the AAS. That put Burke to work to look at their chart recordings to get some data that they could report at the AAS meeting. Burke temporarily set aside all the records that appeared to show the interference so he could concentrate on analyzing the good data. But the chart records without the interference were incomplete and Burke was unable to determine from them alone the nature of a rise and fall in the response that appeared on the good chart records. So, he had to pull out the records affected by the interference and noticed that the interference actually occurred about four minutes earlier each day. As Franklin later remarked, "A strange romance this was turning out to be."[50] Unlike Karl Jansky, Franklin had a strong background in astronomy, and quickly realized that the shift of four minutes a day meant that the interference was not coming from a terrestrial source, but from a direction apparently fixed in space. With further scrutiny, it appeared not to be precisely at a constant position in the sky but was slowly drifting southwest, thus fortuitously remaining in their beam that, following Franklin's earlier decision, was also being continually adjusted to the south. This meant that the source of interference was moving in the sky!

Howard Tatel, a DTM colleague, overheard Burke and Franklin puzzling over this strange moving interference. Just a few nights earlier, Tatel had been using the DTM 7.5 m (25 foot) Würzburg-Riese antenna at 21 cm to try to detect hydrogen from Jupiter. Having this on his mind, he facetiously suggested that their interference might be coming from Jupiter and everyone had a good laugh. Trying to demonstrate that Tatel was wrong, Franklin looked up the position of Jupiter in the American Ephemeris and Nautical Almanac and was surprised to see that both Jupiter and Uranus were about in the same direction as their moving source of interference.

That afternoon, Burke and Franklin drove out to work on their antenna at Seneca and, in the evening twilight, Burke asked about the "exceptional bright object that was almost on the meridian." "That's Jupiter," Franklin explained. Inspired by actually seeing Jupiter in the sky, the next morning Franklin plotted the position of their interference along with the positions of the planets Jupiter and Uranus. Burke later recalled that, as Franklin plotted the position of Jupiter day by day, seeing that it lined up with the position of the radio interference, Franklin called out, "Hey fellows, come over here and look at this."[51] Burke then remembered the strange bursts of noise that he and Graham-Smith had seen a year earlier with their interferometer. He went back to the old chart recording and realized that these bursts were also coming from the same direction in the sky as Jupiter.

Burke and Franklin now had something to report in their scheduled paper at the April AAS meeting just a week later in Princeton, New Jersey. Their abstract appeared with the innocuous title, "High Resolution Radio Astronomy at 1.5 m Wavelength." DTM issued a press release, which was embargoed until after the AAS meeting. But *The New York Times* jumped the gun. While sitting in the audience waiting for Burke to give his talk, Franklin ran across the *Times* headline, "Sound on Jupiter Is Picked Up in the U.S." (Figure 9.5).[52]

In their short published abstract, Burke and Franklin described the calibration of their instrument using a number of strong radio sources and the start of their survey at +22 degrees declination, that included the Crab Nebula, a known galactic nebulosity, and the "very intense burst of radio noise" which they cautiously wrote

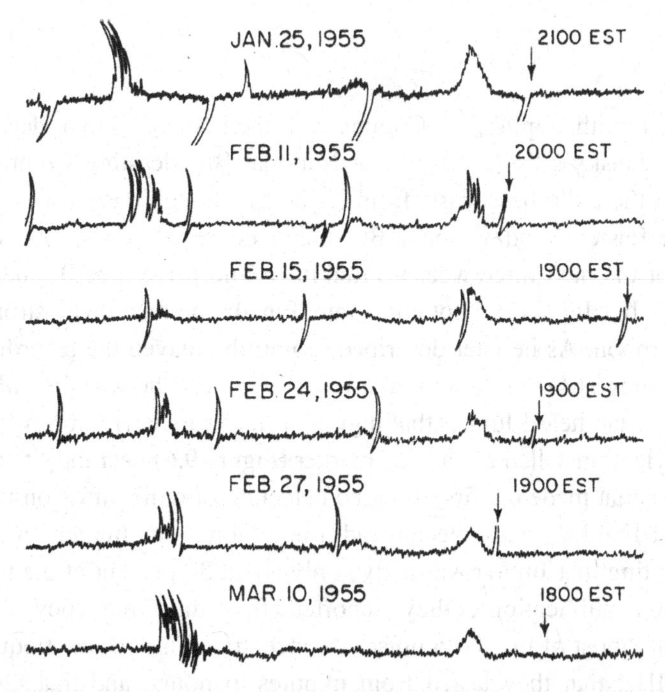

Figure 9.5 DTM chart recording showing Jupiter radio bursts on the left and the Crab Nebula to the right. The changing interval between the two responses reflects the motion of Jupiter. Credit: Franklin (1984).

Figure 9.6 Maryland state road sign marking the location of the DTM array where Burke and Franklin first detected Jupiter radio bursts. Credit: JOVE/NASA.

"is associated with Jupiter."[53] Coming off their success two decades earlier broadcasting Jansky's "star noise," the National Broadcasting Company wanted recordings of the radio broadcasts from Jupiter. Although there was a strong burst of noise on Easter Sunday, the NBC magnetic recorder did not work. Each afternoon, for the next three weeks, Franklin would drive the 20 miles out to the telescope site, but Jupiter did not cooperate. Finally, he recorded a strong event on a Friday afternoon. As he later described it, until he played the recording the next Monday morning at DTM, he was in awe knowing that he was the only person in the world who had heard Jupiter that night.[54] Fifty years later, on 6 April 2005, the state of Maryland installed a roadside marker (Figure 9.6) near the site of the DTM 22 Mc/s cross that made the first reported detection of radio emission from Jupiter.

Burke and Franklin had detected radio emission from Jupiter on 9 of the 31 scans, suggesting that Jupiter was active only about 30 percent of the time. In later more detailed publications, they reported that multifrequency observations indicated that "most of the radio noise is concentrated in the low frequency range below 38 MHz, that they lasted from minutes to hours, and that they occur at slightly different times at different frequencies."[55] Also, they reported a curious "slow drift" in the times of the maximum received signal. Following the AAS meeting in Princeton, Burke and Franklin built a simple interferometer system and quickly discovered that the radio bursts from Jupiter were right-hand circularly polarized, sometimes by as much as 100 percent, demonstrating that they were probably originating in a region with a strong magnetic field.[56] This was the first evidence for magnetic fields in the Solar System, aside from the Earth and the Sun.

Aware that Alex Shain was operating an 18 MHz array in Australia, Burke wrote to Shain informing him about their discovery of 22 MHz radio bursts from Jupiter and asked if Shain could have also detected Jupiter radio bursts. Looking back through his chart recordings going back to 1950, Shain indeed found that his records "showed a series of bursts which had previously been passed over as terrestrial interference," and he was able to confirm the Burke and Franklin result. Especially interesting were records obtained during a six week period when the east–west extension of his antenna system had been fortuitously dismantled, so the remaining north–south antenna dipole elements had a very broad beam in hour angle, enabling observations of Jupiter over an eight hour stretch of time. From these data, Shain was able to show that the radio bursts occurred at intervals close to the approximately 10 hour Jupiter rotation period. By combining these eight hour long observations with data covering the long time span between the Australian and DTM observations, Shain showed that the bursts came primarily from a very localized region on Jupiter and that it rotated with a period of $9^h\ 55^m\ 13^s \pm 5^s$ or close to the rotation period of $9^h\ 55^m\ 40.632^s$ (System II) observed for the atmosphere in Jupiter's south temperate belt. This was about 5 minutes slower than the $9^h\ 50^m\ 30.003^s$ period (System I) associated with the cloud system in the equatorial region of Jupiter. More detailed analysis by Shain and a group at the University of Florida showed that the apparent rotation of the radio burst region was actually $9^h\ 55^m\ 28.8^s$, or about 12 seconds faster than the System II period, and that it was probably related to the rotation of the main body of Jupiter, unaffected by the complex circulation in the active Jovian atmosphere. This System III radio period was later adopted by the International Astronomical Union as the official rotation period of Jupiter.[57]

Burke and Franklin naturally wondered if the Jovian radio bursts were due to electrical storms on Jupiter analogous to terrestrial thunderstorms and set up a simple experiment to observe the radio noise from a local lightning storm. But when Franklin was attaching a cable at the time when a cloud-to-cloud lightning strike occurred, he was abruptly reminded that playing with wire antennas in an open field during a lightning storm was asking for trouble. Franklin survived the incident and showed that "terrestrial lightning strokes generated a strong signal at 22 MHz, but the noise power was at least a million times too weak to explain the radio bursts from Jupiter."[58]

Nearly 10 years later, Keith Bigg in Australia found a $42^h\ 27.4^m$ periodicity in archival Jupiter data, which was close to the time it takes Io, the closest of the four major satellites of Jupiter, to orbit the planet. Somehow, Bigg concluded, Io is able to trigger the radio emission from Jupiter.[59] As later explained by Peter Goldreich and Donald Lynden-Bell, the Jupiter radio bursts are generated by Io as it moves through the planet's intense magnetic field.[60]

Although the intense radio bursts from Jupiter would have sooner or later been recognized by others, and in fact had been detected earlier, at least by Alex Shain in Australia, the discovery by Burke and Franklin was serendipitous in several ways. They were lucky that the DTM array was initially set up to observe the Crab Nebula, that happened to be at nearly the same declination as Jupiter. Franklin later remarked that it had been pure chance that they had decided to survey the sky south of 22 degrees declination in the direction that Jupiter was moving, and not to the north.[61] They were also fortunate that Howard Tatel overheard their discussions, and that he was thinking about Jupiter based on his own research program. Finally, they were lucky that Jupiter was near opposition and so, aside from the Moon, was the brightest object in the evening sky, serving as a beacon waiting to be recognized. Shain was equally fortunate in having the part of his array that gave east–west resolution pulled to pieces. The remaining elements were sensitive to a large swath of sky, enabling Shain to observe Jupiter for nearly a full rotation period, and so he was able to reconstruct the rotation period of the Jovian radio source. But, earlier, he had missed recognizing the multiple occasions during 1950 and 1951 when he had dismissed the Jupiter radio bursts as terrestrial interference in his 18 MHz observations.

Burke and Franklin never understood the cause of the rise and fall of the antenna output that prompted Burke to re-examine the chart records containing the interference that turned out to be from Jupiter. Probably it was a temperature dependent receiver instability. In 1956, Franklin left DTM for New York, where he later became the director of the Hayden Planetarium and a frequent reporter and writer about astronomy for the public. Burke turned his attention to 21 cm studies of interstellar hydrogen. In 1965, he left DTM to join the MIT faculty, where he had a long, distinguished career as one of the most prominent radio astronomers in the United States. The planned 22 MHz sky survey using the DTM Mills Cross was never completed. Studies of the Jovian radio bursts were continued for many years by others, including Alex Smith and Thomas Carr at the University of Florida, James Warwick at the High Altitude Observatory in Boulder, Colorado, as well as Alex Shain in Australia, until his untimely death in 1960 only a few days after his 38th birthday. More recently, the NASA Goddard Space Flight Center, in collaboration with several colleges and universities, has organized the educational and outreach *Radio JOVE* project that allows amateurs and students to interactively monitor and analyze 20 MHz Jupiter radio bursts.[62] Indeed, Jupiter radio bursts are so intense around 20 MHz that they can be easily observed with simple home built antennas and an inexpensive commercial shortwave receiver.

The discovery of strong radio radiation from Jupiter in 1955 not only marked the beginning of planetary radio astronomy, it was also the opening of a new era in planetary astrophysics that had previously been largely confined to celestial mechanics and drawings by hand of fleeting visual observations of planetary surfaces and atmospheres sometimes enhanced by creative imagination.

Intense Radiation Belts around Jupiter When the first American satellite, Explorer 1, was launched on 31 January 1958, it carried a Geiger counter to measure cosmic ray intensity unaffected by absorption in the Earth's atmosphere. University of Iowa Professor James A. Van Allen, who was in charge of the cosmic ray project, was at first bewildered and disappointed that the cosmic ray counts recorded and transmitted to earth were surprisingly low. Further investigation showed that, in fact, the counter was jammed due to the unexpectedly large charged particle density and the finite response time of the Geiger counter.[63] This serendipitous discovery of the intense radiation belts trapped in the Earth's geomagnetic field, which became known as the Van Allen Belts, set the stage for the subsequent discovery of the much more powerful radiation belts surrounding Jupiter.

Plagued by locally generated interference, fed up with the large pointing uncertainties from their alt-azimuth mounted 50 foot dish in Washington, and wanting a larger collecting area, in 1958 NRL built an 84 foot equatorially mounted dish located at a radio quiet site at Maryland Point, Maryland, about 30 miles south of the 50 foot dish at their Laboratory in Washington. Following initial mechanical tests, Russell Sloanaker was given the task of calibrating the antenna pointing, using a new 10.3 cm (2,910 MHz) radiometer that he had recently developed. As part of his gain and pointing calibration process, Sloanaker observed a number of the brighter radio sources, but took the opportunity to observe Jupiter as well. At the 1958 Paris Symposium on Radio Astronomy, National Radio Astronomy Observatory (NRAO) scientist Frank Drake heard NRL radio astronomy head, Edward McClain, report that Sloanaker had found a surprisingly high value of 580 K at 9.2 cm (3.3 GHz), considerably larger than the 145 K that the NRL group had found at 3.2 cm.[64]

Drake was intrigued by this surprisingly high value. Was it really a measurement of hot thermal radiation from deep in Jupiter's gaseous atmosphere or was it some sort of new nonthermal radiation, in which case it should appear even stronger at longer wavelengths? Drake had just joined NRAO in Green Bank, West Virginia, as only the second member of the scientific staff, and was looking for projects using the new 85 foot Howard E. Tatel radio telescope that had just been completed in early 1959.[65] Using the observatory's 22 cm (1,360 MHz) receiver, he pointed the 85 foot telescope to Jupiter and was surprised to find an even higher brightness temperature of about 3,000 K. The only way to tell if this was really a measurement of the temperature deep down in Jupiter's atmosphere, or some nonthermal phenomena, was to observe at an even longer wavelength. With the support of Hein Hvatum, then a visitor from Sweden, who had developed instrumentation for a 440 MHz (68 cm) receiver, Drake observed Jupiter at this relatively long wavelength. Somewhat to their surprise, Drake and Hvatum recorded an unexpectedly strong signal.

Drake quickly realized that the large angular response of the 85 foot telescope at 440 MHz included an area more than 2 degrees across and that, within this large area, there were many cosmic radio sources. Since the position of Jupiter changed from night to night, Drake was able to repeat the observations after Jupiter moved through the telescope beam and thus subtract out the contribution of the confusing radio sources. But still, the measurements of the apparent disk temperature of Jupiter indicated a surprisingly large value of 70,000 K, far too hot to be explained by any thermal process. Indeed, the observed flux density at 68 cm was about twice the value that they had measured at 22 cm, whereas, according to the $(\text{frequency})^2$ dependence expected from the blackbody radiation law, it should have been about a factor of 10 weaker at 68 cm.

The earlier detection of the circularly polarized decameter radio bursts from Jupiter by Burke and Franklin suggested that Jupiter was surrounded by a magnetic field of about 10 Gauss, or about 10 times the value of the Earth's magnetic field. Drake wondered if the excessive brightness of Jupiter might be due to synchrotron radiation from relativistic particles moving in the magnetic field. Lloyd Berkner, the Acting NRAO Director, was anxious to demonstrate a new scientific result from what was perceived by competing radio astronomers as an underperforming and overfunded NRAO, and encouraged Drake and Hvatum to publish their result to get credit and establish priority. In August/September 1959, at the 103rd meeting of the American Astronomical Society in Toronto, Canada, Drake not only reported the large apparent temperature of Jupiter but, encouraged by the recently discovered Earth's Van Allen Belts, he boldly proposed that "the radiation originates as synchrotron radiation from relativistic particles trapped in a Jovian magnetic field."[66]

While visiting NRL and Green Bank from Caltech, John Bolton learned about the high Jovian disk temperatures measured by Sloanaker at NRL and by Drake at Green Bank.[67] Alerted by Bolton, Jim Roberts and Gordon Stanley turned the new Caltech 90 foot radio telescope toward Jupiter and measured a disk temperature of 5,500 K at 31 cm (960 MHz), confirming the surprisingly high values determined at NRL and Green Bank. Roberts and Stanley followed up on Drake's suggestion that the excessive radio emission from Jupiter might be due to synchrotron radiation from relativistic electrons moving in a Jovian magnetic field. They noted

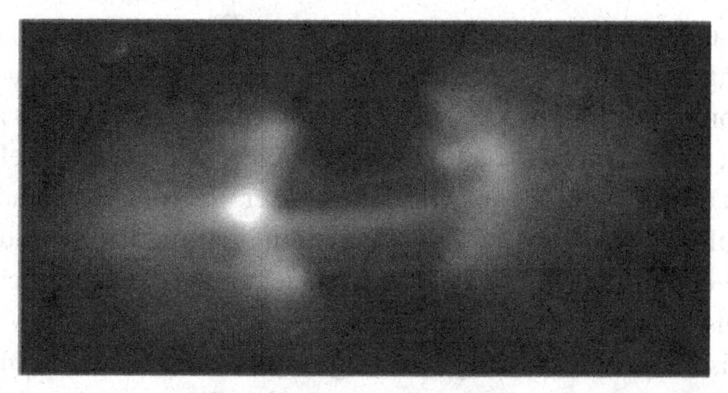

Figure 9.7 20 cm VLA radio image of Jupiter's radiation belts. Courtesy of Imke de Pater.

that the detection of circular polarization in the decametric Jovian radio bursts required a magnetic field of about 7 Gauss but concluded that the required electron density would need to be orders of magnitude greater than in the terrestrial Van Allen Belts.[68]

Shortly after the late 1959 completion of the Caltech Owens Valley Radio Observatory Interferometer, V. Radhakrishnan and Roberts turned the interferometer toward Jupiter and found a strongly linearly polarized radio source several times the size of the planet, that was later imaged by Dave Morris and Glenn Berge[69] – a dramatic confirmation of Drake's conjecture of a Jovian radiation belt and in time to rescue NASA's Pioneer 10 mission to Jupiter from a fatal encounter (Figure 9.7). Later observations, made with better angular resolution, were able to constrain the electron distribution and magnetic field configuration and show the effect of Jupiter's moons on the radiation belts.[70]

Cold War Radars, Venus, and the Astronomical Unit

Unlike the observations discussed in the rest of this book, radar astronomy does deviate from the purely passive observational style discussed in the Introduction. As was described by Gordon Pettengill,

> With radar we're sending out a coherent, completely polarized waveform. We can mark the waveform, we can frequency shift, we can point it in different directions, and that means we have much more information in the echo that comes back. We are halfway in the laboratory: in a sense we're one step beyond the usual astronomy, but not quite as far as the laboratory physicists.[71]

Not surprisingly, the Moon was the obvious first celestial target considered for the powerful radar systems that had been developed during the Second World War by both the Allied and Axis powers, and was apparently discussed at many laboratories and radar stations around the world.[72] As described in Chapter 1, while working as a broadcast engineer in 1940, John DeWitt was one of the first people to confirm Karl Jansky's detection of galactic radio emission. A few months later, DeWitt unsuccessfully tried to detect radio signals reflected from the Moon. DeWitt was an amateur astronomer as well as a radio amateur (W4ERI). Following periods of industrial and military radio and radar development, by the end of the War, DeWitt was a Lieutenant Colonel in charge of the Army's Evans Signal Laboratory at Belmar, New Jersey, with a new opportunity to achieve his dream of receiving radar signals reflected from the surface of the Moon. He called his scheme Project Diana after the Roman goddess of the Moon.

Using a sensitive experimental receiver, an 8 kW 112 MHz transmitter, and a radar antenna that had been used during the War to detect enemy aircraft, DeWitt and his staff began transmitting pulses toward the Moon in September 1946 (Figure 9.8). Since the Moon is about 240,000 miles away, it takes about 2.4 seconds for any reflected radio signal to reach the Earth. The group transmitted a

Figure 9.8 New Jersey radar site where the first radar contact with the moon was made. Credit: NRAO/AUI Archives, Papers of W. T. Sullivan III.

one second long pulse every four seconds and then watched for a return 2.5 seconds after each pulse. Because the antenna was steerable only in azimuth, their experiments were limited to the short periods where the Moon was rising or setting. Finally, after a series of enhancements to their electronics and to the antenna system, at 11:58 a.m. on the morning of 10 January 1946, a distinct "beep" was heard just 2.4 seconds after the transmitted pulse was sent to the Moon as it rose over the Atlantic Ocean.[73] Ironically, DeWitt had gone to a nearby store to buy cigarettes, and was not informed of their success until the next day.

Apparently the concept of contacting the Moon by radio had a big media impact, with multiple stories in *The New York Times* and in *Time* and *Newsweek* magazines.[74] As later described by Harold Webb, one of the Project Diana team members, "Following the announcement on January 25, 1946, that the [Army] Signal Corps experiments were successful in making contact with the moon by radar, the imagination of news reporters and feature story writers went wild with predictions that space ships would soon be a reality."[75] Ironically, Project Diana was never officially endorsed by the military or civilian authorities. DeWitt was acting on his own initiative and a weak justification based on a vague Pentagon directive to investigate ways to track enemy ballistic missiles.

Meanwhile, in Hungary, Zoltán Bay shared DeWitt's interest and enthusiasm in making radar contact with the Moon. Bay was a Professor of Atomic Physics at the Technical University of Budapest and at the same time the director of the United Incandescent Lamp and Electric Co. Ltd., also known as "Tingsram." During the Second World War, he worked on the development of radar, and in early 1944 began a program to attempt to receive lunar radar echoes. Like DeWitt, Bay had a longtime interest in astronomy as well as expertise in electronics. However, as a result of Allied bombing and the advance of the Soviet army into Hungary, Bay was forced to move his laboratory several times and, in the process, much of his equipment for the lunar experiment was either lost or destroyed. Only after a two year delay did Bay finally have all of his apparatus ready.

Bay's transmitter and antenna system were not unlike that used by DeWitt, but Bay did not have access to the sensitive experimental receiver and sophisticated instrumentation that was used by the American team to record the lunar echoes. Knowing that, unlike Project Diana, he would be unable to detect a single pulse reflected from the Moon, Bay devised a very innovative electro-chemical integrator, which he called a "cumulator," that averaged the receiver output in 10 different delay channels from 2.0 to 2.9 seconds, each differing by 0.1 seconds. A true echo would appear only in the channel corresponding to the 2.4 second delay of the round trip to the Moon. As reported by Bay, "The first successful experiment at 2.5 m wavelength (120 MHz) was carried out on the 6th February 1946," just a month after the widely publicized reports of the American experiments.[76] As in the United States, there was great public interest in Bay's achievement, which was announced on Hungarian radio and widely publicized in the newspapers.

Zoltán Bay's cumulator measured the volume of the liberated hydrogen gas from the electrolysis of a solution of potassium hydroxide (KOH) by the amplified and detected lunar echo. In this way, by measuring the gas generated from the sum of hundreds of pulses, Bay was able to improve his sensitivity by more than an order of magnitude. Although primitive by modern techniques, subsequent programs to detect radar echoes from the planets used Bay's scheme of averaging for different delay periods around the expected value. But the modern techniques differ from Bay's in that the accumulation and averaging of the received radar echoes is done using digital sampling and analysis in a digital computer.

The first successful experimental detection of weak radar reflections from the Moon occurred around the same time in 1946 in both the United States and in Hungary. Both projects were driven in part by the desire to demonstrate that radio waves can pass relatively unimpeded through the ionosphere, but also by the ambitions of DeWitt and Bay to achieve a first. Several military and industrial groups also experimented with using the Moon to provide long distance communication. Although none of these attempts proved practical, radio amateurs today, using only modest equipment and antennas, routinely communicate using VHF and UHF signals reflected off the Moon.

The extension of radar astronomy to interplanetary distances was challenging and became an area of Cold War era competition. While the received signal strength of radio transmissions falls off as the square of the distance, in radar experiments the inverse square law comes in twice. The radar signal received at the target depends not only on the square of the distance between the Earth-based transmitting antenna and the target, but the reflected signals received back on the Earth also fall off as the square of the distance between the target and the receiving antenna. So, the ability to detect reflected radar signals falls off as the fourth power of the target's distance from the Earth. In practice, the echo may be further reduced by the irregularities on the planetary surface, which can scatter the incident radiation in different directions.

Approximately every year and a half, Venus makes its closest approach to the Earth, but at each inferior conjunction it is still about 100 times further away than the Moon. Adjusting for its greater diameter still meant that to receive radar echoes from Venus would require an improvement in sensitivity by a factor of ten million over what was needed by DeWitt and Bay to detect echoes from the Moon. Stimulated by Cold War pressures to achieve military superiority and prestige, after only a little over a decade, a number of laboratories in the United States and Europe had developed sufficiently powerful radar systems to attempt to receive radar echoes from Venus, starting with the conjunction of early 1958.

Aside from individual credit and establishing national prestige, a successful detection of radar echoes would have two important scientific consequences. First, the time lapse between the transmitted and received signal would give the distance to Venus, that, knowing the relative distances of the planets from their orbital periods, could be used to more accurately calculate the Astronomical Unit (the average distance between the Earth and the Sun) and thus calibrate the scale of the Solar System. Also, because Venus is covered by clouds, visual observations had been unable to determine the rate of rotation. However, due to the differential Doppler shift from the approaching and receding sides of Venus, the rotation of the planet would spread the spectrum of the returned radar pulse by an amount that depended on the rate of rotation. But the race to be first to detect radar echoes from Venus and the unbridled enthusiasm of the radar scientists would lead to a series of embarrassments before there was a verified successful experiment.[77]

Only a few months after Sputnik had launched the Space Race in 1957, MIT scientists were well poised to use the Lincoln Laboratory's Millstone Hill radar to establish radar contact with Venus. With an 85 foot steerable antenna, a powerful transmitter, and a sensitive maser receiver, Venus appeared to be within reach of their experimental radar, that had originally been built for defense related work. On 10 and 12 February 1958, the Lincoln Lab group made five separate attempts to detect the reflections of their 440 MHz radar from Venus. At the time Venus was about 28 million miles from the Earth and the round trip Earth–Venus–Earth propagation time was about 5 minutes. During each attempt, thousands of pulses were transmitted during an approximate 4.5 minute interval, and the data from the receiver were recorded on magnetic tape for the following 5 minutes. As part of the analysis, to optimize the detection of reflected pulses, the data were carefully filtered to pass only a pulse of the right length, the right frequency corrected for the Doppler shift of the Earth–Venus motion and Earth rotation and around the right time delay resulting from the Earth–Venus round trip propagation time.

After a year of analysis of data from four of the five observing sessions, the Lincoln Lab group became convinced that they had detected echoes from Venus on two of the four runs and concluded that Venus was about 5,000 miles closer than expected from previous visual observations. This led to a revised value for the Astronomical Unit that was announced in a paper published in *Science* magazine and reported on the front page of *The New York Times.* In recognition of their achievement, the group received congratulations from US President Dwight Eisenhower.[78] Meanwhile, Jodrell Bank was unable to get their newly-completed 250 foot diameter antenna ready in time for the 1958 conjunction and was under considerable pressure to detect Venus echoes near the time of the next conjunction in September 1959, when they reported confirmation of the Lincoln Lab revised value for the Astronomical Unit.[79]

However, at the same time, although they had increased their radar power and installed a more reliable receiver, Lincoln Laboratory was unable to confirm their 1958 claim that they had successfully received radar echoes from Venus.[80]

The next opportunity to try to resolve the embarrassing discrepancy was the 1961 conjunction. Both Lincoln Laboratory and Jodrell Bank were poised with enhanced instrumentation, but with new competition from the Caltech Jet Propulsion Laboratory (JPL). The JPL experimenters employed two separate 85 foot antennas, one for transmitting and one, equipped with a sensitive maser amplifier, for receiving. Starting on 10 March 1961, JPL reported the real-time detection of echoes without the need for later computer processing of the data. The new JPL data determined the distance to Venus and established a new value for the Astronomical Unit that was 100,000 km greater than the value claimed by Lincoln Lab in their 1958 experiment and reported as confirmed by Jodrell Bank in 1959.[81] Subsequently, Lincoln Lab, Jodrell Bank, and other laboratories confirmed the JPL result.[82] Lincoln Labs had actually run their experiment a few days before JPL, but did not complete the analysis of their data until after they knew about the JPL detection. Meanwhile, in the USSR, Vladimir Kotelnikov, perhaps driven by Cold War pressures to demonstrate the power of the Soviet radar capability, but apparently unaware of the new JPL results, reported the successful detection of radar echoes from Venus using the tracking station at Yevpatoria, the same antenna that was later used by Sholomitsky to detect the flux density variability in CTA 102 (Chapter 4). The Russian experiment reported a value of the Astronomical Unit 100,000 km smaller than measured at JPL, but close to the incorrect values reported by Lincoln Lab and Jodrell Bank for the 1958 and 1959 conjunctions, respectively.[83] Recognizing the discrepancies after they learned about the new detection at JPL and Lincoln Labs, the Russian group reanalyzed their data and reported agreement with JPL and Lincoln Lab.[84] Later, William Smith reanalyzed the 1959 Lincoln Lab data and found evidence for an echo in agreement with the 1961 results.[85] In 1965, based on the 1961 JPL and Lincoln Lab radar measurements, the International Astronomical Union adopted the new value of the Astronomical Unit.

The 1961 Russian observations initially reported a rotation period between 9 and 11 days while Lincoln Lab found "a relatively smooth rocky surface" that rotated more slowly in about 225 days.[86] It was not until the late 1962 conjunction that the JPL radar was able to peer through the clouds with sufficient accuracy to reveal the long hidden surface and finally show that Venus was only slowly rotating, and that the rotation was in the reverse direction (counter-clockwise) from its revolution around the Sun, the Earth, and most of the other planets.[87] This retrograde motion had been missed by all the previous radar observations and analysis, although it was confirmed by Drake from his passive 10 cm radio observations made between 1961 and 1963, but not recognized until after the JPL radar result was announced.[88] Only three months after its completion, the Arecibo radar detected Venus in February 1964. Observations with the radar near the inferior conjunction of Venus in June 1964 measured a retrograde period of 247 ± 5 days, very close to the current value of 243 days.

Earlier in this chapter, we described how, in 1965, the Arecibo radar succeeded in revealing the unexpected rotation of Mercury. Further advances in radar technology have led to detailed mapping of the Moon, Mercury, Venus, Mars, and some of the satellites of Jupiter and Saturn, from both ground-based radars and in situ mapping of Venus and Titan from orbiting radar systems. The incorrect claims of detected echoes in 1958 and 1959, by Lincoln Labs and Jodrell Bank, respectively, were no doubt due to noise fluctuations filtered by the expected propagation time delay, and perhaps amplified by the optimistic expectations of the experimenters.

Discovering the First Extrasolar Planets

Are there planets surrounding other stars in the Galaxy or is our Solar System unique, the result of an improbable fortuitous event unlikely to be repeated elsewhere? Alternatively, if the formation of planetary systems is a natural consequence of stellar evolution, is it possible to detect other planetary systems? Such questions, which are related to the search for life elsewhere in the Galaxy and theories of planetary formation, have been asked for millennia. The detection of planets around other stars is at best challenging. As seen from elsewhere in the Galaxy, the Sun outshines even Jupiter by a factor of about 10 billion, so until recently it was not possible to detect an extrasolar planet using conventional ground-based telescopes. Numerous observers have wrongly claimed to have discovered other planetary systems using traditional optical techniques,[89] so it is not surprising that the discovery of the first extrasolar planets by two radio astronomers in 1991 came about in a very unforeseen way and was associated with a very unexpected star – a neutron star in the form of a millisecond pulsar.

One typically thinks of planets revolving around their sun, or the Moon revolving around the Earth. But this is not precisely true. According to Newton's laws of gravity, both bodies actually revolve around their common center of mass. In the case of the Sun–Earth system, the center of mass is well within the Sun's diameter. Nevertheless, a distant observer somewhere in the Galaxy could, in principle, observe a small wobble in the position of the Sun over the annual 365 day cycle. The hypothetical distant observer will also see a Doppler shift in the frequency of spectral lines in the observed solar spectrum as the velocity of the Sun also changes. But the velocities are typically slow – of the order of one meter per second, so are difficult to observe in practice.

Just as the small motions of the host star will produce a Doppler shift in the spectral lines, they will also introduce a shift in the apparent frequency of any observable ticking clock. Pulsars provide the accurate ticking clock that can be used to detect planets orbiting a pulsating neutron star. In 1991, a team of radio astronomers from Jodrell Bank noticed that, after correcting for the motion of the Earth, the timing of the pulsar PSR 1829–10 was modulated with an apparent period of six months with an amplitude about 0.008 seconds, an easily measured amount. After considering other possible causes of the observed sinusoidal variation, the team concluded that it was caused by a planet, about 10 times the mass of the Earth, that was orbiting the pulsar with a six-month nearly circular orbit.[90] They then went on to speculate how this discovery would impact theories of both planet and neutron star formation, in particular how the planet could have survived the supernova explosion that created the neutron star. The close agreement of the orbital period with half an Earth year should have, and did, raise a flag, suggesting that the apparent modulation of the pulsar period might be an artifact of the Earth's annual motion around the Sun and not of a planet's motion around the pulsar. However, the observers noted that they did not see such an effect in any of the several hundred other pulsars that they had observed, including one located only two degrees away from 1829–10. The announcement of a planet orbiting PSR 1829–10, reported in the 25 July 1991 issue of *Nature*, created a lot of excitement within the planetary science as well as the radio astronomy communities. It also generated a flurry of theoretical ideas about the origin of planets, including reconsiderations of the probability of extraterrestrial life.[91]

Meanwhile, Alex Wolszczan, a Polish-born scientist working at the Arecibo Observatory in Puerto Rico, was observing pulsars with the 1,000 foot radio telescope. Due to maintenance underway at the Observatory, Wolszczan was unable to scan the telescope or track individual sources and was limited to letting the sky drift through the stationary telescope beam. With this restriction, all he could do was to survey narrow strips of sky looking for new pulsars, one of which was the 6.2 millisecond pulsar, PSR 1257+12. Examining routine timing measurements of PSR 1257+12 following the completion of the telescope maintenance and the restoration of full operations, Wolszczan noticed "an unusual variability superimposed on an annual sinusoidal pattern caused by a small [1 arcmin] error in the assumed pulsar position." After Dale Frail (Figure 9.9) used the NRAO VLA to determine the pulsar position with 0.1 arcsec accuracy, Wolszczan and Frail found that there remained "two strict periodicities" only 0.003 seconds in amplitude, which they interpreted as the result of at least two planets orbiting the pulsar with periods of 66.6 and 98.2 days as well as a possible third planet with a longer period.[92] They calculated the masses of the two planets as 2.8 and 3.4 times the mass of the Earth in nearly circular orbits at 0.47 and 0.36

Figure 9.9 Alex Wolszczan (left) and Dale Frail (right). Credit: American Astronomical Society (AAS), courtesy of AIP Emil Segrè Visual Archives; NRAO/AUI/NSF.

Astronomical Units (AU), respectively. As the authors pointed out, "The characteristics of the 1257+12 planets are not unlike that of the inner Solar System. Both planets circle the pulsar at distances similar to that of Mercury in its orbit around the Sun." Further, they noted that the observed ±0.7 m per second (approximately 2 miles/hr) "pulsar 'wobble' caused by the orbital motions of the planets, ... is entirely inaccessible to the optical methods of planetary detection." The discovery of a planetary system around a pulsar dramatically demonstrated that planet formation could occur under extreme conditions and that the formation of our Solar System was not a unique event.

Wolszczan and Frail (Figure 9.9) submitted their paper to *Nature* on 21 November 1991, four months after the Jodrell Bank paper was published. So, the Wolszczan and Frail paper perhaps received relatively little initial publicity when it was published on 9 January 1992. Further timing observations of PSR 1217+12 by Wolszczan confirmed the existence of the suspected third very low mass planet with a 25.3 day orbit as well as indicating the presence of a possible fourth planet. The new observations also detected the predicted small perturbations of the orbits due to gravitational interaction of the two inner planets, which effectively eliminated any alternative interpretation of the timing observations.[93]

In January 1972, Wolszczan and Frail traveled to Atlanta, Georgia, to report their detection of what they thought was the second known exoplanet system at the 179th meeting of the American Astronomical Society. Andrew Lyne, the leader of the Jodrell Bank team, was invited to the same meeting to describe their dramatic discovery of the first known planet outside the Solar System. Instead, Lyne reported that, following further analysis, they had made a subtle error in their analysis and that, in fact, there was no planet orbiting PSR 1829–10. Because of an initial error in the assumed position of the pulsar combined with their simplifying assumption that the Earth's orbit is circular, the slight ellipticity of the Earth's orbit caused the appearance of a six-month sinusoidal variation in the timing of pulses. Although, as Lyne explained, it is usual to use the pulsar timing data itself to solve for the position of the pulsar. However, that was not done in this case, resulting in an embarrassing false claim to have discovered the first extrasolar planet. Lyne and Bailes reported their retraction in the 16 January 1992 issue of *Nature*.[94]

Nearly four years after the Wolszczan and Frail announcement in *Nature* of the 1257+12 planetary system, Swiss astronomers Michel Mayor and Didier Queloz reported their spectroscopic discovery of a "Jupiter-mass companion to a solar-type star," 51 Pegasi. The 1257+12 planetary system was mentioned only in passing in the last sentence of their submitted paper, that was widely acclaimed by the scientific community as well as in the media.[95] Mayor and Queloz were able to detect the 51 Pegasi planet by measuring the shift of the optical spectral lines due to the relatively large wobble of 51 Pegasi of about 75 m per second resulting from the nearby massive orbiting planet. The 51 Pegasi planet had a reported mass comparable to that of Jupiter and orbited 51 Pegasi every four days at a distance of only 0.08 AU, making it much more massive than the 1257+12 planets and orbiting very much closer to the host star. Further refinement of the Mayor-Queloz technique has resulted in the discovery of more extrasolar planets,[96] but it would be another two decades before the optical methods could reach levels of accuracy needed to detect Earth-mass planets.[97] The more recent NASA missions, Kepler and the Transiting Exoplanet Survey Satellite (TESS), have uncovered thousands of exoplanets.[98] Kepler and TESS were able to detect planets by measuring the small drop in the light when a planet passes in front of the host star. The study of exoplanets is now a major field of astronomy, driven in part by the search for Earth-like planets orbiting solar-type stars in the so-called habitable zone and the hunt for signs of extraterrestrial life.

Both PSR 1257+12 and 51 Pegasi planets are subject to extreme environments. In the 1257+12 system, there is intense radiation from the host neutron star, while the 51 Pegasi planet is exposed to the extreme heat of its nearby host star, indicating that the formation of planets is not uncommon. Wolszczan and Frail discovered the first known extrasolar system containing more than one planet and the first Earth-mass extrasolar planet. Four years after the 1257+12 announcement, Mayor and Queloz discovered a single Jupiter-like planet orbiting a solar-type star. But it would take another decade before any other multi-planet systems would be found.[99] In 2019, Mayor and Queloz received the Nobel Prize in Physics "for the discovery of an exoplanet orbiting a solar-type star."

10

Testing Gravity

Of course there is no hope of observing this phenomenon directly.
Albert Einstein, on the detectability of gravitational lensing[1]

For half a century following the publication of Albert Einstein's General Theory of Relativity (GR)[2] it appeared that GR had few impacts to the real world of observational physics and astronomy. However, starting in the mid-1960s, radio astronomers measured the gravitational bending of radio waves by the Sun with an accuracy reaching several hundred times better than any visual observation, detected the first gravitational lenses, found the first experimental evidence for gravitational waves, provided the most convincing evidence for the existence of dark matter in galaxies, and devised a new test of GR not considered by Einstein. Unlike many other topics discussed in this book, the theoretical framework was established decades before many of the phenomena were observed and the theory tested.

Relativistic Bending

Albert Einstein's (Figure 10.1) Theory of General Relativity (GR) defines the properties of space–time that differ from those associated with Newton's classical laws of physics. The differences are small and at the time bordered on the edge of existing measurement capabilities. The classical differences are:

(1) The precession of the perihelion (point of closest approach to the Sun) by Mercury of 43 arcseconds per century,
(2) The bending of light by gravity that is a factor of two larger than given by Newton's laws, and
(3) The shift in the wavelength of light by gravity.

Figure 10.1 Albert Einstein outside his house at 112 Mercer Street, Princeton, New Jersey on the occasion of his 70th birthday. Credit: Photographer, Alan Richards. From the Shelby White and Leon Levy Archives Center, Institute for Advanced Study, Princeton (NJ).

A fourth, more recently realized test was developed in 1964 and is discussed in the next section.

The first of these was not really a prediction. It had been known since the middle of the nineteenth century that the perihelion of Mercury precessed by an amount that was greater by 43 arcseconds per century than could be explained by Newtonian mechanics and the influence of the other planets. This is a very small angle, about 2 percent of the diameter of the Moon, but it was well determined from astronomical observations. The cause was unknown and led to speculations that the orbit of Mercury was perturbed by the gravitational influence of an unseen planet that was inside the orbit of Mercury. This hypothetical planet even had a name, Vulcan, but extensive photographic searches found no evidence for its existence. General Relativity explained the precession of Mercury as a consequence of the bending of space–time by the Sun and was initially the only demonstrable evidence for the correctness of GR and its relevance to the real world.

The effect of Einstein's third prediction was so small that it would be nearly half a century before it could be tested. During 1959 and 1960 Harvard University Professor Robert Pound and his student Glen Rebka used the then recently discovered Mössbauer Effect of the "recoilless resonant scattering" of gamma rays from solids to demonstrate the predicted gravitational redshift with an accuracy of

10 percent.[3] In their elegant, and now famous, experiment, Pound and Rebka measured the shift in frequency of a narrow gamma-ray absorption line in an isotope of iron of less than one part in 10^{14} over a 74 foot (23 m) height differential obtained by transporting their experiment from the basement to the top floor of Harvard's Jefferson Physics Laboratory building. Finally, 20 years later, Robert Vessot and colleagues at Harvard and NASA flew a hydrogen maser to an altitude of 10,000 km and were able to confirm the expected change in frequency with altitude to better than 0.01 percent.[4]

The bending of light by the gravity of the Sun, however, appeared to offer an immediate opportunity to test the new theory of General Relativity which predicted that the Sun would deflect the light from a star just grazing the limb of the Sun by 1.75 arcseconds. This was twice the amount that Einstein and others had calculated earlier, based on classical Newtonian mechanics and the equivalence of inertial and gravitational mass as demonstrated by Galileo in his famous Tower of Pisa experiment.[5]

Because of the intense optical glare of the Sun, stars located near the Sun can only be observed during the time of a total solar eclipse. But measuring the small position displacement on photographic plates during an eclipse is not straightforward. The expected deflections are distorted by atmospheric turbulence (seeing) and amount to only a few hundredths of a millimeter on a typical astronomical photographic plate. The rapid change of temperature and barometric pressure at the time of an eclipse distorts both the telescope optics and photographic plates, which then need to be compared with control plates of the same field taken during nighttime when there is no bending by the Sun, but subject to different thermal distortions. Moreover, total eclipses of the Sun are rare and eclipse observations generally took place at remote locations that do not provide the same stability of infrastructure as fixed observatories. Also, at the time of a typical eclipse, there are generally too few stars within a few solar radii that are sufficiently bright to adequately measure GR bending and separate it from instrumental distortions.

A very favorable opportunity to test Einstein's new theory occurred at the time of the 29 May 1919 total solar eclipse, that occurred when the Sun was fortuitously in the field of a bright star cluster. The First World War had just ended and international travel was again feasible, but still difficult. Two teams set out from Britain, one to Sobral, Brazil, and one, led by the distinguished astrophysicist Arthur Stanley Eddington, to the Portuguese island colony of Principe, off the west coast of Africa. Both expeditions reported the detection of gravitational bending with a claimed accuracy of about 20 percent that appeared to support the GR and not the classical prediction.[6] The report received broad media attention but, due to weather and equipment failure, the data were poor. There were also charges, which were later refuted, that Eddington had selectively discarded data that did not agree with the GR predictions.[7]

Over nearly the next half century, there was little progress.[8] The precision of photographic astrometry had barely improved. For example, a 1973 expedition to observe an eclipse in Mauritania, in spite of meticulous attention to detail, was able to measure the solar bending with a reported precision of 11 percent. By contrast radio astronomers had developed sophisticated interferometer systems that promised previously unheard of precision (Chapter 11). Moreover, although the Sun is a very strong radio source, it does not swamp the rest of the radio sky, so radio observations can be made at any time and do not require a rare solar eclipse. Fortunately, each October the Sun passes very close to the quasar 3C 279, one of the brightest radio sources in the sky at centimeter wavelengths. Several groups of radio astronomers realized that they could measure the relative position of 3C 279 with respect to another quasar, 3C 273, located only ten degrees away, as the Sun passed close to 3C 279. In October 1969, one team used the Caltech interferometer at 9.6 GHz (3.1 cm) with a baseline of 1.066 km (3,498 ft) over a two week period to measure a deflection of 3C 279 referred to the limb of the Sun of 1.77 arcseconds.[9] Meanwhile, another team used two antennas of the JPL Goldstone Tracking Station to make an independent interferometric measurement of the separation of 3C 279 and 3C 273 at 2,388 MHz (12.5 cm) over a 21.6 km (13.4 mile) baseline, and measured the gravitational bending to be 1.82 arcseconds.[10] Both of the interferometric radio measurements were able to determine the bending with a precision of about 10 percent, comparable to the best visual determination, and were consistent with that expected from General Relativity, but not with the classical prediction of 0.88 arcsec. However, there was a complication.

Radio (but not light) waves passing close to the Sun are bent not only by the Sun's gravity, but also by refraction in the ionized plasma in the solar corona. Because the gravitational bending decreases linearly with the distance from the Sun, while the refraction has a more complex dependence on the radial distance, both groups were able to separate the two effects well enough to determine the gravitational bending.

Richard Sramek further addressed this complication by observing simultaneously at two frequencies, 2,695 MHz (11.1 cm) and 8,085 MHz (3.7 cm) with the 2.7 km NRAO three-element interferometer.[11] Since the amount of coronal refraction depends on the square of the wavelength, while the relativistic bending is independent of wavelength, Sramek was able to better separate the two bending effects to determine the relativistic bending with a claimed precision of 5 percent.

The following years saw increased precision from the use of longer interferometer baselines, higher frequencies, multiple antennas, multiple reference sources, and improved analysis.[12] By 1976, Edward Fomalont and Sramek had extended their NRAO interferometer baseline to 35 km and used two reference sources.[13] With an experimental precision of about 1 percent, they were able to exclude the then competing scalar-tensor gravitational theory at a 99 percent confidence level.[14] In October 2005, Fomalont et al. used the NRAO Very Long Baseline Array (VLBA) to measure the bending of 3C 279 with respect to three nearby calibration sources. By observing at three frequencies, 15, 23, and 43 GHz (1 cm, 1.3 cm, and 7 mm), they were able to effectively eliminate the effect of bending due to refraction in the solar corona and measured the gravitational bending to be 99.98 percent of that expected from General Relativity with an accuracy of 0.04 percent.[15]

With the increasing use of radio interferometers for precision astrometry, the effect of gravitational bending by the Sun is so great that the effect can be seen anywhere in the sky, even far away from the Sun. Indeed, a series of very long baseline interferometer measurements, made to study terrestrial geodesy over periods up to 20 years, led to a determination of the deflection with increasing precision, ultimately rivaling that of shorter experiments expressly designed to test GR bending.[16] Experiments designed to measure gravitational bending itself are no longer of interest; rather General Relativity is assumed to be correct, and the theoretical bending is applied as a matter of routine when analyzing radio interferometer data meant for precision astrometry, measurements of plate tectonics and Earth rotation, or interplanetary navigation.

The Fourth Test of General Relativity

For half a century after Einstein developed his theory of General Relativity, the only experimental tests were the three that he originally described. However, in 1964, MIT professor Irwin Shapiro (Figure 10.2) proposed a new independent test, one that Einstein could not possibly have imagined decades before the first radar systems were developed. Shapiro realized that, because the speed of light (and radio waves) depends on the strength of the gravitational potential along its path, radar reflections from Venus or Mercury that pass near the Sun should be delayed by almost 0.0002 seconds, and he discussed the practical feasibility of conducting the test using the MIT Haystack and Arecibo planetary radar systems.[17]

Following upgrades of the 300 kW Haystack radar system, in 1966 and 1967 Shapiro and his MIT colleagues detected the expected delay in the returned echoes near the time of superior conjunctions of both Venus and Mercury. The measurement was challenging as the 0.0002 second delayed signal was on top of a total Earth–Venus–Earth propagation time of nearly 30 minutes. The received signal was very weak, about 10^{-21} watts. Over the next three years, the group expanded their program to include radar observations from Arecibo as well as adding experiments to detect delayed echoes from Mars. These new experiments of what has become known as the "Fourth Test of General Relativity" or the "Shapiro

Figure 10.2 Irwin Shapiro. Courtesy of Irwin Shapiro.

delay," confirmed that the expected delay in propagation time was within 5 percent of that expected from GR.[18] Later, interplanetary spacecraft were used to demonstrate agreement with GR to better than 0.1 percent.[19]

More recently, radio astronomers have measured the Shapiro delay using pulse time of arrival measurements from binary (neutron star–white dwarf) millisecond pulsar systems with accurately known orbits. From the delay in the pulsar signal time of arrival of about 10 microseconds, induced by passage close to the white dwarf companion, they calculated that these neutron stars have masses more than twice that of the Sun, or significantly more than the canonical neutron star upper limit of 1.4 solar masses. The discovery of these high mass neutron stars forced a reconsideration of the nature of neutron stars and the possibility that they contain a significant non-nuclear component such as hyperons or quarks.[20]

Gravitational Lensing

In a now famous 1936 paper, Albert Einstein pointed out that, due to gravitational bending, if a star was aligned between the observer and a more distant star, the

Figure 10.3 Jeno and Madeleine Barnothy. Credit: AIP Emilio Segrè Visual Archives, John Irwin Slide Collection.

intermediate star could act as a gravitational lens, forming a ring-like image around the lensing star.[21] Apparently Einstein had forgotten that he first worked out the equations for gravitational lensing as early as 1912, but in 1936 he was persuaded by a Czech engineer to publish the analysis of a gravitational lens.[22]

However, aware that the angular size of the ring "will defy the resolving power of our instruments," Einstein concluded that "Of course there is no hope of observing this phenomenon directly," and he remarked that, instead, one would only see an increase in the apparent brightness. He also went on to point out that "if the observer is situated at a small distance, *x*, from the extended central line ... the observer will see ... two point-like light-sources."

Shortly after Einstein published his 1936 paper, the Caltech Swiss astrophysicist, Fritz Zwicky, noted that an entire galaxy, as well as clusters of galaxies, or nebulae as they were then called, not only could, but would act as an observable massive gravitational lens which "becomes practically a *certainty*."[23] With great prescience, in these papers, Zwicky realized that the detection of gravitational lensing would provide an additional test of GR, "would enable us to see nebulae at a greater distance than those reached by even the greatest telescopes," and provide "a direct determination of nebular masses." Zwicky's second paper was apparently motivated by his separate determination, discussed later in this chapter, that the measured mass of galaxy clusters greatly exceeds the visible mass.

Jeno Barnothy and his wife, Madeleine Barnothy (Figure 10.3), did research in cosmic ray physics in their native Hungary before emigrating to the United States in 1948. They both knew about Einstein's short 1936 paper, and during the 1960s and 1970s they wrote many papers and gave many talks, particularly at meetings of the American Astronomical Society, explaining a number of observed phenomena as the result of gravitational lensing. This included the anomalous brightness of quasars, the absence of any magnitude-redshift (Hubble) relation for quasars, the non-uniform redshift distribution of quasars, and the superluminal motion of radio sources, that were all discussed in Chapter 4.[24] Many of their presentations were repetitious and were confused by Jeno's unconventional and controversial "FIB" theory of cosmology.[25] In spite of their warm and friendly personalities, perhaps because of their persistent repetitive talks to decreasingly attentive audiences, or perhaps because their only professional affiliation was with the University of Illinois Medical School Biomagnetic Research Foundation, which was outside the mainstream of American astronomy, their speculations about gravitational lensing were widely ignored.

Quite independent of the Barnothys' long advocacy, Einstein's prediction that the theoretical possibility of gravitational lensing might actually exist in the Universe was dramatically confirmed by the accidental discovery of the double image of the quasar 0957+561.[26] The radio source 0958+56 was catalogued in 1973 by a Jodrell Bank team as part of a 966 MHz (31 cm) survey intended to cover the sky from declination +40 to +70 degrees. They were assigned one month to do this project with the 250 foot (76 m) Jodrell Bank radio telescope (Chapter 11). Due to interference from satellites and other problems, by the end of the month, the survey had only reached +55 degrees. To the dismay of the next scheduled observer, Lovell granted the team another month to complete their survey; otherwise they would not have found 0958+56 at +56 degrees declination. Using an interferometer, the Jodrell Bank radio astronomers were able to determine accurate radio positions, allowing the optical identification of about 70 percent of their radio sources. The other 30 percent, including the source 0958+56, appeared to be too large to observe with the interferometer.

Richard Porcas, who had been part of the Jodrell Bank survey team, joined NRAO after completing his PhD dissertation, and used the Green Bank 300 foot antenna at 6 cm (5 GHz) and 11 cm (2.7 GHz) to re-observe the extended sources with ten times better resolution (1.8 arcmin) than the Jodrell Bank 18 arcmin resolution. He did not know at that time that 0958+561 was in fact two sources, one an unrelated source coincident with the previously identified galaxy NGC 3070 and another weaker one about 14 arcminutes away. Due to his search procedure, he detected only the weaker one and 0958+56 was thus re-designated 0957+561. Megan (Meg) Urry, an NRAO summer student, was tasked by Porcas with searching for optical counterparts of the radio sources. Urry noticed two blue stellar objects (BSOs) about 5 arcseconds apart – both magnitude 17 and near the radio source 0957+561. Apparently this same pair of BSOs had been independently noted by a Jodrell Bank student, Anne Cohen, but the radio position uncertainty at the time precluded any convincing identification with the radio source.[27]

Although there was still an uncomfortably large position difference between the radio and optical positions, Dennis Walsh's attention was drawn to their unusually blue color. Walsh joined up with University of Arizona colleagues to obtain optical spectra and redshifts for the newly-identified radio sources. Spectra taken of the slightly brighter BSO using the 2.1 m (82 inch) telescope at the Kitt Peak National Observatory in Arizona showed two strong emission lines that they identified with carbon IV and carbon III at a redshift of 1.4, supporting its identification as a quasar. Walsh and his colleagues then went on to obtain a spectrum of the other BSO, located 6 arcseconds away. To their surprise, the spectrum was identical, having the same two emission lines at the same redshift. Thinking that they had made some error, they went back to the first BSO to confirm that they had not just repeated the observation. After repeated back and forth observations of the two BSOs, they became convinced that they had identical redshifts, colors, and nearly equal magnitudes.

While Walsh and his colleagues went on to observe the spectra of the other identified radio sources in their list, the following night Ray Weymann used the University of Arizona 2.3 m (90 inch) telescope to obtain high spectral resolution spectra of both quasars. Noting that they had the same absorption lines, Weymann suggested that, due to gravitational lensing by an unseen intervening galaxy, they were seeing two images of the same quasar.[28] Using VLBI, Porcas and colleagues made high resolution radio images and showed that both quasars had the same core-jet structure.[29] Subsequently, MIT radio astronomers used the VLA to better image the radio emission from 0957+561. As shown in Figure 10.4 they found many more radio features than are visible optically.[30] This discovery of the first gravitational lens led to a proliferation of both theoretical and observational studies of gravitational lensing systems. Seven years after Porcas first observed the radio source 0957+561, and more than half a century after Einstein and others first speculated about gravitational lensing, it became a new tool for astronomers to study the Universe and its content.

The lensed images depend on the mass distribution in the lensing galaxy or cluster of galaxies as well as the shape and angular separation of the lensed object from the lensing galaxy.[31] Many kinds of gravitationally lensed images are now observed both at radio and optical wavelengths. Figure 10.5 shows a radio image of the so-called Einstein ring first observed by Jacqueline Hewitt et al. with the VLA.[32]

Fifteen years before the discovery of 0957+561, the Norwegian astrophysicist Sjur Refsdal pointed out that, because different lensed images arrive via different paths, any intrinsic radio or optical variation in the source is seen at different times in the lensed images.[33] Since the propagation delay depends on the lens geometry, including the distance to the source and to the lens, observations of the time delay seen between different images are used to determine the distances to the source and lens. By comparison with the observed redshifts, the discovery of gravitational lensing gave

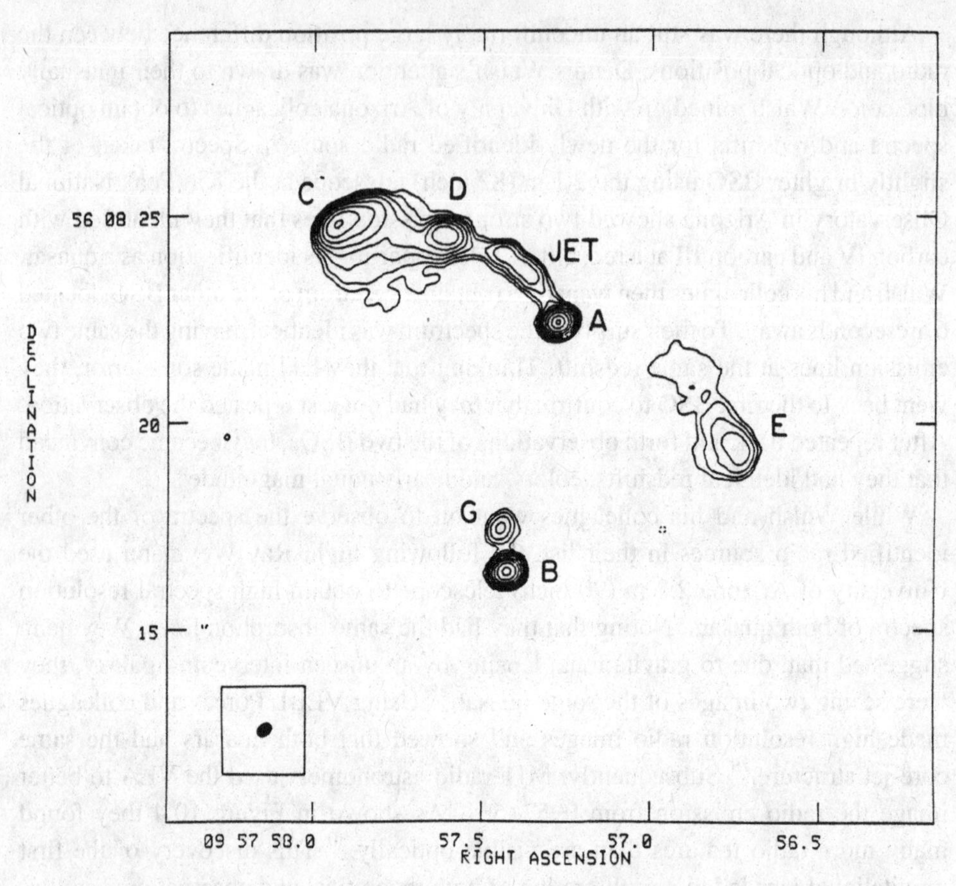

Figure 10.4 VLA contour map of the gravitational lens 0957+561. Credit: Greenfield et al. (1985).

astronomers a new way to determine the Hubble constant, independent of the classical optical techniques.[34] Spurred by the serendipitous discovery of the 0957+561 gravitational lens, Refsdal returned to work on gravitational lensing, which is now a widely used tool for astrophysics with many applications: searches for extrasolar planets, dark matter and dark energy in clusters of galaxies, and determinations of the Hubble constant.[35] Although Refsdal's pioneering work was widely recognized with the flurry of activity following the discovery of 0957+561, the prescient papers of the Barnothys have received relatively little recognition.

Detecting Gravitational Radiation

One of the most profound consequences of GR, as well as other modern theories of gravity, is the prediction of gravitational radiation from any non-symmetric time

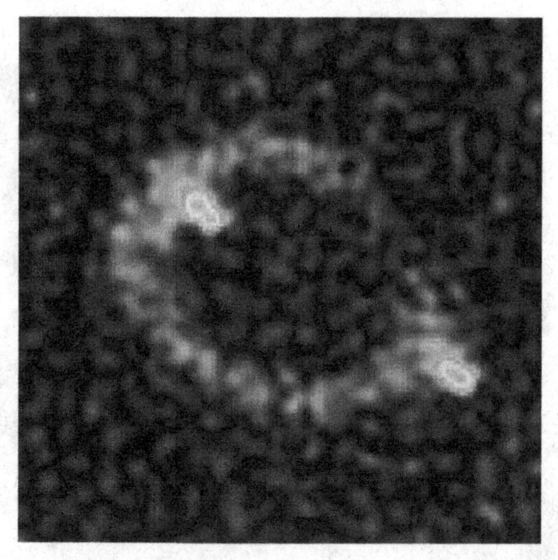

Figure 10.5 VLA image of MG1131+0456 Einstein Ring. Credit: Hewitt et al. (1988).

varying activity involving massive bodies, for example the interaction between two very close massive stars or an asymmetric supernova explosion. However, the expected strength of any gravitational wave radiation is exceedingly small, challenging any experimental verification.

Joe Weber had been a naval officer during the Second World War. Following the War, after a short stint developing electronic countermeasures, Weber joined the University of Maryland faculty, where he directed his skills in microwave electronics toward innovative experiments designed to detect the feeble levels of gravitational radiation that might reach the Earth from cataclysmic celestial events. Over a period of years, Weber built a series of large suspended bars of aluminum, spatially separated by up to 1,000 km, coupled to piezoelectric crystals that he claimed detected the small vibrations induced by passing gravity waves mostly coming from the direction of the Galactic Center (Figure 10.6).[36] Although his claimed detection of gravity waves generated considerable excitement, the implied strength of signals far exceeded anything predicted on theoretical grounds. Meanwhile, others were unable to reproduce Weber's results with experiments that were much more sensitive than Weber's, and Weber's claimed detections of gravitational radiation are now largely discounted.[37] Nevertheless, Weber is credited with having initiated the field of experimental gravitational wave astronomy that stimulated many other workers. Moreover, apparently Weber was the first to suggest the use of a laser interferometer which was used to finally detect gravitational waves nearly half a century later.[38]

However, the first quantitative proof of the existence of gravitational radiation came, not from a designed laboratory experiment, but from two young radio

Figure 10.6 Joseph Weber checking the crystal detectors on his 38 inch bar. Credit: AIP Emilio Segrè Visual Archives.

astronomers who, while searching for new pulsars, discovered the binary pulsar system PSR 1913+16, serendipitously opening a new door to gravitational wave astronomy.[39]

Russell Hulse (Figure 10.7) was a University of Massachusetts graduate student working with his advisor Joseph (Joe) Taylor (Figure 10.8). Hulse had gone to graduate school intending to do research in atomic physics, but instead was persuaded by Taylor to switch to radio astronomy. In the summer of 1973, Hulse and Taylor were given time on the Arecibo 1,000 foot radio telescope to explore the characteristics of different types of pulsars, and Hulse traveled to Puerto Rico to take the data. Using a sophisticated computer algorithm originally devised by Taylor[40] and implemented by Hulse, they searched for pulsed emission over a

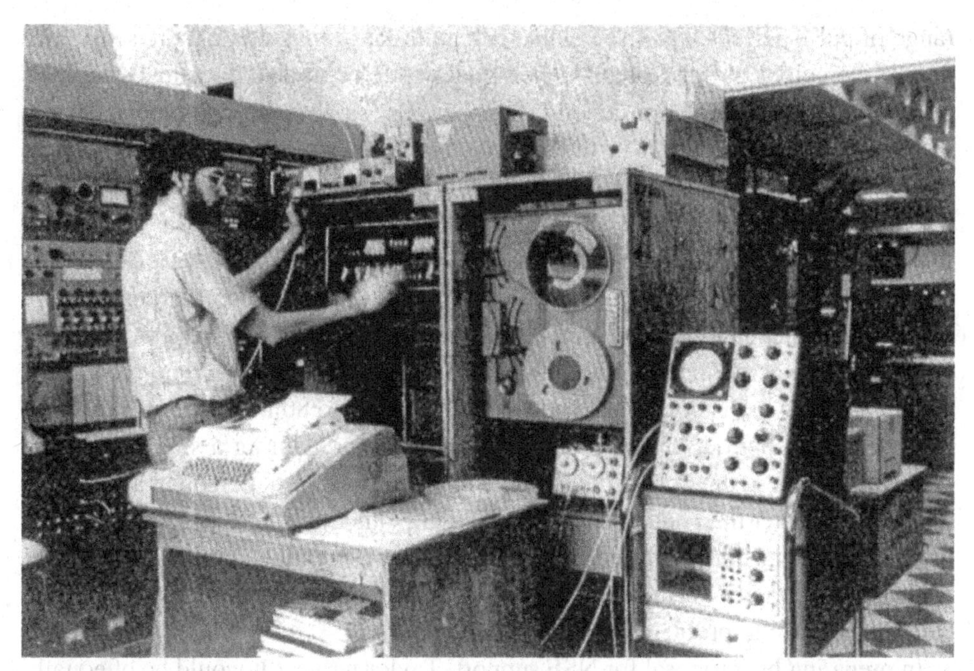

Figure 10.7 Russell Hulse, at the Arecibo Observatory. Courtesy of Joe Taylor.

Figure 10.8 Joe Taylor. Credit: AIP Emilio Segrè Visual Archives, Physics Today.

range of pulse periods and dispersions with an order of magnitude more sensitivity than any previous pulsar search.[41] Their goal was to compile a statistical sample of pulsars to study the relation among their periods, slow-down rates, pulse shape, dispersion, and distribution in the Galaxy. However, among the 40 new pulsars discovered, one, PSR 1913+16, that was only slightly stronger than their search limit, stood out because of its short period of 0.059 seconds. This was the second fastest pulsar known after the 0.0033 second Crab pulsar. Hulse first detected PSR 1913+16 on 2 July 1974, but subsequent observations to refine the period were hard to understand. After correcting for the Doppler shift due to the motion of the Earth, Hulse was perplexed. Each time he observed PSR 1913+16 he recorded a different period in his notebook. The changes were up to 80 microseconds over the observed pulse period, and as much as 8 microseconds during a single 5 minute observation. These unexplained variations in the period were more than an order of magnitude greater than had been observed in any other pulsar. Suspecting a timing error of the kind that confused the Green Bank observations of the Crab pulsar discussed in Chapter 7, Hulse modified the computer program to obtain better time resolution. But the problem remained.

In preparing his proposal for NSF support, Taylor wrote, "It would be of equally high significance to find even one example of a pulsar in a binary system, for measurement of its parameters could yield the pulsar mass, an extremely important number." However, this time they had found about 30 "normal" pulsars, none of which had such strange behavior. So, when they found a pulsar with a varying period, they were not really expecting it.[42]

Finally, on 18 September, Hulse realized what was happening. Pulsar PSR 1913+16 was indeed part of a binary system. At the time telephone communication between this remote part of Puerto Rico and the US mainland was at best marginal. So Hulse quickly dashed off a handwritten letter to Taylor, who was still at the university in Amherst, to report that the period was changing due to the Doppler shift resulting from the pulsar's orbiting motion around a companion with a remarkably short period of only 8 hours. Then, better appreciating the importance of his discovery, Hulse used a shortwave radio system to contact the Arecibo Observatory headquarters at Cornell University in Ithaca, New York, which then patched him by telephone through to Taylor in Amherst. Over the next weeks, Hulse read his measurements to Taylor over the radio-telephone link, from which Taylor calculated the orbital parameters.

Hulse and Taylor reported their discovery of the first binary pulsar in an IAU Telegram and in December at the 144th meeting of the American Astronomical Society (AAS), that was held in Gainesville, Florida.[43] In their AAS presentation, they noted the potential to observe "a number of interesting effects" including, "the relativistic Doppler term, the gravitational redshift [Einstein's third test of GR], and the advance of periastron [analogous to the precession of Mercury's orbit]." Interestingly, they did not mention the possibility of detecting gravitational radiation at the AAS meeting nor in their paper sent to the *Astrophysical Journal* in October.[44]

Hulse and Taylor suggested that the binary companion of the pulsar was probably also a neutron star of comparable mass to the pulsar, but itself did not appear to be a pulsar. They also predicted a GR advance of the highly elliptical orbit by about 4 degrees per year, that was a factor of 3,000 greater than Mercury's 43 arcseconds per century. Ensuing measurements of pulse arrival times established the mass of each of the two binary components to be close to 1.4 solar masses and confirmed the expected precession of the orbit as well as other strong field GR effects. After four years of timing data and correcting the data for these effects, in December 1978, at the 9th Texas Symposium on Relativistic Astrophysics in Munich, Taylor announced what is undoubtedly the most important discovery that came from the binary pulsar – the orbital period was decreasing by precisely the rate expected if the binary system was losing energy by the radiation of gravitational waves.[45] Although most modern theories of gravity imply gravitational radiation, as shown in Figure 10.9 the continued timing observations of the PSR 1913+16 orbital period speed up confirmed the predictions of GR, initially to within 25 percent and later to about 0.2 percent, and provided the first "compelling evidence for the existence of gravitational radiation, as well as new and profound confirmation of the general theory of relativity."[46]

Unlike the earlier laboratory and Solar System tests of GR, the binary pulsar allowed for the first time tests of strong gravitational fields such as those associated with neutron stars and black holes. "For their discovery of a new type of pulsar, a discovery that has opened up new possibilities for the study of gravitation," Russell Hulse and Joseph Taylor received the 1993 Nobel Prize in Physics, thus becoming the fifth and sixth recipients of the Nobel Prize for work in radio astronomy. It would be another four decades after the discovery of the Hulse–Taylor binary pulsar before gravity waves caused by the coalescence of two black holes were directly detected by LIGO.[47]

In preparing for his Arecibo observations, Hulse had set up his computer program to record only those pulsars that appeared to be stronger than seven times the receiver noise. If set lower to search for weaker signals, Hulse would have to deal with many false reports. At the time of the initial discovery observation, PSR 1913+16 was only 7.25 times greater than the noise. Had it been even 3 percent weaker at the time, or had Hulse set his cutoff 3 percent higher, he would not have detected the binary pulsar that gave the first experimental evidence for gravitational radiation.

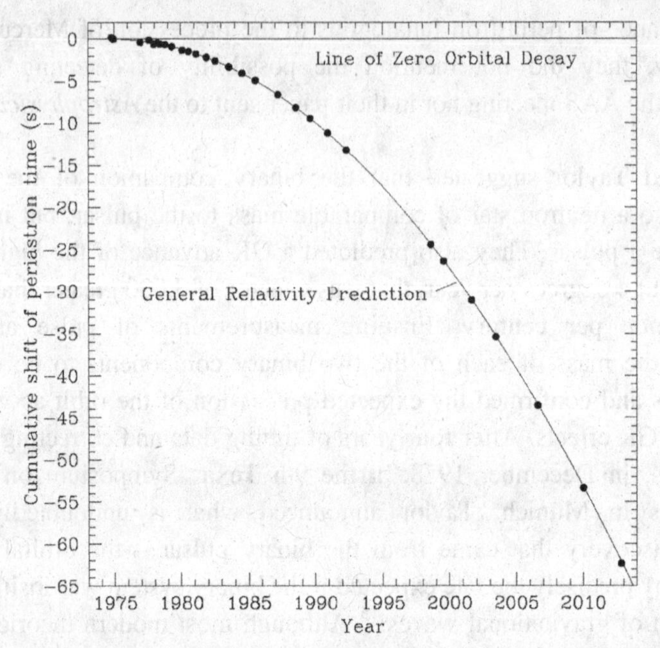

Figure 10.9 Orbital decay of the orbit of PSR B1913+16 as a function of time. The solid line represents the shift from gravitational wave radiation predicted by General Relativity. Credit: Weisberg and Huang (2016).

In 2004, using the Parkes radio telescope, an international team of radio astronomers from Australia, India, Italy, the United Kingdom, and the United States reported the discovery of the first double pulsar system – two pulsars, J0737A and B, in a slightly eccentric 2.4 hour orbit that led to an improvement of nearly another order of magnitude in the change of orbital period due to gravitational wave radiation.[48]

The gravitational radiation subsequently detected from LIGO results from the catastrophic final coalescence that followed the in-spiraling of stellar mass neutron stars or black holes. These events generate gravitational waves with frequencies of about 100 Hertz (periods of 0.01 seconds). In-spiraling supermassive black holes ($M \sim 10^7$ to 10^9 solar masses) can generate gravitational waves with periods of the order of a few years or frequencies measured in nanohertz. Because millisecond pulsars are such precise clocks, timing observations of an array of pulsars can, in principle, detect the small deviations in pulsar arrival times due to the small changes in the position of the Earth from the passage of these incoming gravitational waves. Detecting individual sources of gravitational radiation or the combined stochastic gravitational wave background from these events or from the brief inflationary period following the big-bang is difficult, as it requires a precise understanding of the kinematics of the Earth's motions due to its interaction with the Sun and other planets, as well as other sources of noise. Pulsar timing arrays designed to detect gravitational waves have been in operation in Australia, China, Europe, India, North America, and South Africa for more than a decade, and their combined data forms the International Pulsar Timing Array (IPTA). Recently more than 100 scientists from 13 countries have published IPTA data covering up to nearly three decades of timing of 65 millisecond pulsars that suggests evidence for the possible detection of a gravitational wave background, but more years of precision observations will be needed before there is a definitive result.[49]

Discovering Dark Matter

In a now classic 1933 paper that, until around 1960, was ignored by everyone except Fritz Zwicky himself, Zwicky noted that the large observed velocity dispersion among the 800 Coma Cluster galaxies would exceed the escape velocity of the cluster, unless the cluster contained a lot more mass than was apparent from counting the number of observed galaxies.[50] In order that the cluster of high velocity galaxies be held together by gravity, or not fly apart, Zwicky famously wrote, "Resultat ergeben dass dunkle Materie in sehr viel grösserer Dichte vorhanden ist als leuchtende Materie," or "The results show that the density of dark matter is much greater than luminous matter." Zwicky's suggestion that there might be large amounts of "dunkle Materie" or unseen "dark matter" in galaxies lay dormant until photographic spectroscopic measurements suggested anomalously large rotation velocities in the outer parts of some spiral galaxies.

As early as 1914, Vesto Slipher, at the Lowell Observatory, reported that his spectroscopic observations of the galaxy NGC 4594 showed that the galaxy was rotating. On one side of the galaxy the lines were shifted toward the red indicating recession, while on the other side the blue shift indicated an approaching velocity supporting the long held belief that galaxies are rotating.[51] According to Kepler's third law of planetary motion, later placed on a firm theoretical basis by Isaac Newton, the speed of rotation of the outer parts of a galaxy should depend inversely as the square root of the distance from the center of the galaxy, and is proportional to the square root of the interior mass. Since most of the visible mass in galaxies is concentrated toward the center, it was anticipated that, in the outer parts of galaxies, where the mass increased only slowly, the rate of rotation would decrease with increasing distance from the center. The first indication that there might be something wrong with this simple picture came from spectroscopic measurements of the Andromeda nebula (M31) by Horace Babcock as part of his PhD dissertation. Babcock's spectra surprisingly indicated that the velocity of the rotation increased slowly and then remained constant out to the faintest observable part of the galaxy.[52]

Babcock discussed this discrepancy in terms of the ratio of mass to light (*M*/*L*) as the visible light rapidly decreased with increasing distance from the center of M31, but the constant observed velocity indicated that the interior mass continued to increase even in the outer part of the galaxy. Babcock did not make any connection with Zwicky's earlier suggestion that dark matter is needed to stabilize clusters of galaxies, but instead suggested that the anomalous mass-to-light ratios could be explained by either the absorption of light in the outer part of the galaxy or "perhaps that new dynamical considerations are required." Over the years, other astronomers noted the surprisingly large observed velocities in the outer regions of galaxies, but photographic observations were difficult, requiring many tens of hours of exposure.[53]

Starting around 1970 electronic detectors began to replace photographic plates. Vera Rubin and her colleagues obtained improved spectroscopic rotation curves of galaxies and showed further examples of rotation curves that were flat or decreased only slowly, indicating increasingly large mass-to-light ratios.[54] The problem of the "missing mass" started to receive broader attention, but the measurements were difficult and not entirely convincing. Around the same time, radio astronomers started to use 21 cm observations of neutral hydrogen to map the rotation of spiral galaxies out to distances beyond where there was visible matter and where it was not possible to obtain optical spectra. Already some of the earliest H I measurements of spiral galaxies showed deviations from a simple Keplerian motion, but no one connected the missing mass needed to explain the large rotation rates of galaxies with the dark matter postulated by Zwicky to stabilize clusters of galaxies.[55]

Meanwhile, a series of interferometric studies at Caltech by David Rogstad and graduate student Seth Shostak firmed up the evidence for rotation curves that were "very flat."[56] But still, not all astronomers agreed that there was a "missing mass" problem.[57] An extensive study of M31 by Morton Roberts (Figure 10.10) and visiting NRAO astronomer Robert Whitehurst went well beyond the visible bounds of the galaxy and showed, unambiguously, that the distribution of mass exceeded the distribution of surface brightness, and that there was a large mass distribution outside the visible regions of the galaxy (Figure 10.11).[58] Further evidence for flat rotation curves came from the radio telescopes at Nançay in France, the 100 m Effelsberg dish in Germany, the 1,000 foot Arecibo dish, and especially from the 1978 work of Albert Bosma using the new Westerbork array in the Netherlands.[59] Astronomers were finally convinced that there was a missing mass problem.[60]

The previously discussed observations of gravitational lensing by clusters and the temperature fluctuations in the cosmic microwave background also require a large hidden mass component, but it was the 21 cm rotation curves of galaxies that provided the convincing evidence of dark matter. As Mort Roberts has pointed out,

Figure 10.10 Mort Roberts at the controls of the Green Bank 300 Foot radio telescope. Credit: NRAO/AUI Archives, Records of NRAO.

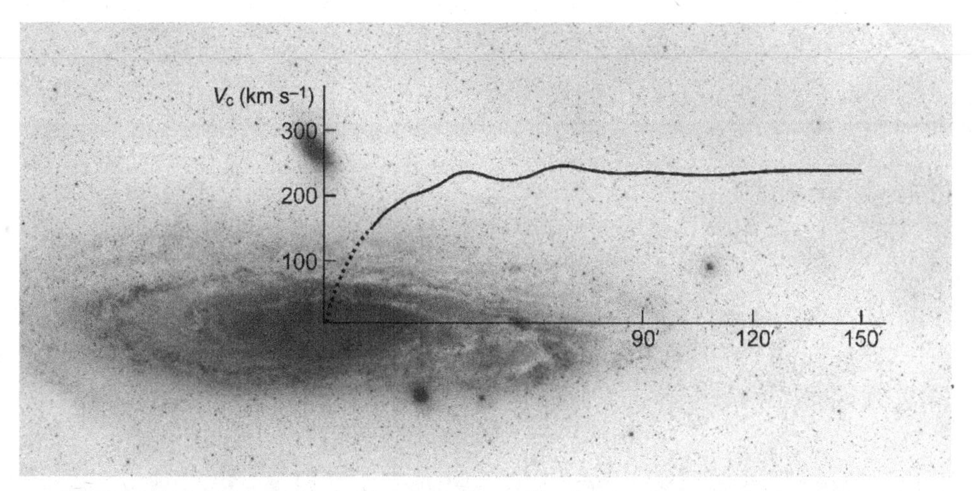

Figure 10.11 21 cm M31 rotation curve showing the rotation velocity as a function of the distance from the center of the galaxy superimposed on an optical image of the galaxy. The near constant velocity, that extends far beyond the optical disk, indicates the presence of dark matter. Credit: Roberts (2008).

it took a half century after Zwicky first discussed the evidence for "dunkle Materie" before astronomers generally recognized that only about 15 percent of the matter in the Universe is ordinary matter composed of neutrons and protons (baryons).[61] The rest, about 85 percent, is in a still unknown form of dark matter. The unsuccessful search for the nature of dark matter has challenged astronomers and physicists now for many decades. There remains, however, a small but vocal minority, led by Mordehai Milgrom's Modified Newtonian Dynamics (MOND) that seeks to explain the dynamics of galaxy clusters and the flat rotation curves of galaxies by a modification of Newton's Laws instead of invoking an unknown dark matter.[62] Although not clearly disproven, these now unconventional ideas remain controversial.

11

If You Build It, They Will Come[1]

> If you are the first person to look under a rock with a new set of tools, you don't even have to be that smart to discover something new.
>
> *Stephen Chu*[2]

The history of radio astronomy is largely defined by the continued development of ever more powerful instruments with increasingly greater sensitivity and vastly better angular resolution. As discussed in preceding chapters, these instrumental developments enabled a wide range of astronomical discoveries – mostly serendipitous. In many cases, the construction of these new more powerful instruments were themselves the result of lucky circumstances, incorrect or naïve scientific justification, or inadequate technical evaluation. In some cases, instruments built for other scientific or military purposes made important astronomical discoveries. In other instances, radio telescopes designed and constructed to address one astronomical area ended up opening whole new and different areas of astronomy research. As the Jodrell Bank radio astronomer Peter Wilkinson has remarked, "What a radio telescope was built for is almost never what it is known for."[3]

Grote Reber's construction of a parabolic dish in a lot adjacent to his home in 1937 was the model for future generations of radio telescopes as well as for terrestrial and space communication systems. Many antenna systems, such as the Bruce Array that allowed Karl Jansky to discover cosmic radio emission, are monochromatic; that is, they work at only one frequency or over a narrow range of frequencies. By contrast, as Reber emphasized, a parabolic dish can be used over a wide range of frequencies by simply changing the feed and radiometer system. At the high frequency (short wavelength) end, dishes are limited by deviations from a perfect parabolic shape due to fabrication errors, as well as from distortions due to gravitational and thermal effects. As a rule of thumb, the shortest operational wavelength is limited to 16 times the rms deviation from the best fitting paraboloid.

At the long wavelength end (low frequency), parabolic dishes are limited by diffraction to wavelengths shorter than about 5 to 10 percent of the dish diameter.

For a long time, Reber's home-built 31.5 foot diameter dish[4] was the largest in the world, even exceeding the ubiquitous German 7.6 m (24 foot) Würzburg-Riese radar antennas which were widely used in postwar radio astronomy. Why did Reber choose 31.5 feet? It was not because he had a specific sensitivity or resolution goal. He used what he had. The backup structure of Reber's dish was constructed from wood. The longest single 2 by 4 piece in the Wheaton lumber yard was 20 feet.[5] Four such pieces fastened together formed the sides of his square backup structure and he thus had a 28 foot diagonal on which to mount the circular dish. Allowing a small overhang around the circumference, probably dictated by the size of the steel plates, Reber used these plates to form the reflecting surface, which resulted in a 31.5 foot diameter dish. Reber's Wheaton dish was moveable only in elevation. Later, at the National Bureau of Standards in Sterling, Virginia, near the site of the current Dulles International Airport, Reber mounted his dish on a turntable to allow for steerability and tracking, but there is no evidence that the Sterling telescope was ever used for radio astronomy. In 1959, Reber had his dish moved to Green Bank, West Virginia, and supervised its re-erection at the entrance to the National Radio Astronomy Observatory, where it remains in use by students nearly a century after Reber first built it in the summer of 1937 (Figure 1.12).

Swords to Ploughshares

The Soviet launch of Sputnik I on 4 October 1957 exacerbated American concerns about the apparent supremacy of Russian military capability and led to the formation by US President Dwight Eisenhower of the National Aeronautics and Space Administration (NASA), as well as the Advanced Research Projects Agency (ARPA) – a well-funded government agency to promote research in science and technology beyond immediate military requirements. Until the early 1970s, when the US Congress adopted the Mansfield Amendment to the 1970 Military Authorization Act that restricted the use of military research funds for university-based scientific research, much of the radio astronomy research in the United States was generously supported by the US Department of Defense (DoD), primarily the Air Force and the Office of Naval Research.

The NRL 50 Foot Radio Telescope[6] The radio astronomy program at the Naval Research Laboratory was an outgrowth of its wartime work in developing short centimeter wave radar. Although the operation of the NRL was led by a Navy

Figure 11.1 Naval Research Laboratory 50 foot radio telescope on top of a laboratory building overlooking the Potomac River and Reagan National Airport. Credit: US Naval Research Laboratory.

Captain, the research program was directed by a civilian scientist. As described in Chapter 2, the early radio astronomy work at NRL concentrated on the Sun, partly because the military was interested in how the Sun affected terrestrial communications and, perhaps, because that was all they could see with their limited postwar instruments. In 1951, NRL scientists John Hagen and Fred Haddock managed to convince the Navy to build a 50 foot radio antenna at NRL on the grounds that it could be used to develop a radio sextant for naval navigation in inclement weather (Figure 11.1).

The NRL antenna was mounted on a surplus five inch gun mount on the roof of the NRL laboratory building overlooking the Potomac River and was, at the time, the largest steerable radio telescope in the world. It had a precise surface that allowed operation down to 1 cm wavelength and was the only large radio telescope in the world that could operate at these short wavelengths. However, typical cosmic radio sources radiate most strongly at longer wavelengths so the research opportunities for NRL were limited. About the only observable radio sources at these wavelengths were the thermal emission from a few planets and a few bright clouds of ionized hydrogen (H II) regions. As Graham-Smith is said to have remarked to Joe Pawsey, "It was the most expensive radio telescope in the world, and all it can see is the Sun and the Moon."[7] Subsequently, however, as discussed in Chapter 9, the NRL 50 foot antenna played a major role in their pioneering observations of the thermal emission from planets as well as the first detection of polarized cosmic radiation (Chapter 4).

The Arecibo 1,000 Foot Radio Telescope[8] The Earth's ionosphere is divided into multiple layers ranging from about 50 to about 1,000 km above the surface of the Earth and is the result of solar ultraviolet radiation that dissociates atoms into electrons and ions. Due to the reflections off the ionosphere, radio waves are able to propagate well beyond the line of sight, thus enabling global radio communications. Prior to the advent of intercontinental cables and later satellite relay systems, shortwave radio transmission via ionospheric reflections played a key role in global civil, government, and military communications. Nearly every government embassy in the world had a large shortwave antenna on its roof to communicate back to the home country.

The US Air Force, in particular, was obsessively interested in any research leading to a better understanding of the ionosphere and how ionospheric disturbances might be used to warn of approaching ICBMs or the passing of a foreign clandestine surveillance satellite. In 1958, Cornell Professor of Electrical Engineering William [Bill] Gordon, a well-known expert in ionospheric research, conceived of an innovative experiment to study the temperature and density of the ionosphere by observing the reflection of radar transmissions off the ionosphere – a technique that came to be known as incoherent backscatter radar. Gordon calculated that, with the then available instrumentation, he would need a radar antenna with an aperture of 1,000 feet (305 m) to probe the ionosphere to heights up to 1,000 km.[9] A key component of this calculation was the thermal agitation of the 1,500 degree ionospheric electrons: due to the Doppler shift of the thermally agitated electrons, the reflected signal from the moving electrons would be spread over hundreds of kilohertz. Therefore, a very large antenna was needed to compensate for the loss of sensitivity due to the width of the reflected pulse.

However, experiments by Kenneth Bowles, one of Gordon's former students, indicated that the returned signal seemed to be narrower than indicated by Gordon's theoretical calculations.[10] Further analysis by Bowles suggested that the electron clouds high in the ionosphere move together with the much heavier ions, hence at slower velocities, and so the thermal broadening is much less than calculated by Gordon. As Frank Drake, who was later Director of the Arecibo Observatory, recalled,[11]

> The spectrum suddenly got about a hundred times narrower than first expected, with the consequence that in fact we needed a dish only about one percent the size in area

> of the dish under consideration. In other words, to achieve the original goal, we didn't need a one thousand foot reflector. A one hundred footer would have done it.

Meanwhile, however, Gordon had convinced ARPA to support the design of his proposed 1,000 foot ionospheric radar system. John Kraus at Ohio State University had reported detecting ionization trails behind the Sputnik satellites, giving some credibility to the idea that Gordon's proposed ionospheric backscattering radar would be able to detect enemy eavesdropping satellites that were beyond normal radar range.

Gordon's publication[12] was issued shortly after Bowles' experiment, but before publication of Bowles' demonstration that the reflected signals covered a narrower band than calculated by Gordon. In addition to Gordon's proposed ionospheric studies, he also noted the possibility of obtaining "radar echoes from the Sun, Venus, and Mars and possibly from Jupiter and Mercury; and receiving from certain parts of remote space hitherto-undetected sources of radiation." Although the large aperture planned by Gordon now appeared unnecessary for his ionospheric investigations, as Marshall Cohen later wrote,[13]

> People then began thinking about a 1000-foot reflector, and the remarkable power of that swamped the original task when the error was discovered less than a year later. Apparently, neither the designers nor the funding agency looked back. No discussion on this point is in any of the Cornell reports available to me. The reports published in 1958 and 1959 do not make reference to Bowles' 1958 paper, although the result was known to the Cornell people and must have been known to ARPA. The entire program remained focused on the 1000-foot dish, with all the enhanced possibilities that entailed. It appears that ARPA, whose interest centered on studying missile wakes, found the 1000-foot dish interesting and worth funding, whereas a 100-foot dish was not interesting.

Or, as Gordon Pettengill described it,[14]

> By the time it was discovered such a large antenna wasn't necessary to meet the original ionospheric objective, it was far too late to kill it. It had just rolled down the slope picking up momentum, and too many people were too heavily involved.

Since the design of the telescope limited its operation to a small region around the zenith, Gordon searched for a site in the tropics where the plane of the ecliptic passes nearly overhead, allowing radar observations of the Moon and planets as well as studies of the ionosphere. After an intensive search, Gordon found a natural bowl-shaped depression in the Puerto Rican karst region near the town of Arecibo. Gordon's original concept was for a 1,000 foot parabolic reflector looking only at the zenith, but the growing interest in planetary radar and radio astronomy drove a design that allowed moving the beam up to 20 degrees from the zenith, requiring movement of the feed structure by the same amount. In order to avoid the loss of sensitivity when a parabolic dish is used off axis, the design was modified to form a spherical reflector. Unlike a parabolic reflector, a spherical reflector focuses incoming celestial radiation along a line rather than at a single point, an effect known as spherical aberration. The decision to build a spherical antenna required the design and fabrication of a sophisticated 105 foot (32 m) long line feed that could handle the anticipated 2.5 megawatt peak power of the radar transmitter. But, unfortunately, there was a fundamental flaw in the design of the line feed, not discovered until after the construction was completed, which caused a large side lobe response that resulted in a loss of sensitivity.

Driven by the Cold War specter of urgency, excavation started in the summer of 1960, the antenna, feed, and radar system were rapidly fabricated, and the Arecibo Ionospheric Observatory was dedicated on 1 November 1963. Four years later the Air Force lost interest; funding to operate the Observatory was transferred from ARPA to the NSF, and in 1971 the Arecibo Observatory became part of the Cornell National Astronomy and Ionospheric Center (Figure 11.2). It never detected any ionospheric disturbances due to missiles or satellites.

Figure 11.2 Arecibo Observatory 1,000 foot spherical reflector. Courtesy of Arecibo Observatory and AIP Emilio Segrè Visual Archives.

In 1974, the old wire mesh surface was replaced by 38,778 aluminum panels to allow operation at higher frequencies, and in 1997 most of the line feeds were replaced by a subreflector system that corrected for the spherical aberration and could be used over a wide range of wavelengths. The telescope was damaged by Hurricane Maria in September 2017, but continued operation after repairs. In August 2020, one of the cables supporting the feed structure became disconnected from the supporting tower and crashed through the surface. Three months later another cable broke under the additional strain. Before engineers could implement a planned controlled "decommissioning," on 1 December 2020 other cables and one of the support towers broke and the entire feed structure collapsed, destroying the telescope.

The scientific justification for the large size of the Arecibo 1,000 foot dish was based on calculations known to be wrong. The proposed investigation could have been achieved with a much more modest antenna with only about 1 percent of the area of the Arecibo reflector. By ignoring the evidence against constructing this huge facility, radio astronomers had access to the largest and, in many respects, the most sensitive radio telescope in the world.

Although probably best known to the public for its prominent role in the James Bond movie *GoldenEye*, the Arecibo radio telescope had a huge impact on late twentieth century astronomy. With the important investigations of pulsars, including the discovery of millisecond pulsars, the discovery of the binary pulsar that gave the first evidence for the existence of gravitational radiation, the first observational evidence for large scale structures in the Universe, planetary radar including the unexpected discovery of the rotation of Mercury, the discovery of the first extrasolar planets, and the most sensitive searches for signals from extraterrestrial intelligence, not to mention decades of ionospheric studies that went well beyond the design goals, the Arecibo radio telescope was another serendipitous success story.

The Haystack Observatory[15] During the Cold War period, the US Department of Defense operated a number of radar antennas in Arctic regions. In support of various defense-related programs, the Lincoln Laboratory, north of Boston, Massachusetts, was operated by MIT. It was built as part of the process to develop a large protective radome that could itself withstand the effects of snow, wind, and ice common to the Arctic environment. With funding from the Air Force and ARPA, Lincoln Laboratory built what was then the largest radome ever constructed, a 150 foot diameter fabric covered metal structure. To evaluate the effect of the radome on radar systems, they installed a 120 foot diameter dish, the largest that could fit inside the 150 foot diameter radome (Figure 11.3).

Figure 11.3 Haystack Observatory 120 foot radio telescope near Westford, MA. Reprinted with permission, courtesy of MIT Lincoln Laboratory, Lexington MA.

The Haystack facility was apparently described by a newspaper reporter who pronounced that the antenna has the capability of finding a needle in a haystack, and the name stuck.[16] When completed in 1964, the Haystack antenna pioneered the use of digital computers to control the antenna and data analysis systems. While the Haystack antenna was initially used primarily for space communications and surveillance, radar astronomers also used the Haystack facility for a series of radar observations of the Moon and planets, as described in Chapter 9, as well as for tests of General Relativity (Chapter 10). Although not designed for passive radio astronomy, with its alt-azimuth configuration and digital computer assisted operation, it was years ahead of the NRAO 140 Foot radio telescope in Green Bank.

The Green Bank 140 Foot telescope, by contrast, had, at the insistence of astronomers, a mechanically complex equatorial mount designed to track celestial objects without the need for coordinate conversion. When finally completed in 1965, after years of design and fabrication problems, contractual issues, and cost overruns, the 140 Foot radio telescope suffered from inconsistent pointing and uncertain gain issues. However, with its outstanding receivers and other state of the art instrumentation, the Green Bank 140 Foot radio telescope dominated American radio astronomy for the next decade (Figure 11.4).[17]

Figure 11.4 NRAO 140 Foot radio telescope in Green Bank, WV. To the left of the telescope is the service tower used to access the focus area to replace and repair receivers. Credit: NRAO/AUI/NSF.

With the decreasing demand for military applications, the Haystack antenna became more available for radio astronomy. In addition to the Solar System radar experiments, it enabled some of the pioneering discoveries of interstellar molecular and atomic gas (Chapter 8) and was involved in some of the earliest VLBI observations. Later upgrades to both the dish and the radome allowed operation at short millimeter wavelengths for both radio astronomy and continued space surveillance programs. Following the Mansfield Amendment, the antenna was transferred to MIT to operate for the North East Radio Observatory Corporation (NEROC) with financial support from the NSF, although the classified military research programs continued.

The Sugar Grove Debacle In the late 1950s, US Navy scientists developed an ill-conceived plan to detect incoming Russian missiles by means of observing their radio telemetry and radar reflected off the Moon. There were two problems with this plan. First, it required a very large antenna in order to have sufficient sensitivity to receive the weak reflected radio signals. Second, the Moon is simultaneously visible above the horizon in both Russia and the United States for only a few hours each day, and there seemed to be no discussion of what would happen if Russia launched their attack during the remaining time.

Undaunted, Naval Research Laboratory scientists laid out a plan to build a 600 foot diameter (the length of two American football fields) fully steerable radio telescope at a remote site near Sugar Grove, West Virginia. The proposed antenna was to be built from one million pounds of aluminum (450,000 kg), 60 million pounds (27 million kg) of steel, and 14,000 cubic yards (10,700 m^3) of concrete, at a cost of $20 million. Military expediency called for operation by mid-1962, so construction by the US Navy Bureau of Yards and Docks (BuDocks) began in 1959 concurrently with completing the engineering design. Although apparently built for highly classified covert operations, both the open scientific literature and the popular news media widely discussed the project's applications for both radio astronomy and for clandestine spying on Russian military operations.[18] If built according to specifications, the Sugar Grove 600 foot antenna would have been the most powerful radio telescope in the world. Adjacent to the 600 foot antenna, the Navy planned an underground 12,000 square-foot two-floor operations building lined with two foot thick concrete walls. Curiously, a 1960 memo received by President Eisenhower's Science Advisor, George Kistiakowski, acknowledged that "the justification and possible uses for this facility have been gradually changing."[19]

By July 1962 the estimated cost to complete the antenna construction had risen to $235 million. Secretary of Defense Robert McNamara canceled the project, claiming that satellites were now able to provide the same intelligence planned for the 600 foot antenna; he made no reference to any radio astronomy applications. One can only speculate on the actual intended purpose of the Sugar Grove project.

From Cosmic Rays to Radio Astronomy

The 250 foot antenna at Jodrell Bank was, for many years, the largest fully steerable radio telescope in the world and a widely recognized icon for radio astronomy. Prior to the start of the Second World War, Bernard Lovell was doing research on cosmic ray showers using cloud chambers to detect incoming cosmic ray events. On Sunday morning, 3 September 1939, he happened to be with a colleague at the Saxton Wold station of the Chain Home defense radar station when the British Prime Minister, Neville Chamberlain, announced that they were at war with Germany. When Lovell saw obvious echoes on the radar monitor, he asked the Women's Auxiliary Air Force operator why she did not report it to Fighter Command Headquarters. "Oh," she responded, "those are not enemy aircraft, they're ionosphere." Lovell and his colleagues immediately speculated, "Radar echoes from cosmic ray showers," and they wondered whether powerful radar systems could be used to study them.[20] Together with his University of Manchester physics department head Patrick M. S. Blackett, who would later win a Nobel Prize for his studies of cosmic rays, Lovell published a paper discussing the possibility of detecting radar echoes from ionization trails left by cosmic ray showers.[21] Two months later, the ionospheric physicist T. L. Eckersley wrote to Blackett, asking if Lovell had considered the effect of damping from the interaction of ionospheric electrons with other molecules. Apparently, Blackett never showed this letter to Lovell until after Lovell had established his research facility at Jodrell Bank and had begun plans to develop his cosmic ray radar program.

Early experiments quickly showed that electrical interference precluded any serious attempt to detect cosmic ray echoes on University grounds. Lovell was driven to find a more radio-quiet location, which he found at the University's Jodrell Bank biological research station. Also, as he later lamented, "A factor of a million [sensitivity] was involved. If I had done the calculations correctly, Jodrell Bank would never have started."[22] Lovell was aware of the work of Jansky and Reber, but as he remarked, "It wasn't my line of work." However, he realized that the high sky brightness due to galactic radio emission at meter wavelengths would limit the sensitivity of radio receivers. The only recourse to improve the sensitivity to radar reflections from cosmic ray ionization trails was to build a bigger antenna. With a grant of 1,000 pounds (about 2,500 US dollars) from the UK Department of Scientific and Industrial Research, Lovell built a fixed 218 foot wire parabolic reflector. But, by 1947, he realized that the sporadic radar echoes he first witnessed in 1939 were not from cosmic ray trails, but from meteor trails, and he turned his attention to studying the radar echoes from them. Within a year he was already beginning to develop ambitious plans to construct a large fully steerable parabolic dish reflector to study radar reflections from meteors, cosmic ray showers, the aurorae, the Moon and planets, as well as for use in the newly-emerging field of radio astronomy.

In his books, *Voice of the Universe: Building the Jodrell Bank Telescope* and *Astronomer by Chance*,[23] Lovell has told of his nearly decade long odyssey, including multiple setbacks, to fund, design, and construct the 250 foot diameter radio telescope at Jodrell Bank. As originally conceived, Lovell anticipated that he would use the antenna at relatively long meter wavelengths, so the reflecting surface could be fabricated from coarse 2 inch (5 cm) square wire mesh, and the required surface precision was only about 5 inches (12 cm). However, following the 1951 discovery by Ewen and Purcell of the 21 cm hydrogen line discussed in Chapter 8, Lovell upgraded his specs to allow for operation at 21 cm without appreciating the impact to the design and cost of constructing his antenna. Further specifications creep, contractual conflicts, re-engineering, legal disputes, multiple delays, and corresponding cost increases nearly led to disaster. Lovell's initial estimate of £50,000 to £60,000 steadily rose to £400,000 and then £660,000, leading to a net £254,000 deficit. The University appealed to the British public for support. Although school children and disabled pensioners wrote to Lovell offering their pocket money, Lovell was faced with a serious problem. He had committed funds he didn't have and faced, at best, personal financial ruin and, at worst, a lengthy prison term resulting from numerous claims and counterclaims amounting to one million pounds. Only when NASA began to reimburse the University of Manchester for use of the Jodrell Bank radio telescope to support the US space program and additional funding was received from the Nuffield Foundation was Lovell absolved of his liability.[24] Unlike the NRL, Haystack, and Arecibo radio telescopes, the Jodrell Bank telescope was not justified on military grounds, but it was Cold War tensions that bailed out Lovell and the University of Manchester from their huge financial debt (Figure 11.5).

Lovell's Jodrell Bank antenna has had a remarkable career lasting more than six decades. During the first years the telescope was used for meteor research and Lovell spent thousands of hours unsuccessfully trying to detect radio flares from stars, but he never went back to pursue his original goal of detecting echoes from cosmic ray showers. It was the first radio telescope outside Cambridge to follow up on the Cambridge discovery of pulsars discussed in Chapter 7; it was responsible for many important pulsar investigations and, as described in Chapter 10, Jodrell Bank discovered the first gravitational lens. Since 1958, the 250 foot antenna has been used as part of the Jodrell Bank long baseline interferometer systems (see following section). Ironically, however, the Jodrell Bank antenna is probably best known by the public, not for its astronomical research, but for a very different accomplishment.

Even before the telescope was completely operational, with great publicity, on 12 October 1957, using their lunar radar, Jodrell Bank was the only facility in the world able to detect and track the Sputnik I carrier rocket launched a week earlier. A year later, in what was supposed to be a secret arrangement with the US Air Force, the Jodrell Bank radio telescope was used, also with great publicity, to track the US Pioneer rocket on its 221,000 mile (356,000 km) ill-fated voyage toward the Moon.[25] In February 1966, the Russian spacecraft Luna 9 made the first controlled landing on the Moon and sent back the first photographs taken from the lunar surface. Surprisingly, the transmissions were unencoded and used

Figure 11.5 The 250 foot Jodrell Bank Radio telescope, ca. 1960. Credit: University of Manchester.

the commercial fax format to transmit the photographic images. Jodrell Bank scientists, who were monitoring the lunar mission, recognized the familiar fax transmission. Using a borrowed fax machine from a local newspaper, they obtained the first images of the Moon's surface, that, to the chagrin of Russia, appeared the next day in London's *Daily Express* newspaper.

Like Arecibo, the Jodrell Bank radio telescope was justified on assumptions that were known to be in error before the telescope was actually built. In the case of Arecibo, the required size was overstated by a factor of about one hundred; the proposed scientific program for the Jodrell Bank dish was not even feasible. Fortunately these errors were overlooked and both facilities went on to make new discoveries that have had a huge impact on radio astronomy and our understanding of the Universe.

The Power of Politics[26]

In 1960–1961, the National Radio Astronomy Observatory (NRAO) constructed a simple 300 foot (91 m) transit radio telescope in Green Bank, West Virginia, at a cost of less than one million dollars. The 300 foot antenna was built with off-the-shelf components, including a crude bicycle-like chain that moved the antenna in elevation. With a chicken wire surface, barely usable at 21 cm wavelengths, the 300 foot radio telescope was intended to serve only as a temporary facility for perhaps five years, until a proper fully steerable precision telescope could be constructed at the new national radio observatory. Although the construction of a large steerable radio telescope was a high priority for every review of the needs of US radio astronomy, it was never the first priority and, following a series of repairs and upgrades, the 300 foot telescope remained very productive for nearly three decades. However, in 1988, an NSF advisory committee, concerned about the need to find money for the operation of several new telescopes then under construction, recommended that the NSF cease funding the operation of the NRAO 300 foot telescope, along with the Arecibo and Haystack Observatories.

During November 1988, NRAO scientist James (Jim) Condon and his colleagues were using the 300 foot telescope to survey the radio sky for new radio sources.[27] They were rapidly scanning the nearly 30 year old telescope up and down in elevation, an observing mode not envisioned by the original telescope designers. On the evening of 15 November, due to metal fatigue, one of the gusset plates that joined members of the backup structure cracked. This added additional stress to the attached steel elements, which then cracked, spreading additional forces to the surrounding structure. Within a few minutes the whole structure collapsed. Only five months after the NSF had decided to terminate funding for the telescope, NRAO and the NSF were no longer faced with the difficult burden of closing down the telescope. All that apparently remained to do was to clean up the mess and sell the 500 tons of aluminum and steel for its scrap value. Unfortunately, however, the steel girders were under considerable tension, and like the game of "pick-up-sticks," releasing one member might release a dangerous spring that could cause serious, if not fatal, injury. All but one potential contractor wanted to get paid to remove the debris, but one local company agreed to do it at no charge (Figure 11.6).

Following the collapse, the media converged on Green Bank, as they often do with any disaster, and dramatically reported loss of "one of the most powerful telescopes in the world"[28] as a devastating setback and a major blow to world astronomy. No matter that only a few months earlier a committee of experts had concluded that it was no longer competitive with the newer telescopes that were becoming available. West Virginia Senator Robert C. Byrd, who was well-known

Figure 11.6 Debris of the 300 Foot telescope as seen the day after the collapse. Credit: Richard Porcas NRAO/AUI/NSF.

for bringing federal money to West Virginia, pointedly asked the NSF Director, Erich Bloch, how he was going to replace the 300 foot telescope, but was told that the NSF did not plan to do so. Moreover, explained Bloch, the NSF had a formal procedure for new projects that required a detailed proposal, peer review, etc. which would need to be evaluated in the context of other priorities. Byrd was enraged by Bloch's stubborn refusal to replace the telescope. Red-faced, he told Bloch that in all his years in Washington, he had never encountered such an uncooperative agency head. Jay Rockefeller, the junior senator from West Virginia, in a not too hidden threat to the NSF budget, calmly explained to Bloch that the "Leader [Byrd] is about to become Chair of [Senate] Appropriations," and "he will have his finger on every dime in the Federal budget. Now are you prepared to let us help you?" (Figure 11.7).[29]

Fortunately, even before the 300 foot telescope collapse, NRAO had begun to develop plans for building a new fully steerable radio telescope, with possible funding from NASA to support their space missions. However, the first priority of NRAO and the US radio astronomy community was to upgrade the VLA and to construct a large array of antennas to operate at short millimeter wavelengths. NRAO did not want another NSF funded project to stand in the way of the planned Millimeter Array. Nevertheless, following the unexpected 300 foot collapse and recognizing the potential prospects of support from the West Virginia senators, in order to present a credible plan to the NSF and to consolidate community support, NRAO convened a series of meetings to discuss options for replacing the 300 foot telescope.

Figure 11.7 West Virginia Senator Robert C. Byrd. Credit: US GPO.

There was little agreement among the meeting participants on the basic design goals of the proposed replacement, for example the tradeoffs between antenna size and precision. Observers interested in short millimeter wavelength molecular spectroscopy argued for a precision structure. Others, who were more interested in the longer wavelengths, such as 21 cm hydrogen spectroscopy, claimed that, since the weather in Green Bank was not optimal for millimeter spectroscopy, the money would be better spent on building a larger but less precise dish. Some radio astronomers, who did not fully appreciate the political consequences, argued for a new array of small dishes instead of a single large antenna, and many pointed out the weather and atmospheric restrictions of a West Virginia site. A particular point of contention was whether or not to build an asymmetric unblocked antenna with the feed structure located off axis or a larger conventional symmetric antenna that could be built for the same price.[30] Others worried that a new NSF funded radio telescope would take money away from whatever was their pet project.

Following considerable debate, NRAO converged on a tentative design for a 70 m (230 feet) diameter antenna that could be built in Green Bank for a proposed 50 million dollars. Apparently Byrd misunderstood or deliberately chose to misunderstand that 70 m referred to meters and not millions of dollars and quietly included 75 million dollars for the replacement of the NRAO 300 foot radio telescope in the FY 1989 Senate Dire Emergency Supplemental Appropriations

Act.[31] In an unusual display of Congressional efficiency and bipartisanship the Senate appropriations bill was passed by voice vote in the full Senate, and in the House of Representatives by an overwhelming vote of 316 to 8, then quickly signed into law by President George H. W. Bush.[32]

However, with the funding too easily in hand, the problems were only beginning. A hastily prepared proposal was quickly approved by the NSF, and in June 1990 NRAO optimistically solicited a request for bids to build a 100 m asymmetric unblocked aperture antenna with a precision active surface controlled by motor driven screws to continually adjust the surface to compensate for deformations due to gravity and thermal differentials. Three bids were received, but only one, from Radiation Systems Inc. (RSI), was even close to the available budget. After a series of negotiations, in December 1990 NRAO signed a firm fixed price contract with RSI for $55 million to build a 100 m off-axis antenna with an active surface, with a contracted completion date of August 1994.[33] However, neither NRAO nor RSI fully understood the added complexities of building a radio telescope with an unblocked aperture and off-axis feed support structure, and it would take 10 years before the telescope was completed and even longer before it was fully instrumented and operational.

To meet the ambitious construction schedule, and to minimize the impact of inflation on the fixed price contract, RSI began construction even before the design was finalized. It was soon clear that the RSI design would not meet the contracted performance goals, and NRAO offered design changes leading to delays and additional costs, accompanied by charges and countercharges of who was responsible for the changes. Faced with safety issues, including one fatal accident, poor workmanship, schedule slippage, and increased costs, RSI was first sold to the Communications Satellite Corporation (COMSAT), which was later acquired by the Lockheed-Martin Company. COMSAT lodged a claim against NRAO for an additional $29 million, charging that NRAO was responsible for the delays and increased costs. NRAO countered with a claim of $12 million for increased management costs, as well as the negative impact to science and to NRAO's reputation associated with the delays. Following an extensive legal battle that itself cost millions of dollars, COMSAT was awarded $6.6 million and NRAO $2.5 million, leaving the manufacturer with a net award of only $4 million out of a $55 million contract, although it was agreed by all that the actual cost to build the new Green Bank Telescope (GBT) was close to $120 million.[34]

The GBT was finally dedicated in August 2000, six years after the contracted completion date and, at 17 million pounds, it was the largest movable land structure anywhere on Earth.[35] With further refinements over the next few years, it has more than met the optimistic design goals with precision pointing and good performance even at short millimeter wavelengths. It is a key part of the international pulsar timing

Figure 11.8 100 × 110 m unblocked aperture Green Bank Telescope. Credit: NRAO/AUI/NSF.

array (Chapter 10), has made extensive contributions to millimeter spectroscopy, and is by far the most powerful fully steerable radio telescope in the world (Figure 11.8).

Naturally, following the collapse of the 300 foot antenna, the NSF initiated an investigation to determine the cause of the accident, but they were unable to find anyone to blame. Rather, the investigation committee complimented NRAO on the innovative low cost design and concluded that there was no human error involved in the collapse. However, months later, after NRAO had completed the long delayed data reduction software upgrade, Jim Condon was finally able to analyze his survey data and found that, during the month before the 15 November collapse, there was an increasing difference in the apparent position of radio sources depending on whether the telescope was moving up or down in elevation. Moreover, "during the final week before the collapse, the north–south beamwidth had increased from 3 arcmin to 4 or 5 arcmin."[36] This was a clear indication that the structure was dangerously bending. Had the telescope software not been undergoing revision and the data been inspected on a daily basis, surely the changes in performance would have been noticed; for safety reasons, the observing program would have been stopped; the 300 foot telescope, that was already earmarked for closure, would probably not have collapsed; Senator Byrd would not have insisted that the NSF replace the telescope; and the GBT would never have been built.

Imaging the Radio Sky[37]

The ability to see fine detail, or the angular resolution of any radio or optical telescope, depends on the ratio of wavelength to the size of the telescope, and is given approximately by $\Theta \sim 70\lambda/D$ degrees, where Θ is the angular resolution, λ is the wavelength of observation, and D is the linear size of the instrument or, in the case of a circular aperture, the diameter. Since radio waves are longer than light waves by a factor of about 100,000, for many years it was assumed that the resolution of radio telescopes was hopelessly limited when compared to that of even modest sized optical telescopes. Indeed, in his pioneering 1945 paper discussing radio spectral lines, Henk van de Hulst remarked, "Without telescopes of enormous aperture the radio waves will never furnish a detailed picture of the heavens."[38] A few years later, Otto Struve, in reviewing the dramatic discoveries made by radio astronomers, commented, "Unfortunately, it has not yet been possible to perfect radio telescopes to such an extent as to match the actual resolving power of the human eye or of a real optical telescope. This is due mainly to the greater length of the radio waves."[39]

In fact, for several reasons the opposite is true, and modern radio telescope systems can image the sky with hundreds of times better resolution than the best optical telescopes. How is this possible? Firstly, in practice, the resolution of optical telescopes is limited not by the size of the aperture, but by turbulence in the terrestrial atmosphere, which causes the familiar twinkling of stars or what astronomers call "seeing." In practice this limits the resolution to a few tenths of an arcsecond, even for telescopes located on the best mountain sites. Secondly, because radio telescopes operate at relatively long wavelengths it is possible to build instruments of essentially unlimited extent and still keep the mechanical tolerances small compared to a wavelength.

In Chapter 3 we described how the sea interferometer was used to obtain increased angular resolution that led to the discovery of radio galaxies. As we discussed in Chapter 2, even earlier, an Australian team used a sea interferometer to study the brightness distribution across the Sun and were able to demonstrate that the enhanced radio emission came from the direction of prominent sunspots. As first pointed out by Albert Michelson in his classic 1890 paper on the astronomical use of interferometry, by combining observations made at different interferometer spacings, even if made at different times, it is possible to mathematically reconstruct an image of the celestial brightness distribution.[40] Michelson was thinking only about optical interferometry and its application to measuring the diameter of stars, and it would be another half century before Lindsay McCready, Joseph Pawsey, and Ruby Payne-Scott in Australia remarked on the potential applications of radio interferometry to obtain high resolution images of the radio sky. In their published paper on sea interferometer observations of the Sun, they noted, "it is possible in principle to determine the actual form of the distribution in a complex case by Fourier synthesis," but cautioned that varying the height of the cliff interferometer "would be feasible but clumsy."[41]

The previous year, Martin Ryle and Derek Vonberg at the Cambridge Cavendish Laboratory used surplus German radar equipment to build a simple two element interferometer.[42] As discussed in Chapter 2, they used their interferometer to pinpoint the location and size of solar radio emission and at the same time rejected the broad radiation from the Galaxy. At the time, Ryle and Vonberg did not comment on the broader applications of interferometers for imaging the radio sky, but subsequently Ryle and his Cavendish colleagues went on to develop a series of interferometer arrays of ever increasing sensitivity and angular resolution. Starting with the 2C, 3C, and 4C radio source surveys discussed in Chapters 3 and 5, followed by the development of the powerful One-Mile Radio Telescope and then the 5-km Radio Telescope, the innovative Cambridge work led to the award of the 1974 Nobel Prize in Physics to Martin Ryle "for his observations and inventions, in particular of the aperture synthesis technique."

Martin Ryle feared competition from the better funded observatories, and the Cambridge radio group played their cards close to their chest. Somewhat paranoid, Ryle was concerned that while he had all the ideas, others, mainly in the United States, had all the money. He was afraid that if he discussed his plans, the Cambridge research program might be scooped by the better financed radio astronomy groups. He even went so far as to suggest that if his PhD graduates took postdoctoral positions in the United States that this might adversely affect their professional future in the United Kingdom.

The power of Fourier synthesis was not confined to Cambridge, and was understood elsewhere, especially in Australia, where Christiansen and his colleagues, following up on the ideas formulated earlier by McCready et al., implemented a form of Fourier synthesis to study the Sun. However, because Ryle and his Cavendish colleagues were not imaging the rapidly varying Sun, they could use a small number of interferometer elements at different spacings to accumulate data over longer periods. Also, the Cavendish Laboratory had one of the then most powerful computers in the world, the EDSAC I, and later the EDSAC II, that facilitated the lengthy calculations needed to created sky images from multiple interferometer pairs.

Ryle's fears were soon realized when a former Cambridge student, Swedish born radio astronomer Jan Högbom, moved to the Netherlands and led the design of the 12 (later 14) element Westerbork Synthesis Radio Telescope (WSRT) completed in 1972.[43] In 1980, American radio astronomers completed the construction of the NRAO 27 element Very Large Array (VLA) in New Mexico. The VLA, for the first time, allowed astronomers to image the radio sky with angular resolution comparable to that of optical telescopes, and for more than four decades it has been the most powerful radio telescope in the world.[44] Later sophisticated synthesis telescopes were built in Australia, India, and South Africa but, by this time, the radio astronomy program at Cambridge had migrated to other areas and Cambridge no longer played a major role in synthesis imaging.

Recently, optical astronomers have devised ways to minimize the effects of atmospheric seeing by rapidly adjusting the shape of the telescope reflecting mirror to compensate for the atmospheric turbulence. So far, this adaptive optics technique has been applied only at relatively long infrared wavelengths and is effective only for bright targets. Space-based observatories which are above the Earth's atmosphere, such as the Hubble Space Telescope (HST) and the new James Webb Space Telescope (JWST), can reach their theoretical resolution limit, but the size of space-borne mirrors is limited by the practical constraints of placing large precision instruments in space.[45] By contrast, radio astronomers are able to record and then adjust their data during the computer imaging process, which is equivalent to mechanically distorting the mirror of an optical telescope. The process used by radio astronomers, known as self-calibration, is much more powerful, as it is done after the data are obtained and can be done iteratively until the best image is formed. By using the target source itself as a calibrator of instrumental and atmospheric instabilities, interferometric radio telescope arrays are able to make high resolution images, limited only by their overall dimensions and essentially free of the effects of atmospheric turbulence and instrumental instabilities.[46]

In principle, interferometry is also possible at optical wavelengths but, in order to form optical interferometer pairs, the incoming light must be split and divided up among the interferometer mirrors. Since there is no technology at optical wavelengths for coherent amplification, each time the light is split, there is at least a factor of two loss of sensitivity. By contrast, radio waves, unlike light waves, can be amplified, divided, and distributed over large distances without any loss of sensitivity.

Although it was Ryle's group in Cambridge that led the development of synthesis imaging, the drive for higher angular resolution in radio astronomy was led by the radio astronomers at the nearby Jodrell Bank Observatory. As discussed in Chapter 3, it was initially widely thought that the discrete radio sources were galactic stars with stellar dimensions, so it appeared that intercontinental baselines would be needed to resolve their images.[47] In order to extend their radio interferometer baselines, Robert Hanbury Brown and Richard Twiss developed a new type of interferometer that compared only the intensity of the received signal at the two ends of the interferometer. Since no phase information was needed, the intensity interferometer could operate over unlimited distances by simply recording the intensity on separate tape recorders located at each end of the interferometer for later playback and comparison.[48]

In a series of observations, the Jodrell Bank radio astronomers observed the two strong northern radio sources Cas A and Cygnus A using a number of different interferometer baselines extending out to 12 km antenna spacing.[49] To their surprise the sources were resolved even with these modest antenna separations. They found the Cas A source to be about 4 arcmin in extent, but the radio emission from Cygnus A appeared to be coming from two separate regions that straddle the optical galaxy identified by Baade and Minkowski.[50] Around the same time, conventional Michelson interferometer observations by Bernie Mills in Australia and Graham-Smith in Cambridge confirmed the finite size of the two sources, but they did not have sufficient resolution to disclose the double nature of Cygnus A. The concept of "radio stars" with very small stellar dimensions gradually faded from the literature, although, as discussed in Chapter 3, not without some resistance (Figure 11.9).

It appeared, therefore, that interferometer baselines longer than a few tens of kilometers might be unnecessary. Shorter baselines could be implemented with conventional Michelson interferometers. Moreover, the Hanbury Brown intensity interferometer had greatly reduced sensitivity compared with conventional interferometers. Hanbury Brown and Twiss used their intensity interferometer over baselines up to 9.2 m to measure the optical diameter of the star Sirius as 0.006 arcseconds.[51] Based on this success, Hanbury Brown went on to build the optical Narrabri Stellar Intensity Interferometer in Australia where it was used to measure the optical diameter of stars, but the intensity interferometer played almost no further role in radio astronomy.[52]

In order to extend their Michelson interferometer separations to greater distances than could be practically connected by cables, the Jodrell Bank radio astronomers implemented a series of 158 MHz (1.9 m) interferometer experiments using radio links to connect antenna elements separated by up to 115 km (71 miles or 10,700 wavelengths).[53] The radio link carried both a reference local oscillator signal from Jodrell Bank to the remote antenna as well as the intermediate frequency signal back for correlation with the 250 foot Jodrell Bank antenna. To provide logistical support, each of the four remote observing sites was set up near a convenient pub with whimsical names such as the *Cat and the Fiddle*. Only six of the 384 3C radio sources investigated were found to be smaller than 1 arcsecond, all of which later turned out to be quasars (Chapter 4). Since only about 1 percent of the observed radio sources showed interference fringes at the longest spacings, it appeared that there was little reason to extend radio interferometers to even longer baselines. Little did anyone

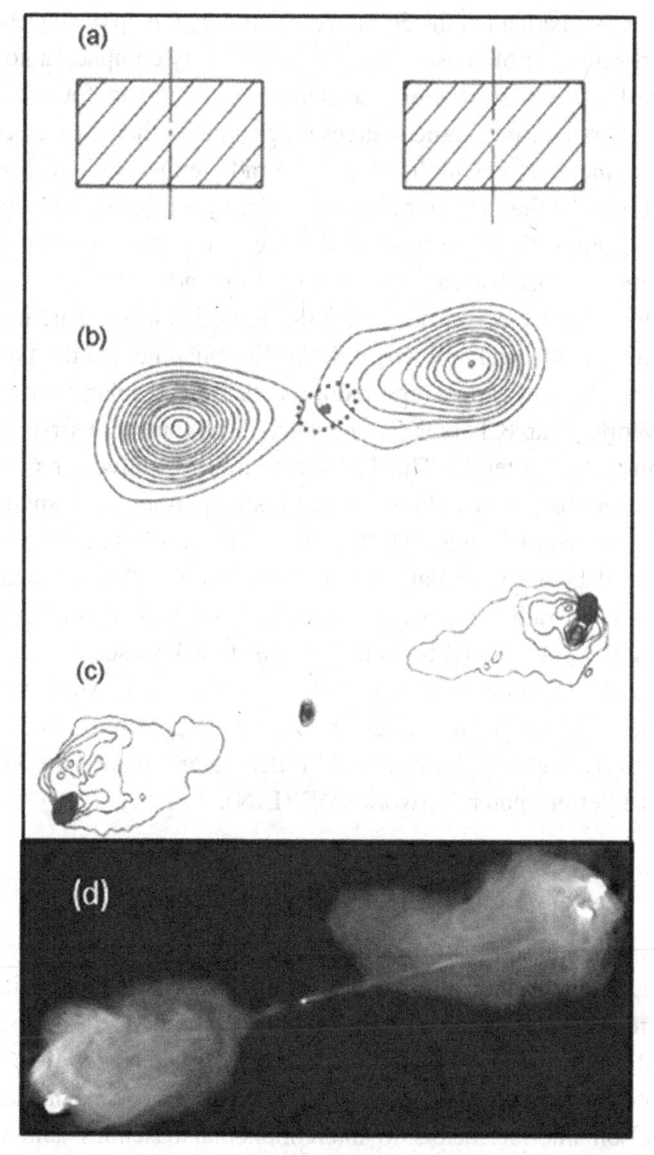

Figure 11.9 Images of the radio source Cygnus A made at nearly decade intervals, demonstrating the dramatic improvement in resolution and image quality from instruments of steadily increasing sophistication. (a) The double nature reconstructed from the Jodrell Bank intensity interferometer data. (Jennison and Das Gupta, 1953). (b) Contour map observed at 1.4 GHz with the Cambridge One-Mile Radio Telescope at an angular resolution of 23 by 30 arcsec. The dotted region represents the optical counterpart (Ryle et al., 1965). (c) Image obtained with the Cambridge 5-km Radio Telescope at 5 GHz with an angular resolution of 2 by 3 arcsec. This image showed, for the first time, a compact flat spectrum source coincident with the optical counterpart in addition to the extended radio lobes (Hargrave and Ryle, 1974). (d) 5 GHz image obtained with the NRAO Very Large Array with 0.4 arcsec resolution showing the thin jets emanating from the central component feeding energy to the outer lobes (Perley, Dreher, and Cowan, 1984). Images (a–d) adapted from Thompson, Moran, and Swenson (2017).

appreciate that this 1 percent of the 3C sources was only the tip of an iceberg, and that there was a large number of quasars associated with very compact radio sources that radiate primarily at much shorter wavelengths than those found in the meter wavelength 3C survey. Since radio sources much smaller than an arcsecond become self-absorbed at meter wavelengths, that is the radio emission is reabsorbed by the dense concentration of the same population of relativistic electrons that generated the synchrotron radiation, they are readily observed only at very short wavelengths where the relativistic electrons are transparent to the radio waves.

This was dramatically confirmed when the Jodrell Bank researchers teamed up with the scientists at the Malvern Royal Radar Establishment under the direction of J. Stanley Hey, who had earlier reported the detection of solar radio bursts during the Second World War (Chapter 2) and later discovered the first discrete radio source, Cygnus A (Chapter 3). The 127 km Jodrell to Malvern baseline at 6, 11, and 21 cm showed that many radio sources, mostly quasars, were smaller than 0.05 arcsec, and a few even less than 0.025 arcsec.[54] In the meantime, as discussed in Chapter 4, interplanetary scintillations, the observations of variable radio sources, and the discovery of radio sources with spectral cutoffs at decimeter wavelengths all suggested radio source structures as small as 0.001 arcsec.

Motivated by the exciting results from the Jodrell Bank to Malvern radio-linked interferometers, the Jodrell Bank radio astronomers went on to develop the Multi-Telescope Radio Linked Interferometer (MTRLI), later renamed the Multi-Element Radio Linked Interferometer Network (MERLIN).

Very Long Baseline Interferometry (VLBI)

The Jodrell Bank radio-linked interferometers opened the door to try even longer interferometer baselines, but already the Jodrell Bank to Malvern interferometer involved two repeater stations, and there was a practical limit to distances that could be covered by radio links without the use of expensive intermediate repeaters. Fortunately, by the mid-1960s several new technologies had been developed that made it possible to extend Michelson interferometers to intercontinental baselines, and ultimately to baselines nearly the distance to the Moon, without the need for any physical connections between the interferometer elements. The common local oscillator was replaced at each end of the interferometer by separate very stable atomic frequency standards that were becoming commercially available. Also, the television and computer industries had developed tape recorders that permitted radio astronomers to record broad bandwidth data at each end of the interferometer. Two teams, one in the United States and one in Canada, raced to demonstrate that very long baseline interferometry was possible using independent atomic local oscillators and tape recorders. Following a series of false starts, by mid-1967 both groups were able to demonstrate the successful operation of transcontinental interferometer baselines, a technique that has become known as very long baseline interferometry or VLBI.[55]

VLBI baselines were quickly extended to intercontinental baselines and to shorter wavelengths to make unprecedented high angular resolution images of radio galaxies and quasars, as described in Chapters 3 and 4, and molecular masers, as discussed in Chapter 8, reaching better than 1 milliarcsec (0.001 arcsec) resolution. Particularly challenging were the series of VLBI observations between antennas in the United States and in the former Soviet Union. Since a by-product of VLBI is the accurate location of the antennas, the intelligence agencies on both sides were wary of the possible consequences of any collaboration between Russian and American radio astronomers, but finally acceded to the idea. Although these non-military radio astronomy observations involved the shipment of sensitive recording equipment and atomic frequency standards to the USSR, they continued throughout the Cold War period and resulted in lasting collaborations and friendships between Russian and American radio astronomers. Only years later did the Americans learn from one of their Russian colleagues who had emigrated to the United States that the Soviets had carefully inspected and photographed the imported American technology.

Even at the longest terrestrial baselines there were still unresolved sources, both quasars and water masers, and for several decades European and US radio astronomers unsuccessfully proposed building a dedicated VLBI satellite that could extend interferometer baselines beyond terrestrial limits. Although none of these radio astronomy satellite programs were ever funded, the first demonstration of an Earth-to-space radio interferometer came from using the NASA Tracking and Data Relay Satellite System (TDRSS). Although designed to provide communications links between orbiting NASA satellites and US ground stations, in 1986 and 1987 the TDRSS 4.9 m antenna was turned instead toward several bright quasars to form radio interferometer baselines extending out to more than 27,000 km (16,000 miles).[56] Then, in 1997, as part of an internationally supported mission, Japanese radio astronomers launched the HALCA VLBI satellite into low Earth orbit, extending interferometer baselines to several times the length of the longest terrestrial baselines.[57] Finally, in 2011, following a development program that lasted many decades, through the fall of the Former Soviet Union and the subsequent economic collapse, Russian radio astronomers launched the sophisticated RadioAstron VLBI satellite into a highly elliptical orbit with interferometer baselines extending up to 340,000 km.[58] RadioAstron continued in operation until the failure of on-board critical instrumentation in 2019, and was able to measure quasar dimensions as small as 20 microarcseconds.

Meanwhile, with improvements in instrumentation and sensitivity, especially at millimeter wavelengths, radio astronomers were also imaging radio galaxies,

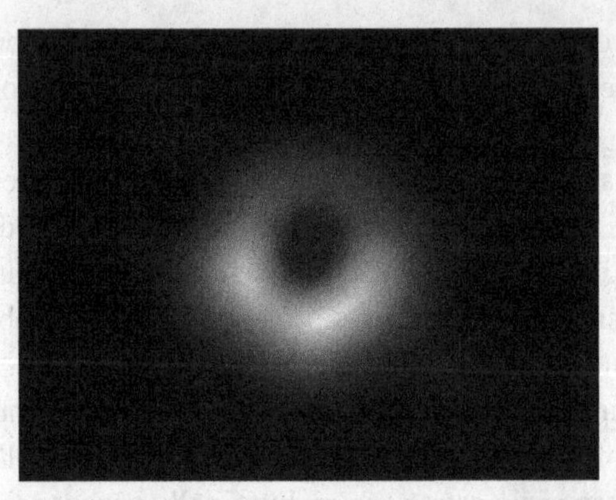

Figure 11.10 Image of the shadow of the supermassive black hole in M87 reconstructed from 1.2 mm data taken with the Event Horizon Telescope. The diameter of the ring is only about 60 microarcseconds. Credit: EHT Collaboration.

quasars, and interstellar masers with angular resolutions as fine as 20 microarcs, more than 10,000 times better than any images from optical telescopes in space or on the ground. The VLBI image of the shadow of the black hole in M87 was reportedly seen by over a billion people (Figure 11.10).[59] During the same time, VLBI astrometric position measurements accurate to about 10 microarcs, exceeding the visual space astrometric mission *Gaia*, revealed the true spiral structure, size, and rotation of the Milky Way.[60]

Although the first implementation of independent-oscillator-tape-recording interferometry occurred in the United States and Canada in 1967, the first seeds of this powerful new technique were sowed some five years earlier in the Soviet Union. In 1962, Russian radio astronomers were apparently the first to develop the ideas that would lead to VLBI, but their application for a patent and permission to publish were delayed by Soviet security concerns until 1965.[61] Unfortunately, a technical error in their analysis led them to believe that their technique would not have adequate sensitivity on the longer baselines, so they did not pursue a practical realization of the technique.

The Square Kilometre Array: Too Big to Fail[62]

As the end of the twentieth century was approaching, radio astronomers in several countries reflected on the remarkable discoveries discussed in previous chapters and began to develop ambitious plans for a next generation radio telescope with greatly improved sensitivity.[63] These discussions came to a focus in 1990 following an unscheduled presentation at a New Mexico conference celebrating the 10th anniversary of the VLA dedication, where Peter Wilkinson from Jodrell Bank proposed building what he called the "Hydrogen Array." Wilkinson argued that to observe 21 cm radiation from neutral hydrogen in the very early Universe, the Hydrogen Array would need to have a collecting area of one million square meters.

Starting in 1993, radio astronomers, primarily from Australia, Canada, China, the Netherlands, and the United States, began to meet informally to prepare a scientific justification and to discuss the technology and organizational structure needed to construct what they later called the Square Kilometre Array (SKA).[64] In 2000, scientists from nine countries signed a Memorandum of Understanding establishing the International SKA Steering Committee (ISSC) to provide broad leadership and coordination, with a planned completion of construction by about 2015. Less than three years later, they established the International SKA Project Office (ISPO) in the Netherlands with Richard Schilizzi, who had been Director of the Joint Institute for VLBI in Europe, as the Director. In 2008, a new Memorandum of Agreement replaced the ISSC with the SKA Science and Engineering Committee (SSEC) and the ISPO by the SKA Program Development Office (SPDO). Funding for the ISPO/SPDO came from a combination of voluntary national and institutional entities as well as grants from the European Commission.

The ISSC/SSEC and ISPO/SPDO considered a number of possible technical implementations of the SKA, including an array of a large number of small dishes (LNSD) promoted by the United States; an array of about 60 Arecibo-like dishes suggested by China; an array of fixed reflectors fed by balloon-supported feeds promoted by Canada; and a so-called Aperture Array consisting of thousands of simple dipole antennas, each with its own receiver, advocated by the European radio astronomers led by the Dutch. The Australian radio astronomers first suggested an array of dielectric Luneberg lenses, then an array of cylindrical paraboloids, and finally an innovative concept of an array of parabolic dishes, each with multiple feed/receivers known as phased-array-feeds (PAFs) at their foci to provide an unprecedented wide field of view. The SKA did not adopt the Australian suggestion to use parabolic cylinders as the antenna element, but the idea resurfaced years later with the very productive Canadian Hydrogen Intensity Mapping Experiment (CHIME) radio telescope.

Although the SKA was initially conceived to study neutral hydrogen in the distant Universe, it was soon apparent that such a powerful radio telescope would have applications to almost all areas of radio astronomy.[65] But increased interest in other scientific areas meant expanding the frequency range of the SKA, which in turn introduced constraints on the site selection and the engineering design.[66]

Initial cost estimates of the SKA were only 100 to 200 million US dollars, but as reality crept in the cost estimates began to increase. Meanwhile, with the improved sensitivity of receivers, it was agreed that a collecting area of only 400,000 m^2 was needed, although the name "Square Kilometre Array" was retained. By 2005, the estimated cost had increased to a billion Euros. Recognizing the competition from other planned large astronomy projects in both Europe and the United States, the ISSC/ISPO proposed to start with a Phase I that would comprise only 5–10 percent of the full SKA. Moreover, since it was not possible to cover the full range of frequencies needed to cover all of the anticipated science with a single technology, the SKA was split into three instruments: SKA-low, SKA-mid, and a possible future SKA-high, covering meter, decimeter/centimeter, and millimeter wavelengths, respectively.

In many countries and regions, interested scientists formed committees and consortia to promote the SKA project and secure national funding to support their SKA planning. In 1998, NRAO convened a meeting in Green Bank to discuss possible US roles in planning for the SKA. However, concerned about possible NRAO conflicts with their planned Millimeter Array and VLA upgrade, NRAO never enthusiastically supported the SKA. Meanwhile, wanting to keep control within the university community, a few months later a small group met to form the US SKA Consortium (USSKAC) to coordinate US participation in the SKA. Over the next decade, under leadership from Cornell University, the USSKAC received $13.5 million from the NSF to support the participation of US radio astronomers in international SKA meetings and to fund engineering design studies at various universities as part of the global effort.

In 2000, European radio astronomers from eight countries banded together to form the European SKA Consortium. Grants from the European Commission and national funding agencies over the next decade totaling tens of millions of Euros (dollars) supported major global SKA studies headed by ASTRON in the Netherlands (Square Kilometre Array Design Study or SKADS), and by the University of Manchester in the United Kingdom (PrepSKA). The later study, that was more comprehensive and better focused than the US funded studies, and that included the active participation of representatives from national funding agencies or government departments, had more impact on the SKA design and the future political structure than did the US effort.

Many large international science projects start as national programs but become international in order to pool resources and to minimize the cost for each participating country. The SKA, however, was different. From the beginning the SKA was planned to be an international project and adopted the theme "born global," anticipating that the United States, Europe, and the rest of the world each would contribute about one-third of the costs. In the hope of facilitating future government funding, the ISSC encouraged representatives from the national funding agencies to become more active in the SKA planning. Soon, the funding agencies formed their own Informal Funding Agencies Group (IFAG) and began to hold their own meetings. In 2009, they formalized their role and created the Agencies SKA Group

(ASG) to coordinate the planning and funding of the SKA and gradually they assumed control of the project. In 2011, the funding agencies formed a new non-profit legal entity called the SKA Organization (SKAO) to plan for the construction of the SKA, and the ASG, SSEC, and SPDO were dissolved. Phil Diamond, who had been Chief of the Australian CSIRO Astronomy and Space Science Division, became the Director-General of the new SKA Observatory.

A major question for the SKA, as for all astronomical telescope projects, was where to locate it. Requirements included a clear dry climate, stable ionosphere, freedom from RFI, availability of a suitably large site, presence of a strong technical infrastructure, and political and economic stability. Following expressions of interest from Argentina, Australia, China, South Africa, and the United States, the ISSC selected Australia and South Africa as the two best prospects to host the SKA. The Argentine and Chinese sites were considered to be geographically too limited and, in the case of Argentina, subject to excessive ionospheric instabilities. The United States never submitted a formal proposal for hosting the SKA, in part due to concerns that the United States was too susceptible to RFI and in part due to the lack of any organization prepared to sponsor the effort.

To better establish their credentials for hosting the SKA, and to demonstrate technical feasibility of design concepts, both Australia and South Africa began to build elaborate prototypes. South Africa constructed an array of sixty-four 13.5 m dishes known as MeerKAT. After a longer than anticipated development effort, Australia succeeded in demonstrating the feasibility of phased array feeds and completed the 36 element Australian SKA Pathfinder (ASKAP) with 47 dual polarized phased array feeds, on each of thirty-six 12 m diameter antenna elements that formed up to 36 separate beams in the sky.[67] In 2011, both countries submitted detailed proposals to host the SKA. In an attempt to broaden the global participation, South Africa, which had little background in radio astronomy and a fragile political and social structure, had been encouraged by the SKA leadership to consider hosting the SKA, but many considered South Africa a dark horse. Most of the SKA community felt that the Australian proposal, with its strong technical base, extensive experience in building and operating radio telescopes, and solid economic and political stability was more attractive to host the SKA.

Following extensive evaluation of the two proposals by the SSEC, the SPDO, an external SKA Site Selection Advisory Committee, unanimously concluded that southern Africa was the preferred site for the SKA.[68] The report of the Site Selection Committee was not well received by the Australian government, which claimed that the evaluation was flawed.[69] Concerned that Australia might withdraw their funding support for the SKA, the new SKAO Board decided that the SKA should be split between Australia and South Africa. A 512 station SKA-low aperture array with each station containing 256 log-periodic elements plus a mid-frequency survey telescope based on the ASKAP concept was to be built in Western Australia. SKA-mid, consisting of an array of 250 15 m diameter dishes, would be built in South Africa near MeerKAT. But by 2015, it became clear that even the already descoped Phase I SKA could not be built for the 690 million Euro budget cap set by the SKAO Board. The Australian survey telescope was deferred and instead of building 250 new antennas for SKA-mid along with 512 SKA-low stations, it was agreed the SKA would add only 133 new SKA 15 m dishes to the existing 64 MeerKAT 13.5 m dishes, while SKA-low would consist of only 256 stations.[70]

The competition to host the SKA headquarters was equally contentious. Following 2007 and 2011 contested decisions to host the SPDO, and then the SKA Project Office in Manchester, United Kingdom, in 2015 an outside advisory panel was appointed to recommend the site of the permanent SKA headquarters. To the surprise of many, the advisory panel chose Padua, Italy over Manchester. But, faced with the possibility that the United Kingdom might withdraw from the project, and under pressure from the United Kingdom, the SKAO Board chose to locate the SKA Global Headquarters on a site at the Jodrell Bank Observatory near Manchester.

In 2019, an international treaty signed by seven countries established the SKA Observatory as an intergovernmental organization. The United States, which played an important role during the first decade of SKA planning, was no longer involved. Although the 2010 US National Academy of Sciences Decade Review of Astronomy noted the substantial scientific opportunities of the SKA, concerned about the unrealistic budget estimates and aggressive time scale proposed, they did not recommend any further funding from the NSF for SKA design and construction.[71] However, by the middle of the decade, with the VLA upgrade and the enormously successful international Atacama Large Millimeter/submillimeter Array (ALMA) having been completed, NRAO Director Tony Beasley proposed that the United States construct what he called the next generation VLA (ngVLA). The ngVLA was strongly endorsed by the US *Astro 2020* Decade Review, and is planned to include 244 dishes, each 18 m in diameter spaced throughout North America, as well as a Short Baseline Array of nineteen 6 m antennas. Operating from 1.2 to 116 GHz, the ngVLA will provide the capabilities originally anticipated for SKA high, as well as better sensitivity than the Phase I SKA-mid at centimeter wavelengths. The ngVLA is planned for completion by the mid-2030s at a projected cost of around $2.5 billion. Meanwhile, SKA Phase 1 is scheduled to be completed in 2028 at a cost of 1.3 billion Euros (approximately $1.4 billion), about five to ten times the original cost estimate of the full SKA, with less than one tenth of the planned collecting area, and 10 years past the earlier planned completion date.

Although construction of SKA Phase I itself is only beginning, the SKA prototypes, precursors, and pathfinders have themselves already proven to be powerful new radio astronomy facilities that are having unanticipated scientific impacts. MeerKAT is one of the most sensitive radio telescopes in the world, and certainly the most powerful in the Southern Hemisphere, with applications to pulsars, radio galaxies, star formation, high redshift H I, and low-surface brightness Galactic structures. Although their proposal for using an array of large spherical reflectors was not adopted for the SKA, Chinese radio astronomers went ahead to build a single Five-hundred-meter Aperture Spherical Telescope (FAST) in the remote Guizhou province. With an effective collecting area of 70,000 m^2, FAST is the most sensitive radio telescope in the world for pulsars as well as for H I in distant galaxies. As discussed in Chapter 7, ASKAP, with its unprecedented combination of high angular resolution and wide-field imaging, has found and localized new FRBs. Although unrelated to the SKA, CHIME, which was originally conceived in part as a result of SKA discussions to map the large-scale distribution of hydrogen at high redshift, is an even more powerful FRB search instrument.[72] Fortuitously the designed frequency range of 400 to 800 MHz for redshifted ($z = 0.75$ to 2.5) hydrogen is also the sweet spot for observing FRBs, and that has already enabled CHIME to have discovered more than 1,000 new FRBs. Like ASKAP, with its large fields of view, CHIME has had an unexpected impact on the study of FRBs, a field of research that was not even known when these instruments were conceived.

Although radio astronomy began with the pioneering work of Karl Jansky and Grote Reber at low radio frequencies (meter to dekameter wavelengths), due to the poor angular resolution, the impact of ionospheric disturbances, and the drive to study interstellar molecules, for the next half century radio astronomers gradually migrated to shorter and shorter wavelengths. However, motivated by the improvements in signal processing, toward the end of the twentieth century radio astronomers again turned their attention to the longer wavelengths, and there were two separate proposals to build what was intended to become SKA-low. Western Australia, with its low population density, was considered to be the best interference free site for SKA-low, followed by the southwestern United States, and last, due to high levels of radio interference, the Netherlands. Nevertheless, Dutch radio astronomers were able to obtain national funding to build a low frequency array, called LOFAR, provided that it was located in an economically disadvantaged area in the northern part of the country. Later, supplementing the construction of 38 LOFAR stations in the Netherlands, in order to obtain improved angular resolution, additional stations were built in locations throughout Europe, extending across a 2,000 km region. With a total of about 20,000 individual log-periodic elements, LOFAR has become a powerful meter-wavelength radio telescope with sub-arcsecond resolution.

Following the breakup of the Australia–US–Netherlands LOFAR collaboration, American and Australian radio astronomers formed a new collaboration to build a separate low frequency array in Australia. However, the US-based group, which initially included MIT, NRL, and the University of New Mexico, subsequently dissolved. The University of New Mexico went on to build its own Long Wavelength Array (LWA) on a site near the VLA. The MIT group, in the meantime, continued to work with an Australian group to establish the Murchison Widefield Array (MWA) near the ASKAP site in Western Australia. LOFAR, the LWA, and the MWA have each resurrected low frequency radio astronomy and are providing new opportunities for a wide range of research programs, including radio galaxies, quasars, cosmology, and pulsars, as well as solar-terrestrial physics. Curiously, the LWA's wide field of view has also re-energized the study of meteor ionospheric radio afterglows. None of these three instruments has had an impact on the detection of the Epoch of Reionization, which was the main science driver for the low frequency component of the SKA.

12

Expecting the Unexpected

> The history of research does not necessarily follow the logic of plans by expert committees or central administration ...[1]

Radio astronomy is a technique-oriented science. Starting with Karl Jansky, the major discoveries discussed in previous chapters were made primarily by skilled scientists who often built, or at least understood, their instruments. They were able to recognize and interpret their results, which were often unexpected and unrelated to what they were looking for. Many of these pioneering radio astronomers were trained as physicists or radio engineers. They had no background or training in astronomy and lacked any advanced degrees, and so their research was not constrained by astrophysical theory and what they could or could not expect to see. Many brought with them to the academic environment the practical electronics training they received in military service.

In this chapter we review the circumstances surrounding the discoveries made by radio astronomers over nearly the past century, the impact of technological innovation, of peer review, and of the role of theoretical predictions. We conclude with speculations about what new discoveries may come from future radio astronomy investigations.

Enabling Discovery

In his books, *Cosmic Discovery*, *In Search of the True Universe*, and *Cosmic Messengers*, Martin Harwit has studied the history of astronomical discoveries, including some of the radio astronomy discoveries discussed in preceding chapters.[2] Harwit suggested that perhaps as much as one-third of all cosmic phenomena has already been discovered, and he outlined those "traits common to many discoveries" in astronomy. These include the importance of technological innovation and the hands-on involvement of researchers in observational

Table 12.1 *Major radio astronomy discoveries*

Year	Discovery	Investigator(s)	Chapter	Key Instrument Involved
1933	Cosmic Radio Emission	Jansky[a]	1	Bruce Array
1943	Solar Radio Bursts	Hey	2	Radar Antennas
1944	Milky Way Radio Map	Reber[a]	1	Parabolic Antenna/ Receiver
1944	The Solar Corona[b]	Reber[a]	2	Parabolic Antenna/ Receiver
1946	First Lunar Radar Detection	DeWitt, Bay	9	Radar/Integrator
1949	First Supernova Remnant – The Crab Nebula	Bolton, Stanley, & Slee	3	Sea Interferometer
1949	Radio galaxies	Bolton, Stanley, & Slee	3	Sea Interferometer
1951	HI	Ewen	8	Horn/Instrumentation
1953	Double Radio Sources	Jennison & Das Gupta	3	Intensity interferometry
1954	Galactic Nucleus – Sgr A	McGee & Bolton[c]	4	Hole-in-Ground Reflector
1955	Jupiter Dekametric Radio Bursts	Burke & Franklin	9	Mills Cross
1956	Evolving Universe	Ryle & Scheuer	5	Synthesis Arrays
1960	Jupiter Radiation Belts[d]	Multiple observers	9	OVRO Interferometer
1961	AU and Venus Rotation	MIT and JPL Teams	9	Powerful Radar Facility
1962	Synthesis Imaging	Ryle[e]	11	EDSAC computer
1963	Quasars	Schmidt, Matthews, Hazard	4	Lunar Occultation
1963	First Interstellar Molecule	Weinreb & Barrett	8	Autocorrelation Spectrometer
1964	4th Test of GR	Shapiro	10	Radar
1964	Interplanetary Scintillations/Solar Wind	Clarke & Hewish	7	Short Integration Time
1964	Radio Recombination Lines	Dravskikh & Sorochenko[f]	8	Instrumentation
1965	Mercury Rotation	Pettengill & Dyce	9	Radar Instrumentation
1965	CMB	Penzias & Wilson[g]	6	Horn Reflector/Receiver
1965	Cosmic Masers (OH)	Gundermann[h]	8	New Receiver
1967	Pulsars – Neutron Stars	Bell & Hewish[i]	7	New Array/Short Integration Time
1968	Water Masers	Cheung & Townes	8	New Receiver
1970	CO and Interstellar Molecules	Wilson, Jefferts, Penzias	8	New Receiver/ Spectrometer
1970	GR Solar Deflection	Sramek & Fomalont[j]	10	Interferometry/VLBI
1971	Superluminal Motion	MIT and NRAO teams	4	VLBI
1974	Binary Pulsar	Hulse & Taylor[k]	10	De-dispersing Software
1974	SgrA*	Balick & Brown	4	GBI
1979	Gravitational lensing	Walsh	10	Optical Spectroscopy
1991	Exoplanets	Wolzczan & Frail	9	De-dispersing Software
1995	First Supermassive Black Hole (NGC 4258)	US-Japanese Team	8	VLBA
1996	CMB Spectrum and Anisotropy	Mather & Smoot[l]	6	Dedicated Designed Satellite
2007	Fast Radio Bursts	Lorimer	7	De-dispersing Software/13 feeds
2019	SMBH image	Large International Team	11	mm VLBI

Notes: Column (1) Year of discovery; (2) Short description of each discovery; (3) Key scientist(s) involved; (4) Chapter where discovery is discussed; (5) Key instrument(s) that facilitated the discovery. [a] Both Jansky and astronomy, the educational background of scientists who made new astronomical discoveries, the role of the military establishment in astronomy, and the role of chance or serendipity.

In tabulating and discussing scientific discoveries, one must have a clear definition of what defines a discovery. Here we adopt Harwit's criterion that a discovery should have been given a name. Nevertheless, it is sometimes difficult to pin down a specific time or specific individuals responsible for a given discovery that occurred as a result of different observations made over a period of time, especially when the initial investigators did not fully understand or appreciate the importance of their discovery. With that caveat in mind, in Table 12.1 we have listed the major discoveries made by radio astronomers. Others might come up with a slightly different list that contains a few different items, and perhaps does not include some of those in Table 12.1, but there would surely be general agreement on most of the entries. We have not included the important discovery of dark matter in Table 12.1 since this came about slowly, over a period of nearly half a century, and as a result of many investigations at both radio and optical wavelengths.

Some of the entries in Table 12.1 reflected a real Eureka moment. The 1946 detections of lunar radar echoes; Ewen's detection of 21 cm galactic hydrogen; Burke and Franklin's discovery of electrical storms on Jupiter; Maarten Schmidt's realization of the high redshift of 3C 273; the radar detection of the non-synchronous rotation of Mercury; the discovery of interstellar, OH, CO, and H_2O masers; and finding the first extrasolar planets were all Eureka events.

Reber were nominated for Nobel Prizes in 1948 and 1950, respectively. [b] Although Hey and Southworth had previously detected solar radio bursts and thermal radio emission, their work remained classified until after the War. Reber was the first to publish the detection of radio emission from the Sun, and the first to recognize the enhanced radio emission from the solar corona. [c] Although reported earlier, McGee and Bolton (1954) were the first to associate the radio source with the nucleus of the Milky Way Galaxy. [d] The OVRO interferometer measurements that showed the presence of radiation belts followed the NRL, Green Bank, and OVRO determinations that the Jupiter radio emission was nonthermal, and the speculation by Frank Drake that the Jovian radio emission was due to relativistic particles moving in an intense magnetic field. [e] Although Martin Ryle is generally credited with the development of Fourier or Aperture Synthesis and shared the 1974 Nobel Physics Prize, some years earlier, Chris Christiansen, in Australia, developed alternate techniques that he called "Earth Rotation Synthesis." [f] The reported detection of RRLs by two independent Russian teams in 1964 was somewhat marginal, but was confirmed the following year by Hoglund and Mezger (1965a, 1965b). [g] Penzias and Wilson shared the 1978 Nobel Physics Prize. [h] Gundermann was probably the first to recognize the anomalous nature of OH emission, but delayed publication. Following confirmation at Berkeley and MIT, the MIT group was the first to point out that they were seeing "anomalously excited OH" (Weinreb et al., 1965). [i] Anthony Hewish received the 1974 Nobel Prize "for his decisive role in the discovery of pulsars." [j] Earlier radio interferometer experiments at Caltech and JPL had a precision comparable to that of the optical measurements. Sramek's experiment was the first to use dual frequencies to separate out the effect of refraction and was the first to clearly prefer the GR prediction over the Newtonian value. Later experiments by Fomalont and others greatly improved the precision. [k] Awarded the 1993 Nobel Physics Prize. [l] Awarded the 2006 Nobel Physics Prize.

By contrast, Jansky's realization that he had discovered radio noise from the Galaxy came about incrementally over a period of several years. Other things, such as the discoveries of cosmic masers, the nucleus of our Galaxy, and radio recombination lines, along with the confirmation of solar GR bending, also came slowly, and in some cases were the result of work at multiple institutions. At the other extreme, John Bolton rejected his discovery of the first radio galaxies, and later of quasars, because he was not able to accept the implied paradigm change. The CMB was independently discovered by multiple scientists, but neither they nor Penzias and Wilson understood the implications of their discovery until it was explained to them.

We have not discussed the discovery of cosmic magnetic fields. That came about slowly as the result of many radio observations, including the interpretation of the strong nonthermal galactic radio emission as synchrotron radiation from ultrarealistic electrons moving in weak magnetic fields, the near simultaneous detection of polarized radio emission at both Parkes and NRL, the discovery of synchrotron emission from Jupiter's radiation belts, and the detection of the Zeeman splitting of the 21 cm line by interstellar magnetic fields.[3]

We elaborate here on the circumstances surrounding these discoveries that have redefined our understanding of our Universe and its contents.

Sensitivity In Figure 12.1 we show the dramatic improvement of radio telescope sensitivity from Jansky's simple Bruce Array to the present day by a factor of about 10^{11}. Ronald Ekers has called attention to the similarity of this plot with the energy increase of particle accelerators with time, known as a Livingston curve.[4] As M. Stanley Livingston first pointed out, the historical increases in accelerator energy resulted from the introduction of a series of qualitatively new technologies starting with electrostatic accelerators, then cyclotrons, bevatrons, synchrotrons, alternating gradient synchrotrons, and particle colliders. In radio astronomy the improvements in sensitivity have come from increased antenna collecting areas and cryogenically cooled low noise receivers. Improvements in digital autocorrelation spectrometers made possible by the explosion in digital signal processing have greatly increased the capability of spectroscopic systems. Continuum sensitivity has exploited the increased bandwidth of modern radiometers. Also, the advances in high angular resolution synthesis have increased the maximum obtainable integration times, enabling sub-microjansky sensitivity without the systematic effects characteristic of filled aperture observations.

Radio Imaging and Angular Resolution Figure 12.2 shows the improvement with time in the angular resolution of radio telescopes. Jansky's antenna

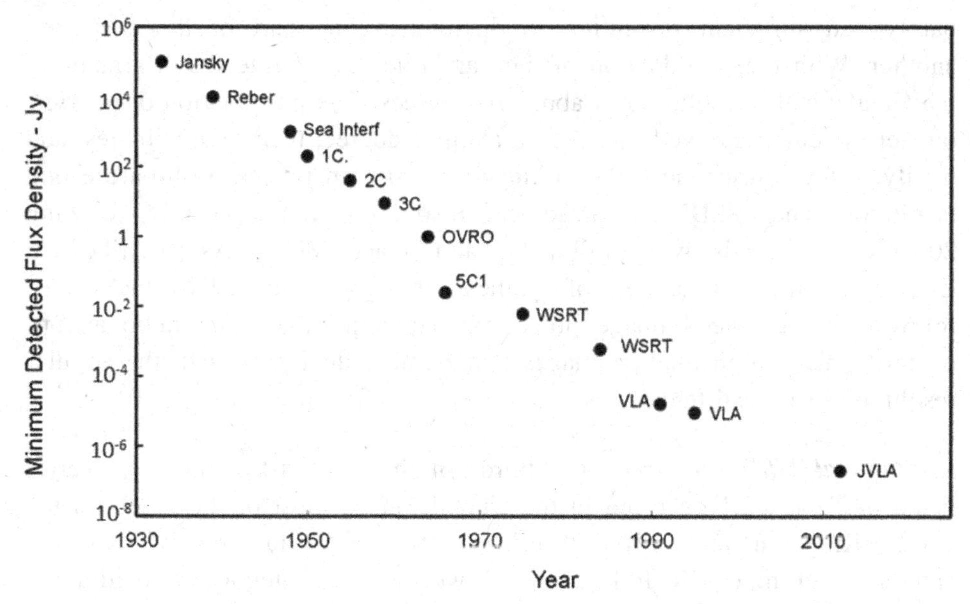

Figure 12.1 Sensitivity of radio telescopes versus time. The sensitivity is expressed in terms of the weakest continuum radio detected at the time.

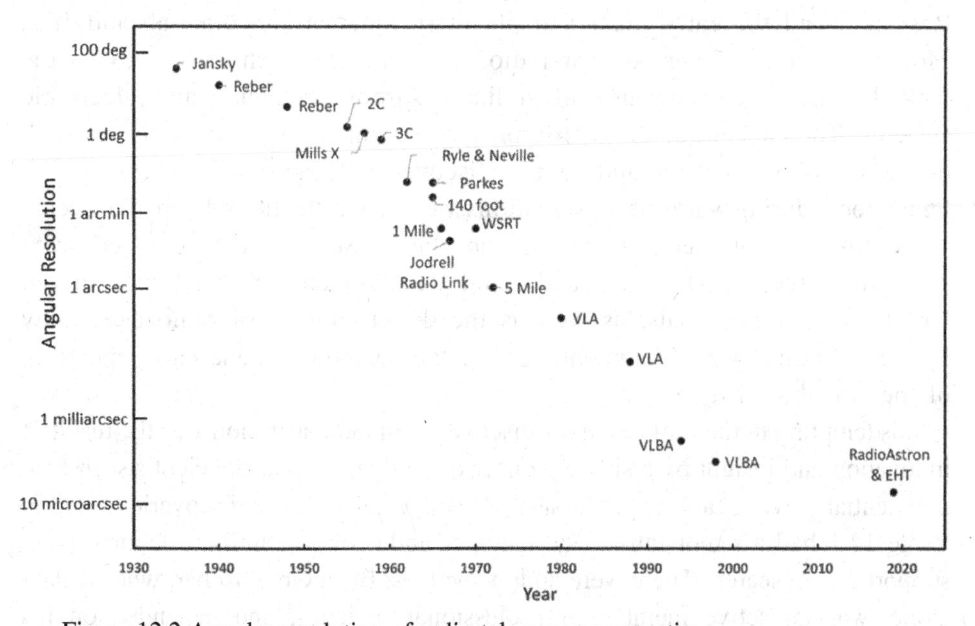

Figure 12.2 Angular resolution of radio telescopes versus time.

barely had sufficient resolution to distinguish one part of the sky from another. With the introduction of the parabolic dish, Grote Reber's maps of the Galaxy had a resolution of about 10 degrees. The introduction of interferometer systems improved this first to about 1 degree, then to arcminutes, and finally a few arcseconds. Radio linked interferometry gave sub-arcsecond resolution, and VLBI milliarcsecond resolution, most recently reaching 20 microarcseconds with millimeter and space VLBI. As described in Chapter 11, the development of synthesis imaging in the 1950s and 1960s, followed by advanced image processing techniques, allowed radio astronomers to make high quality images that rivaled, then exceeded, the angular resolution of optical telescopes.

Institutional Affiliation About two-thirds of the radio astronomy discoveries listed in Table 12.1 occurred in the United States, most of the others in the United Kingdom, and only a few in the rest of world, notably Australia, Hungary, and the USSR. In Figure 12.3, we show the category of institutional affiliation of the scientists responsible for each discovery listed in Table 12.1. While many of these discoveries happened in a university environment, we note the contribution of militarily motivated research in the independent discoveries of solar radio bursts by Alexander, Hey, and Slee, the independent discovery of pulsars by Charles Schisler, the first radio measurements of the temperature of the planets at NRL, funding of OVRO by the Office of Naval Research, and the early lunar and planetary radar at Lincoln Lab and JPL. Moreover, much of the postwar radio astronomy research, especially in the United Kingdom, made use of military surplus antennas and electronic systems. The concept of sea interferometry, so important in the early research on solar radio emission and in the discovery of radio galaxies, was first understood during wartime observations of radar reflections from the ocean. Also, radio astronomers, at least in the United States, have benefited since 1940 from the well-financed and rapid development of electronics driven largely by military goals, as well as the direct support of radio astronomy by the US military establishment, such as the construction and early operation of the Arecibo Observatory.

It is tempting to think of the many discoveries in radio astronomy as the result of innovation and insight by a single pioneer or, perhaps, a pair of scientists. In fact, in essentially every case, the radio astronomers who made the discoveries listed in Table 12.1 had a major university, national, industrial, or military laboratory to support their research. There were no lone wolves. Even Grote Reber, who worked alone, was an active member of professional societies, and depended on his

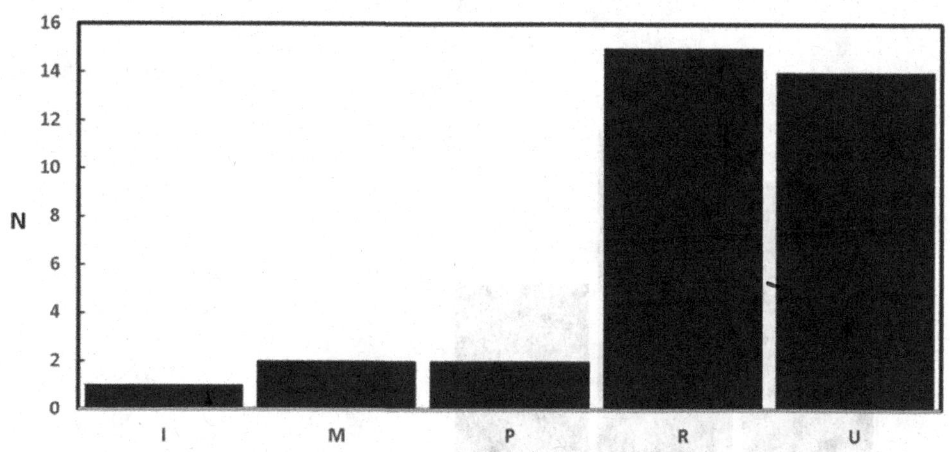

Figure 12.3 Histogram showing the distribution of institutional affiliations of scientists responsible for the major radio astronomy discoveries.
Note: I = Industrial, M = Military, P = Private, R = Research Organization, U = University.

connections to the electronics industry for low noise tubes. Karl Jansky, who perhaps best exemplifies the image of the single researcher doggedly pursuing a perplexing puzzle resulting in a series of well thought out publications, was surrounded by some of the greatest minds in the electronics industry. Doc Ewen was a graduate student working alone building his own equipment, but critical to his experiment was the equipment he borrowed on weekends from the Harvard Cyclotron Lab, without which he would not have detected Galactic 21 cm hydrogen radiation.

Age of the Investigators It is well known that scientists generally do their best work as young men and women and, as shown in Figure 12.4, this has certainly been true of radio astronomers. Many of the workers we have discussed, starting with Jansky and Reber, were in their 20s when they did their pioneering work. Most of the others were in their 30s. Bruce Slee was only 21 when he detected solar radio bursts while operating a wartime radar station on the north coast of Australia. Slee published his last paper in 2015, at age 91, one year before he died. The oldest entry in Table 12.1 is for the COBE determination of the CMB blackbody spectrum and anisotropy. Both Mather and Smoot were about 50 years old at the time the results of their experiment were announced.

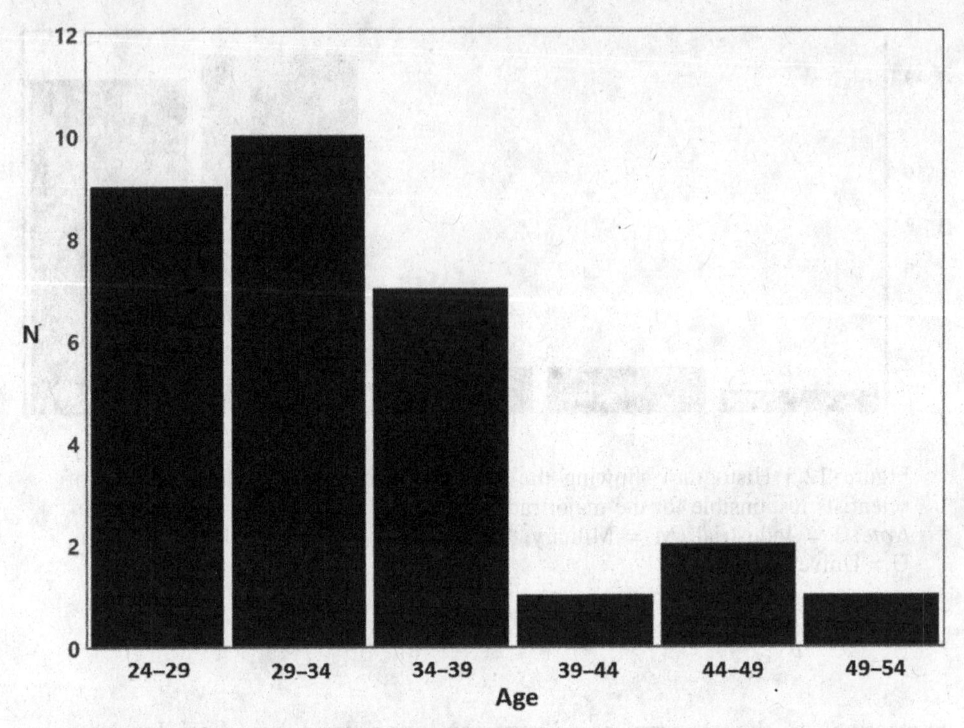

Figure 12.4 Histogram showing the distribution of the ages of scientists at the time of their discoveries. Average age was used when two people were equally involved. The age of the lead author was used when there were three or more scientists involved.

The Role of Luck As eloquently exclaimed by Louis Pasteur in 1854, "Dans les champs de l'observation le hasard ne favorise que les esprits préparés," or "In the fields of observation chance favors only the prepared mind." Indeed, this has been true in radio astronomy. At the 1983 conference marking the 50th anniversary of Karl Jansky's discovery of cosmic radiation, the participants agreed that, although the path toward the correct answer was often circuitous, the major discoveries involved "the right man, in the right place, at the right time, doing the right thing," although Frank Drake offered that "serendipity requires the right person in the right place, at the right time doing the wrong thing." Several speakers also noted that sometimes it helped not to know too much.

In nearly all of the discoveries listed in Table 12.1, the investigators either had designed and built their own specialized instrumentation, had developed specific software, or at least had a deep understanding of their equipment. Karl Jansky was lucky that his older brother Curtis was able to get him a job at Bell Labs. He was lucky that he was assigned to work at the New Jersey field station rather than in the main Bell Labs office in New York City. He was also lucky that he happened to be working during a minimum in the 11 year sunspot cycle. Had it been a few years earlier or a few years later when the Sun was more active, the ionosphere would have been opaque to the galactic radio emission. Alexander, Hey, Reber, and Slee were all lucky that the mid-1940s was a period of great solar activity. Burke and Franklin were lucky that they decided to survey the sky south instead of north of the Crab Nebula, and so stumbled on the intense decametric bursts from Jupiter. Alex Shain was lucky that the East–West arm of his array had been dismantled, so he was left with a broad beam in hour angle that could observe Jupiter for 8 hours each day, allowing him to pin down the rotation period. Russell Hulse was lucky that, at the time he was observing at Arecibo, the binary pulsar just happened to be a few percent above his detection limit. Had he set his limit a few percent higher, he would not have detected the binary pulsar that led to the first detection of gravitational radiation.

The Japanese radio astronomers who discovered the high velocity maser wings of NGC 4258, later used to detect the first black hole and to determine the Hubble constant, were very lucky. They were observing the strong central maser feature NGC 4258 to test out a new ultra-wide band spectrometer, not to study NGC 4258. Only months later did one of the scientists bother to also look at the high velocity frequency channels to discover the previously unknown maser emission that made the VLBA discoveries possible. Wolszczan and Frail were lucky that the Arecibo radio telescope was undergoing maintenance and repair, so they were unable to move the telescope to look at different regions of the sky. Instead, they were confined to observing a region near the zenith as the rotation of the Earth swept their fixed antenna beam across the sky and they found PSR 1257+12 being orbited by the first known planets outside our Solar System. As Frail later remarked, "The exoplanet field was launched by two clueless radio astronomers who stumbled on the discovery using a broken radio telescope."[5] Maarten Schmidt was lucky that 3C 273 happened to be the only quasar having the simple Balmer spectral line series so that he could unambiguously determine the redshift. The radio astronomers who made these and other discoveries discussed in this book were in the right place, at the right time, and mostly doing the right thing. They were lucky but they all had more than just good luck.

Radio Astronomy, Technology, and Computers In his three books, Martin Harwit has emphasized the role that new technologies had in astronomical discoveries. This has been particularly true in radio astronomy. As noted, nearly all of the discoveries listed in Table 12.1 have resulted from exploiting a new technology – large antenna structure, lower noise receivers, digital signal processing, interferometry, including VLBI, and aperture synthesis.

Probably no other area of technology has developed more rapidly than the high speed digital computing used to control the operation of modern radio telescopes and the handling of their huge data output. Not surprisingly, the use of computers has transformed radio astronomy, along with essentially all other areas of scientific research. The high resolution images of radio galaxies and other cosmic radio sources would not be possible without fast digital computers and the development of sophisticated image processing software. Indeed, it was, in part, the availability of the state-of-the-art EDSAC computer at the Cambridge Cavendish Laboratory that facilitated the two-dimensional synthesis imaging developed by Ryle and his colleagues. VLBI would not be possible without high speed computers. Russ Hulse would not have found the binary pulsar and Duncan Lorimer would not have found the first FRB without computer-aided searches for signals with unknown dispersion rates.

On the other hand, we recall that, in her thesis, Jocelyn Bell remarked,

> It is worth emphasizing that if the aerial output had been digitized and fed directly into a computer, these sources might well not have been discovered, because the computer would not have been programmed to search for unexpected objects.[6]

A similar caution about using computers was raised by John Bolton in 1964, when visiting the new National Radio Astronomy Observatory in Green Bank. Concerned about the impact of the apparent pervasiveness of computers and involvement of the scientists in their observations, Bolton wrote back to his colleagues in Australia,[7]

> If you make your observation by writing a set of instructions for a telescope operator to carry out, and then write a set of instructions for a computer to extract some data from the results, then it is rather unlikely that you are going to find anything other than what you are looking for.

Robert Wilson, who had been a student of Bolton's at Caltech, expressed it more broadly:[8]

> Anytime you make a perfectly matched filter to your problem, you are going to get the optimum signal to noise ratio in your problem, and learn nothing about anything else. The computer merely allows one to do that very easily, most of the time.

Beware of Theoreticians

In the classical description of experimental science, frequently taught in high schools, the scientist starts with a theory or hypothesis, then does an experiment or observation that either proves or disproves the theory. In practice it is often more complex. As we have seen, theory played little or no role in most of the discoveries listed in Table 12.1, and in some cases delayed the actual discovery by discouraging observational confirmation.

For only a few of the entries in Table 12.1 did a theoretical prediction actually anticipate a subsequent discovery or confirmation: the bending of radio waves by the Sun, Shapiro's fourth test of General Relativity, the blackbody spectrum and anisotropy of the CMB, and the shadow of the M87 black hole observed by the Event Horizon Telescope (EHT). Interestingly three of the four were directly related to General Relativity. We also note that, following the discovery of the CMB, there were actually a succession of theoretical predicted levels of anisotropy that were, in succession, shown by observations with improved sensitivity to be wrong. The COBE measurement confirmed the most recent prediction.

Henk van de Hulst's prediction of the 21 cm hydrogen line is often considered to be the one textbook example in radio astronomy where the observation and detection followed from a theoretical prediction.[9] In fact, as we saw in Chapter 8, van de Hulst was quite skeptical about the prospects for detecting the line, concluding that it "does not appear hopeless." Shklovsky was more optimistic, but a prominent Soviet physicist ridiculed the idea and discouraged any observational attempt to detect the 21 cm line.

More generally, many of the discoveries made by radio astronomers were serendipitous or unexpected. Following Jansky's discovery of cosmic radio emission, the discoveries of solar and Jupiter radio bursts, radio galaxies, double radio sources, Jupiter's radiation belts, quasars, pulsars, cosmic masers, and FRBs were all unpredicted. In some other cases, such as the 21 cm detection of neutral hydrogen, or radio recombination lines and interstellar molecules, the theory was wrong, and in the case of interstellar water, theory prevented an earlier detection. In other instances, such as the solar corona, the solar wind, the CMB and superluminal radio sources, the phenomena were predicted, but the predictions played no role in the experimental discovery. Nor did Franco Pacini's speculation on electromagnetic radiation from rapidly rotating neutron stars play any role in the discovery of pulsars. Martin Rees predicted superluminal motion to explain the rapid time variability of quasars, but it played no role in the 1971 discovery by two VLBI teams.

In cases such as the radio bursts from the Sun and Jupiter, cosmic masers, and interstellar carbon monoxide, the radio emission is so strong that they could have been discovered much earlier. In fact, some of these phenomena were observed earlier, but not recognized. The CMB was not only predicted and had been marginally detected by multiple radio workers but was ignored or misunderstood. Moreover, as was later realized, the CMB was responsible for the long known anomalous excitation of interstellar CN. Jupiter bursts were observed by others before Burke and Franklin, but were unrecognized because they did not fit into the generally accepted paradigm. John Bolton initially rejected his discovery of radio galaxies because it seemed too extreme. A decade later, Bolton, as well as Greenstein, Sandage, and others, rejected the recognition of the first quasar, 3C 48, also because it did not fit into their preconceptions. Two years later, because he was not thinking of a large shift for what he thought was a galactic star, it took Maarten Schmidt nearly six weeks to recognize the Balmer series spectrum of 3C 273. Moreover, until alerted by Bolton's report of the (incorrect) occultation position, both Caltech radio and optical astronomers had ignored 3C 273 because they thought it was too big to have a large redshift, even though it was one of the strongest radio sources in the sky, and the bright 13 magnitude quasar was almost begging to be identified with the radio source.

Although theory has had little impact in anticipating the discoveries in Table 12.1, the new observational data made possible by radio telescopes of increasing sophistication has served to constrain theoretical models, thus leading to a new and better understanding of our Universe and its contents. In some cases, the theoretical understanding was remarkably rapid. As discussed in Chapter 9, although observers and theoreticians had been in agreement for more than half a century that Mercury did not rotate with respect to the Sun, a theoretical "prediction" was published in the same issue of *Nature* as the results of the Arecibo radar determination that showed that the 2/3 resonance was a natural consequence of Mercury's highly elliptical orbit. Indeed, Gordon Pettengill later remarked "Any theoretician who had looked at this fact, even at this stupidly simple level, would have immediately realized that Mercury had to be rotating with a period of 59 days. I think this has to be another form of oversight."[10]

Unconfirmed Discoveries Along with all the successes of radio astronomy, we need to mention the radio discoveries that later turned out to be wrong. Perhaps the most prominent of these were the erroneous detection of the redshift of Cygnus A and the 1958 Lincoln Laboratory claimed detection of radar echoes from Venus. In the former case, the observers, having made an error in calculating the redshift, weren't even looking at the right frequency. In the latter case, the false detection, including the wrong value for the distance and the wrong direction of rotation, were later "confirmed" by radar scientists at both Jodrell Bank and in the USSR. These unfortunate events illustrate that the narrow search filters described by Robert Wilson can not only optimize the signal-to-noise ratio and reject unexpected results, but they can also fool observers into finding something that is not there.

The Evolution of Radio Astronomy

In Figure 12.5, we show the cumulative number of discoveries as a function of time. Following a modest start in the years after 1933, in the early 1960s there was an explosive growth in the rate of new discoveries lasting for about a decade. These were clearly the Golden Years of radio astronomy. Surprisingly, however, since about 1980, the rate of new discoveries has dramatically decreased, despite the introduction during this period of powerful new radio telescopes in Australia, Chile, Germany, India, the Netherlands, the United States, and most recently China. Harwit has made a similar analysis that covers all discoveries in astronomy and astrophysics dating from the time of antiquity.[11] Interestingly, Harwit's plot does not show the same decrease in the rate of new astronomical discoveries as we see in radio astronomy. Harwit attributes the recent growth in new astronomical discoveries to the introduction of the new messengers of X-ray, gamma-ray, infrared, neutrino, and most recently gravitational wave research.

Along with the steady improvement in the sensitivity and resolution of radio telescopes has been an increase in their cost and their complexity. Jansky's and Reber's antenna systems each cost only a few thousand dollars; ALMA cost somewhat over a billion dollars, an increase by a factor of a million. (Over this period, inflation accounts for only about a factor of 20). Not surprisingly, researchers are increasingly driven toward these large user facilities, in part because of their improved capabilities, but also because of the perception that they are where the resources, including students, are.

No longer do individual scientists, or even a small group of scientists, build and use their radio telescopes. As a result of the complexity of modern radio telescopes, it often takes a large group of skilled scientists, engineers, and computer experts to implement and interpret new radio astronomy investigations. The development of the so-called user facilities such as those at the NRAO and Arecibo, and later in Australia, Germany, India, and the Netherlands, in principle provide the user with sufficient technical support so that a single individual or small group of scientists can propose, carry out, interpret, and publish significant new observational results at only modest cost to them, and without having to have detailed technical involvement with the instrument or

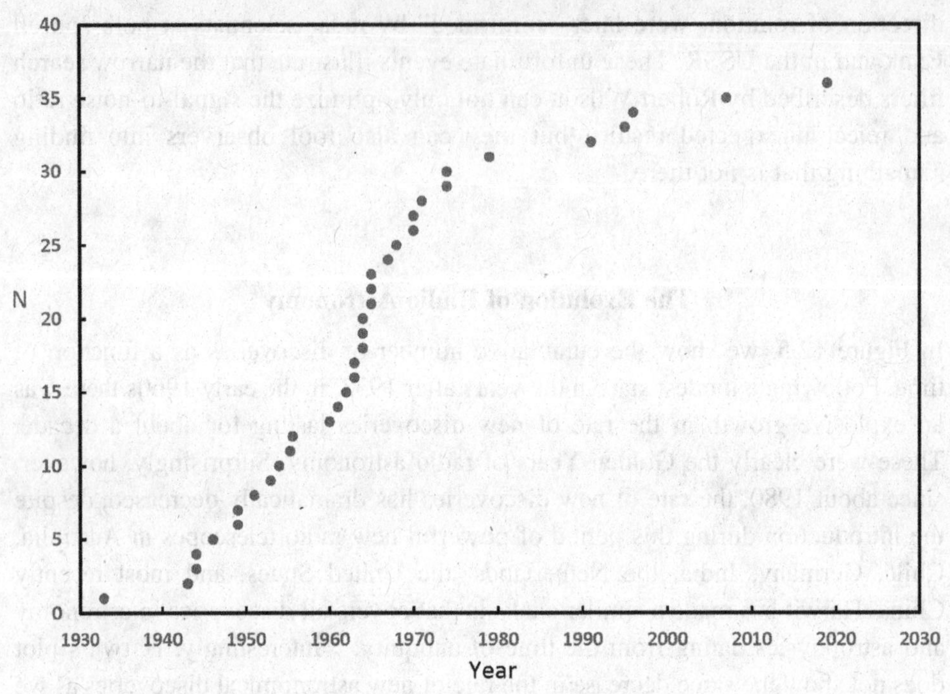

Figure 12.5 Plot showing the cumulative number of discoveries as a function of time.

even with their data. Scientists no longer develop hardware. That is a task mostly implemented by professional engineers who are rarely involved in observation and analysis. The current generation of radio astronomers are instead more likely to develop the sophisticated new algorithms needed for imaging, pulsar timing and searching, and for spectroscopy, although many of these tasks are also being taken over by professional software developers.

Does Figure 12.5 indicate that radio astronomers have run out of things to discover, or is it a consequence of the growth of radio astronomy into "big science" and the lack of opportunity for innovative research?

Peer Review Until the middle of the twentieth century, it was not unusual for the editors of scientific journals to assume the responsibility of accepting papers for publication. Starting in the postwar era, it became more common for scientific journals to seek the opinion of expert reviewers, a practice that has come to be known as "peer review." Referees generally provide valuable advice to both the editor and the author in suggesting improvements to the presentation, even correcting author errors or misunderstandings, although they can and have rejected or delayed publication of important papers. There have been claims that papers were withheld or delayed because the proposed publication may appear to be competing with the referee's own research program, or because a referee fails to understand the importance of a paper being reviewed.

Duncan Lorimer's Fast Radio Burst discovery paper has had over 1,100 citations and has spawned a whole new research industry, as well as motivating the construction of expensive new facilities devoted to discovering and locating new Fast Radio Bursts. When Lorimer submitted his paper to a well-known prestigious journal, the editor dismissed the paper writing "the paper as it stands is rather 'thin'," and "we are unable to conclude that the paper – as it stands – provides the sort of fundamental advances in scientific understanding that we generally look for."[12] Fortunately, this important paper was quickly accepted by a different journal. As another example of the possible adverse impact of peer review, we recall the discussion in Chapter 3 surrounding the controversy over the delayed publication of Ron Bracewell's paper reporting the polarization of Centaurus A.

Peer review has become pervasive and extends well beyond the review of submitted publications. With the increasing domination of large, user-oriented radio astronomy facilities, the competition for observing time is so great that the observatories seek referees to help decide which proposals to accept and which to reject. Inevitably, this procedure favors the safe proposals, those that are mostly likely to succeed, while the more innovative and often more risky proposals, those that might result in major breakthroughs, are difficult to get through the refereeing system. John Broderick and Thomas Gold have cautioned against what they call, respectively, the "buffalo syndrome" or "herd instinct" that constrains new research that does not fit clearly into the prevailing mainstream.[13]

As Gold warned,

> The enemy of new ideas is the "herd instinct." This instinct, which has no doubt a great deal of sociological significance, is a disaster in science. If a large proportion of the scientific community in one field are guided by this instinct, then they cannot adopt a novel viewpoint, since they cannot imagine that the rest of the herd would all adopt it at the same time. The herd stays together if it trots along the same road, and anyone who departs from it may find himself alone. If support from peers, moral, or financial, is of importance, then staying with the herd is the successful strategy – successful, that is, for this limited purpose, but not for the pursuit of science.
>
> Staying with the herd also has the advantage that one need not fear ridicule or fear being exposed if one does not know how to justify the case. If it is the commonly accepted viewpoint, then that seems sufficient justification for adopting it. The sheep on the interior of the herd are well protected from a bite in the ankle by the sheepdog. To run with the crowd seems to give many people the feeling of security and of

> comfort that they seek, and they are not going to give that up for such an insecure thing as scientific progress.
>
> The tendency to herd behavior may be greatly aggravated by the structure of support for science. If for any line of work, support is dependent on widespread judgement that it is on the right lines, then of course one will comb out all those who do not wish to join the herd.[14]

Noting the extensive use of peer review not only by journals and observatory telescope time assignment committees, but in applying for grants to support one's research and by committees charged with choosing job applicants or awarding prestigious prizes, Gold then went on to claim,

> The peer review system, regarded as the only fair way of distributing funds and recognition, is in fact a disastrously unstable device. The more concentrated and "efficient" such a peer review system is, the less noise there is in it, the more firmly and quickly will it produce its disastrous end result of a herd stampeding in a well-defined direction – a direction which was determined perhaps only by an insignificant fluctuation.

As described in Chapter 8, probably the most notorious result of peer review in radio astronomy was the rejection of a proposal to NRAO to use the 140 Foot radio telescope to search for interstellar water vapor at 22.2 GHz (1.3 cm). On the advice of a distinguished theoretical referee, NRAO declined the proposal, only to later learn that Al Cheung, a Berkeley graduate student, using a small 6 m (20 foot) antenna found the extraordinarily strong 22.3 GHz signal from interstellar water vapor. Indeed, the observed emission was so strong that it could easily have been detected years, if not decades, earlier, perhaps even before the hydrogen line. But no one thought to try. Also, we recall the story, also related in Chapter 8, of Iosef Shklovsky, who was unable to search for the 21 cm hydrogen line because Lev Landau, the great Soviet theoretician, declared that "this was clear pathology."

Not only does the herd instinct limit the opportunity to pursue new ideas and experimental approaches, but it may also limit the acceptance of new ideas and results that do not fit into currently popular concepts. Recall in Chapter 7, how the radio astronomy community was skeptical until FRBs were finally detected in data from a different telescope than Parkes, the delay in accepting the discovery of radio recombination lines, or rapid quasar time variability, because they didn't fit neatly to then popular ideas. Perhaps the best example of the herd mentality is the repeated reports of the locked rotation of Mercury by generations of astronomers, all apparently based on the original erroneous claim of a single visual nineteenth century astronomer.

At the 1998 dedication of a monument erected by Bell Labs to recognize the discovery of cosmic radio waves by Karl Jansky, John Pierce, a contemporary of Jansky's at Bell Labs and recognized as the father of communication satellites, wrote, "Jansky's work shows that very important phenomena can be disregarded when they don't find a niche in the science of their times." Pierce went on to caution, "Let today's men of science gaze on our monument to Jansky and his work, and let them ask themselves if in their research they may not sometimes disregard the new and significant in pursuing the current and popular."[15] On the same occasion, Charles Townes wrote that, Jansky's "work was such a radical break-through that its value to astronomy was for many years unrecognized," while Hanbury Brown added, "It is a powerful reminder of the vital importance of experiment to the progress of science. No one would have predicted Jansky's results theoretically."

Radio Astronomy and the Nobel Prize

There is no Nobel Prize in astronomy; nevertheless, no fewer than 21 scientists have been honored with the Nobel Prize in Physics for their work in observational astronomy, along with five others for their contributions to theoretical astrophysics. A widely repeated account for the lack of a Nobel Prize in astronomy is that Alfred Nobel's wife had run off with an astronomer, although there is no evidence that Nobel was ever married.[16]

Prior to 1974, there had not been any Nobel Prizes awarded for any research work in observational astronomy. Then, as mentioned in the Introduction, the first three Nobel Prizes in observational astronomy, that went to six individuals, were all for discoveries in radio astronomy. In 1974, Martin Ryle and Tony Hewish became the first observational astronomers, ever, to be recognized by a Nobel Prize "for their pioneering work in radio astrophysics." Then, in 1978 Arno Penzias and Robert Wilson received the Nobel Prize for their 1965 detection of the cosmic microwave background, followed by Joe Taylor and Russell Hulse in 1993 for their discovery of the binary pulsar system "that opened up new possibilities for the study of gravitation." Later, in 2006, John Mather and George Smoot were additionally recognized for their COBE determination of the CMB blackbody spectrum and anisotropy, respectively. Both Karl Jansky (in 1948) and Grote Reber (in 1950) were nominated for Nobel Prizes but radio astronomy had not yet made the impact that followed in the 1950s and 1960s, and it seems that their nominations were not seriously considered. Other radio astronomers who were nominated for the Nobel Prize include Bernard Lovell with 14 nominations between 1963 and 1970, and Hanbury Brown who was nominated in 1965 and 1966 along with Richard Twiss, probably for their invention of intensity interferometry.

Not until 2002, when Raymond Davis, Masatoshi Koshiba, and Riccardo Giacconi were recognized for the detection of cosmic neutrinos and cosmic X-ray sources, was there a Nobel Prize for observational or experimental astronomy outside the field of radio astronomy. This was followed by the 2016 Prize to Rainer Weiss, Barry Barish, and Kip Thorne for their LIGO detection of gravitational waves. The first Prize received for work in classical optical observational astronomy was not until the 2011 Prize to Saul Perlmutter, Brian Schmidt, and Adam Riess for their "discovery of the accelerating expansion of the universe." This was quickly followed in 2019 when Michel Mayor and Didier Queloz were recognized for their "discovery of an exoplanet orbiting a solar-type star." The following year Rheinhard Genzel and Andrea Ghez, were the co-recipients of the 2020 Prize for their infrared wavelength "discovery of a supermassive compact object at the center of our galaxy."[17]

Although between 1974 and 2006 eight scientists were recognized for their work in radio astronomy, there were none after 2006, apparently reflecting the explosive growth of radio astronomy during the period from the late 1950s to the late 1970s, that was followed by the dramatic decrease in the rate of new radio astronomy discoveries after the mid-1980s, as displayed in Figure 12.5. Starting in 2002, seven astronomers were recognized for their work involving observational techniques outside the radio window, reflecting the dramatic development of new techniques in optical astronomy as well as the emergence of other new fields of X-ray, neutrino, and gravitational wave astronomy, all based on newly-developed technologies.

We note that there were also a number of Nobel Prizes awarded for contributions to theoretical astrophysics, starting with Hans Bethe in 1967 and Subrahmanyan Chandrasekhar in 1983, for their contributions to understanding stellar energy production and evolution. Chandrasekhar shared the 1983 Prize with Willie Fowler for Fowler's "theoretical and experimental studies of the nuclear reactions of importance in the formation of the chemical elements in the universe." Later James Peebles and Roger Penrose, respectively, shared the 2019 and 2020 Prizes for their contributions to "understanding the evolution of the universe," and "the discovery that black hole formation is a robust prediction of the general theory of relativity." As early as 1936, Victor Hess was recognized "for his discovery of cosmic radiation." However, until the 1950s, when the Russian scientists Iosef Shklovsky and Vitaly Ginzburg, and Geoffrey Burbidge in the United Kingdom, became interested in cosmic radiation, cosmic rays were primarily a source of high energy particles for physics experiments rather than a means of probing the Universe.

Although the most contentious Nobel Prizes have been the Peace and Literature Prizes, the Nobel Prizes recognizing work in astronomy and astrophysics have not been without controversy. The 1974 award to Tony Hewish "for his decisive role in the discovery of pulsars" has probably been the most controversial. Much has been written and more has been said about Jocelyn Bell's omission from the 1974 Prize. Among the most vocal critics was Fred Hoyle, who faced potential legal action for his criticism of Hewish for allegedly "stealing Bell's results," and for his criticism of the Nobel Committee for "not understanding the true facts."[18] Was Jocelyn Bell passed by because of her young age and because she was only a student at the time of her discovery? Probably not. Others have received the Nobel Physics Prize based on their PhD dissertation research, including Russell Hulse, a PhD student at the University of Massachusetts, who two decades later was recognized together with his supervisor, Joseph Taylor, with the 1993 Nobel Physics Prize for their "discovery of a new type of pulsar."[19] Brian Josephson won the 1973 Prize for work he did in 1962 as a 22 year old PhD student at the Cambridge Cavendish Laboratory, just five years before Jocelyn Bell discovered pulsars. Or, as some have suggested, was Bell overlooked because she was a woman? Only four women have won the Nobel Physics Prize, including most recently Andrea Ghez in 2020 for her role in the discovery of the black hole at the center of the Milky Way.[20] While her status as a young female student probably did not exclude Jocelyn Bell from being considered, the bar for evaluating her contribution may have been placed higher than it was for Hewish. It is perhaps relevant that Russell Hulse appeared as the first author of a two author paper, whereas Bell was the second author of a five author paper with Hewish as the lead author, but one would expect that the Nobel Committee would have looked beyond the publications to learn who did what.

In 1977, Bell offered her own perspective on the issue:[21]

> First, demarcation disputes between supervisor and student are always difficult, probably impossible to resolve. Secondly, it is the supervisor who has the final responsibility for the success or failure of the project ... I believe it would demean the Nobel Prizes if they were awarded to research students, except in very exceptional cases, and I do not believe this is one of them. Finally, I am not myself upset about it – after all, I am in good company, am I not!

Hewish, who designed the array used by Bell and initiated the IPS survey project, dismissed the omission of Bell by comparing his role to that of a ship's captain who makes all the important decisions, implying that Bell's role was more akin to "somebody up on the masthead [who] says land ho."[22]

Somewhat less contentious, but perhaps more perplexing, was the 2019 Prize awarded to Mayor and Queloz for their 1995 announcement of "an exoplanet orbiting a solar-type star" that followed the discovery, four years earlier, by Wolszczan and Frail of a planetary system consisting of three to four Earth-like planets orbiting a pulsar. Apparently, the Nobel Committee put more weight on "a solar type star" than on "Earth-like planets," although it was the Wolszczan and Frail radio observations that showed, for the first time, that the formation of planetary systems was not unique to our Solar System. Possibly, the earlier false claim by Lyne and Bailes that they had detected the first extrasolar planet using essentially the same technique as Wolszczan and Frail left the Nobel Committee skeptical about the validity of the pulsar result.[23]

Known Knowns, Known Unknowns, and Unknown Unknowns

At a 2002 press conference, in response to a question about Iraqi weapons of mass destruction, the US Defense Secretary Donald Rumsfeld famously remarked,[24]

> as we know, there are known knowns; there are things we know we know. We also know there are known unknowns; that is to say we know there are some things we do not know. But there are also unknown unknowns – the ones we don't know we don't know.

Following Rumsfeld's terminology, the items listed in Table 12.1 can now be considered as "known knowns." However, there remain important "known unknowns" in the radio sky. These include:

The Hubble Parameter, H_0 For many decades after its first determination in 1929, the value of the Hubble parameter, H_0, describing the rate of expansion of the Universe and consequently the age and scale of the Universe, was uncertain by as much as an order of magnitude. By the end of the twentieth century, the observational measurements of H_0 were becoming more precise, and the agreement among different methods appeared to be converging. However, as noted in Chapter 8, the value of H_0 determined by VLBA observations of H_2O megamasers, while in good agreement with the Hubble Key Project and other measurements in the local Universe, differs from the value as estimated from the CMB anisotropy and the distribution of distant galaxy clusters. If confirmed by future observations, the tension between these measurements of the Hubble parameter may require a change in our understanding of the early Universe.

Dark Matter and Dark Energy As described in Chapter 10, assuming that Newton's laws are valid, galaxy rotation curves give convincing evidence that 85 percent of the matter in the Universe is composed of dark matter. Observations of gravitational lensing and the anisotropy in the cosmic microwave background radiation (Chapter 6) provide further evidence for the existence of dark matter.

Optical observations of distant supernovae have convincingly shown that the rate of expansion of the Universe is accelerating due to an unknown form of dark energy.[25] The evidence for dark energy is supported by the anisotropy of the cosmic microwave background, the clustering of galaxies, and radio observations of gravitational lensing.[26] Dark energy comprises 68 percent of the total mass-energy density of the Universe, dark matter 26 percent, and ordinary (baryonic) matter only 5 percent. Experimental and observational efforts to identify the nature of dark matter and dark energy have so far been elusive and remain a challenge for both astronomers and physicists.

Gravitational Background Radiation In Chapter 10, we briefly discussed the difficult pulsar timing array programs aimed toward detecting the gravitational background radiation resulting from the collective inspiraling of supermassive black hole binaries during the final few thousand years prior to their catastrophic merging. The detection of the predicted gravitational wave background would be exceedingly interesting but has also remained elusive. Thousands of hours of observing time using the world's largest radio telescopes have hinted at a detection, but the data are subject to a variety of poorly understood uncertainties.[27] Longer observations with more sensitive antennas are needed to reduce the noise, but even then it may be unclear if the pulsar timing arrays have detected the background gravitational radiation, or if the results are dominated by an unknown systematic error such as timing errors due to uncertainties in the motion of the Earth or to variations in pulsar spin rates.

The Epoch of Reionization (EoR) Less than a million years after the big-bang, the Universe became sufficiently cool that all the gas existed in its neutral or un-ionized state. When the first stars and galaxies were formed, their ultraviolet radiation reionized the intergalactic medium, ending the so-called cosmic "dark ages." It is thought that this Epoch of Reionization (EoR) occurred when the Universe was about one billion years old, or at a redshift of about 10 to 20, where the 1.4 GHz (21 cm) hydrogen line is shifted to about 100 MHz. However, it remains unclear if the epoch of reionization was brief, in which case there should be a narrow signature in the radio spectrum of the background radiation, or if it happened more gradually over a period of time, resulting in a weaker signal spread over a larger bandwidth. The detection of the EoR signal is a very difficult experiment. Separating a real signal from spectral distortions due to instrumental effects and foreground radio emission from the Galaxy, radio galaxies, and quasars, as well as the proliferation of RFI in this frequency band, is challenging.[28]

In a carefully designed experiment that was carried out in 2018 in a radio quiet region of Western Australia, a team of radio astronomers announced the apparent detection of a surprisingly strong EoR signal centered at 78 MHz spread over 19 MHz. If real, the reported signal suggested that either the primordial hydrogen gas was much colder than expected, or the background temperature after reionization was hotter than expected.[29] This important finding was quickly challenged by the Cambridge astronomers on the grounds that the claimed temperature increment was unrealistically large and not justified by the data.[30] The status of the EoR remains controversial. If verified by other experimenters, the next step will be the even more difficult challenge to map the EoR across the sky with one of the many radio telescopes being built for this purpose.[31]

The Nonthermal Cosmic Background Radiation The apparent existence of a nonthermal cosmic background radiation was first disclosed in centimeter wavelength balloon observations and is supported by older dekameter wavelength ground-based observations.[32] The excess temperature found from the balloon observations appears to be statistically significant, although a systematic error in this one-time experiment cannot be excluded, while the dekameter data depend on observations that are more than 50 years old. So far, the interpretation of the nonthermal cosmic background in terms of known or postulated phenomena remains a challenge.[33] If confirmed, the nature of the nonthermal background radiation will be difficult to understand in terms of the known constituents of the Universe.

The Search for Extraterrestrial Intelligence (SETI) Probably no discovery in science would have a more profound impact on society than the detection of radio signals received from an intelligent extraterrestrial civilization.

The first serious SETI observation in modern times was by 30 year old Frank Drake in 1960 using a new 85 foot radio telescope at the National Radio Astronomy Observatory in Green Bank, West Virginia. Drake observed two nearby stars, Tau CETI and Epsilon Eridani, over a narrow range of frequencies near the 1,420.4 MHz hydrogen line. He chose this frequency, not because of any astronomical reason, but because NRAO was building instrumentation for this frequency to study the radio emission from galactic hydrogen.[34] After 150 hours of observation, including one false alarm, Drake did not find any evidence for intelligent extraterrestrial civilizations. Since Drake's pioneering project, the sensitivity of radio telescopes and the number of frequency channels that can be simultaneously observed has increased so greatly that modern SETI studies, can in one second, reach the detection sensitivity that would have taken Drake 100 million years. There have as yet been no claimed successful detections of extraterrestrial civilization. However, there is probably no other scientific topic devoid of any positive results that has been the subject of more publications and conferences than SETI.

While Drake was still preparing for what he called Project Ozma, two Cornell physicists, Giuseppe Cocconi and Philip Morrison, published their now famous paper, "Searching for Interstellar Communication."[35] Cocconi and Morrison not only speculated on the probability of intelligent life elsewhere in the Universe, but also suggested that, since hydrogen was the most abundant element in the Universe, any extraterrestrial would recognize the hydrogen radio line at 1,420.4 MHz as a unique frequency for interstellar communication. While that might make sense if the extraterrestrials are theoreticians, if they are radio astronomers they would probably recognize that this is the most inappropriate frequency to use for communication. Here on Earth, there is only a single small band of frequencies around 1,420.4 MHz that is reserved by international agreement for radio astronomy, which means no radio transmissions are permitted anywhere in the world at any time in that band. Surely, any advanced extraterrestrial civilization would similarly reserve the 1,420.4 MHz band for radio astronomy and not pollute it with powerful radio transmissions intended for unknown recipients.

Indeed, soon after Project Ozma and the Cocconi and Morrison paper, and following the discovery of intense galactic OH emission near 18 cm wavelength (1.7 GHz) discussed in Chapter 8, Hewlett–Packard Vice President Barney Oliver suggested that the band between the 18 cm OH lines and the 21 cm hydrogen line, that he dubbed the water-hole ($H + OH = H_2O$), was the obvious place that extraterrestrials would transmit just as traditional water-holes were the place for ancient societies to gather. But Charlie Townes, the inventor of the maser and co-inventor of the laser, argued that, instead of looking in the radio spectrum, we should instead look for infrared beacons from powerful extraterrestrial lasers.[36] Moreover, Townes has argued that if lasers had been invented before vacuum tubes and radio technology, the history of SETI might well have developed along different lines.

Over the next half century, suggestions ranged from using neutrinos for interstellar communication, to looking for matter–antimatter annihilations from alien spacecraft propulsion systems, to suggestions that the Fast Radio Bursts discussed in Chapter 7 are really beams used to propel alien intergalactic light sails, to use by aliens of quantum communications. No one has so far suggested looking for gravity wave signals, but they surely will. During the same period, Princeton physicist Freeman Dyson suggested that alien civilizations would leave infrared signatures of their technology, initially called "Dyson spheres," and more broadly recently referred to as "technosignatures."[37]

Over the past half century alone, the "optimum" technique for interstellar communication has apparently changed on a time scale of a decade. As we have discussed in preceding chapters, during this same period, astronomers have not had a good record in predicting newly-discovered astrophysical phenomena. Quasars, pulsars, FRBs, interstellar masers, solar and planetary radio bursts, the rotation of Mercury, superluminal motion, the cosmic microwave background, and dark matter were all discovered serendipitously, largely as a result of newly-developed technology. Why then do we think that we might understand the technology, not to mention the sociology, culture, and motivation, of civilizations many hundreds or thousands of years more advanced than ourselves?

At the 1961 Green Bank *Conference on Extraterrestrial Intelligent Life*, a small group of invited scientists met for a few days to consider the outstanding issues involved in detecting intelligent life elsewhere in the Universe. It was at this meeting that Frank Drake presented his now famous Drake Equation that provided a useful framework to isolate and discuss the issues. Drake estimated the number of communicative civilizations in the Galaxy based on the rate of star formation, the fraction of stars with planetary systems, the number of planets in each system with conditions appropriate for life, the fraction of life-bearing planets that host intelligent life possessing manipulative capabilities, the fraction of planets with intelligence that develops the capability and interest for interstellar communication, and finally the average lifetime of intelligent civilizations.

At the time of the 1961 conference little was known about any of these quantities. We now understand the rate at which new stars are formed during the history of the Galaxy, the number of extrasolar planets, and the number of planets with an Earth-like environment. We can make an intelligent guess about the emergence of technical civilizations. What remains unknown is the lifetime of technical civilizations. If intelligent civilizations are long lived, then Drake's Equation suggests that there should be many technically developed extraterrestrial civilizations capable of interstellar communications. But if the lifetime of technical civilizations is limited by nuclear warfare, disease, or global environmental contamination, that might explain the lack of any successful detection of other intelligent civilizations. Indeed, as early as 1950, the Nobel Prize winning physicist Enrico Fermi famously asked, "Where is everybody?" Fermi's Paradox, as it has become known, has a variety of possible explanations, but with the increasing number of negative SETI searches made over the past 70 years with vastly improved sensitivity, there is increasing speculation asking "Are we alone?"[38]

Perhaps not surprisingly, there have been numerous false alarms generated by the multitude of SETI investigations. The first false alarm came during the first night of Project Ozma. After observing the star Tau Ceti for several hours, Frank Drake turned his telescope toward Epsilon Eridani. Immediately, there was a strong pulsating signal that repeated every 8 seconds – just what he was looking for. According to Drake, "It lasted a few minutes and then disappeared." We were "dumbfounded. Could it be this easy? . . . We were so surprised and unprepared for it, we didn't know what to do."[39] Subsequent observations when the pulsating signal returned a few nights later showed that it was terrestrial interference, probably from an aircraft radar or an orbiting satellite.

Perhaps the best known of the false alarms resulted from the original Cambridge discovery of pulsars. As recounted in Chapter 7, once they had determined that the mysterious pulses were not terrestrial but coming from somewhere in the cosmos, it was very difficult for Jocelyn Bell and her Cambridge colleagues to understand how such narrow pulses could be generated with unprecedented regularity by any natural process. Instead, they speculated that the pulses might be artificial interstellar navigational beacons, much akin to the terrestrial LORAN system operated by the US Navy in the 1960s for marine navigation.

As described in Chapter 4, the surprising detection of the approximately 100-day time variations in the quasar CTA 102 was difficult to comprehend in terms of conventional synchrotron radiation theory. Although later understood to be the result of highly beamed relativistic jets, the speculation that CTA 102 was an extraterrestrial beacon signal was widely reported in the global media and was the subject of the whimsical lyric by the American rock band, *The Byrds*.[40]

Perhaps the SETI account that had the biggest impact on the scientific community was the famous WOW signal reported by Jerry Ehman, an enthusiastic SETI researcher at the Ohio State University's Big Ear radio telescope. Having completed an all sky survey for radio galaxies and quasars, radio astronomy pioneer John Kraus and his former student Robert Dixon led a 25 year SETI program with the Big Ear transit radio telescope. On 15 August 1977, the Big Ear detected a short burst of noise that was confined to a single 10 kHz wide frequency channel near the 1,420.4 MHz hydrogen line. The signal lasted for only about one minute, or the length of time that a celestial body passes through the narrow stationary antenna beam, and Ehman impulsively scribbled "WOW!" on the paper chart record display.[41] John Kraus enthusiastically generated lots of publicity that inspired numerous follow-up observations, none of which showed any evidence for repetition of the WOW signal, which was almost surely instrumental or terrestrial interference.[42]

One of the most interesting SETI results came from the five year search by Paul Horowitz and Carl Sagan using an 8.4 million channel spectrometer. After eliminating sources of terrestrial interference, they recorded 37 events above their detection threshold with "a rough but suggestive correlation of candidate signals with the Galactic plane."[43]

The most powerful SETI program to date is the ten year $100 million *Breakthrough Listen* project, funded by the billionaire Yuri Milner, that is using essentially all of the world's most powerful radio telescopes. Like all the other SETI programs, *Breakthrough Listen* has not detected any signs of extraterrestrial intelligence, but it has been enormously successful in raising broader awareness of SETI and bringing a new generation of students and scientists to the field.

Although the imagination of SETI researchers has been almost unlimited, thinking has been largely constrained by what might be called an expanded anthropomorphic principle. While it is generally accepted that extraterrestrials may not be humanoid in appearance, it is tacitly assumed by most that they are biological. However, in 1960, Stanford professor Ron Bracewell made the innovative suggestion that rather than send one-way signals from their home planet, alien civilizations will more likely send self-replicating probes to other solar systems, thus facilitating practical two-way communication without the long time delay that would be involved in communicating between galactic planetary systems separated by many light years.[44] It is not much of an extrapolation of current technology to anticipate that an advanced computer on the interstellar probe will contain all of the knowledge of the parent civilization – sort of an interstellar Wikipedia – and that the terrestrial-based and probe-based computers will both have a level of artificial intelligence that will develop a common language, and exchange information about their respective histories, science, technology, medicine, economics, and culture. Will humans be involved? Will humans still even exist when the first contact with extra-terrestrials is made? Even if they do, how much of what the terrestrial computers learn from their extraterrestrial counterparts will they share with their human creators?

The Unknown Unknowns As should now be clear to the reader, the discoveries that changed our understanding of the Universe made over the past almost century by radio astronomers were nearly all unpredicted. In Rumsfield's terminology these were the unknown unknowns. In Figure 12.5 we have plotted the cumulative number of discoveries listed in Table 12.1 as a function of time. The rapid rate of discovery during the 1950s and 1960s is apparent, as is the small number of new discoveries during the past three decades. Indeed, during recent decades there have been only four significant discoveries by radio astronomers: exoplanets, the discovery of an extragalactic supermassive black hole, the CMB anisotropy, and Fast Radio Bursts, plus the demonstration by the EHT of the predicted shadow surrounding the black hole at the center of the radio galaxy M87. Are we running out of new phenomena to discover, as Harwit suggests? "Is there anything left for radio astronomers to discover?" Or has the development of the new technology that played such an important role in the earlier discoveries stagnated?[45] Perhaps radio astronomy techniques have become so sophisticated and targeted that they focus on narrowly defined targets, making it difficult to discover anything new or unexpected.

The new radio telescopes that have come on line during the past decade or so, ALMA, ASKAP, CHIME, LOFAR, the LWA, and the MWA, have opened the new windows of millimeter/submillimeter, long wavelength, and transient radio astronomy. Meanwhile the new MeerKAT and the recently upgraded GMRT, and the VLA, along with old stalwarts such as the Jodrell Bank MK I and Parkes 210 foot antennas, equipped with new instrumentation, continue to make new inroads into sensitivity, sky coverage, and spectroscopic capabilities. But if history serves as an example, the excitement of these instruments and, more importantly, the next generation of radio telescopes such as the SKA, ngVLA, and DSA-2000, will not be in the old questions which they answer, but the new questions they will raise and the yet newer instruments that will be built to answer them.

Notes

Introduction

1 Found on a Capetown South Africa restaurant placemat.
2 1 nanometer is 10^{-9} m and is equal to 10 Angstroms. The range 350 to 700 nanometers (3,500 to 7,000 Angstroms) roughly corresponds to the range of radiation visible to the human eye.
3 Merton and Barber (2004), pp. 1–2.
4 William Shakespeare, *The Tragedie of Cymbeline*, Act 4, Scene 3.
5 Translation from an 1854 lecture by Louis Pasteur at the University of Lille, "Dans les champs de l'observation le hasard ne favorise que les esprits préparés."
6 Merton and Barber (2004), pp. 233–241.

Chapter 1

1 Parts of this chapter are taken from Kellermann (1999, 2003, 2004) and Kellermann, Bouton, and Brandt (2020), pp. 1–33. More detailed accounts of the events described in this chapter are found in C. M. Jansky, Jr. (1958), Reber (1958, 1984), and Sullivan (1984), Sullivan (2009), pp. 29–79.
2 Taken from Jansky (1932).
3 Bown (1927). Throughout this book we give the frequency of radio waves in kilohertz (kHz), Megahertz (MHz), or Gigahertz (GHz), with the corresponding wavelength in meters, centimeters (cm), or millimeters (mm) given in parentheses or, where appropriate, in the reverse order. The relation between frequency, ν, and wavelength, λ, is given by $\nu = c/\lambda$, where c is the speed of light. With c = 300,000,000 m/second, and a wavelength of 10 m, ν = 30 MHz. Radio astronomers and radio engineers have the bad habit of using frequency and wavelength interchangeably. Prior to the adoption of the Hertz in the International System of Units (SI) in 1960, the units of frequency were expressed in cycles per second or, as appropriate, kilocycles per second (kc/s) and Megacycles per second (Mc/s), although these units were often carelessly written as just "kc" and "Mc."
4 The use of shortwave radio communication was pioneered by the amateur radio community during the mid-1920s. See, e.g., DeSoto (1936). Southworth (1962) gives a good account of the development of radio telephony at AT&T and the migration from long waves to shortwaves. See also Beck (1984).
5 Beck (1984). Similar effects were apparently widely known to radio amateurs.
6 Karl's older brother, Cyril Moreau Jansky, Jr. (known as C. M. Jansky and to his friends and to family as "Moreau") was an influential leader in the electronics industry.
7 AT&T's R. Kestenbaum interview with Bell Labs colleagues Harald Friis, Al Beck, John Schelleng, and Arthur Crawford, 10 February 1965, NAA-Jansky Papers, Oral Histories.
8 We are indebted to former Bell Labs scientists, Tony Tyson and Robert Wilson, and to University of Washington's Woodruff T. Sullivan III, for providing us with copies of Jansky's work reports, to the University of Wisconsin Archivist Bernard Schermetzler for allowing us to make copies of Jansky's letters, and to Karl's late wife, Alice, as well as his son and daughter, David and Ann

Moreau, who have kindly shared with us other family correspondence. The original letters are preserved at the University of Wisconsin Archives. Fortunately, Karl's family kept his letters, which were found in the attic of the family home after the house was sold, and the new owner, recognizing their significance, donated the letters to the University Archives. Copies of Jansky's correspondence, work reports, and notebook entries are held in the NRAO/AUI Historical Archives in Charlottesville, Virginia: NAA-Jansky Papers, NAA-Kellermann Papers, and NAA-Sullivan Papers, which also contains Sullivan's Working Files and his interviews with Jansky's Bell Labs colleagues.

9 Southworth (1956).
10 KGJ to family, 23 September 1928.
11 KGJ to family, 5 December 1930.
12 KGJ to family, 18 January 1932.
13 The International Scientific Radio Union is one of many international scientific unions. Members resident in the United States are organized as a formal Committee of the National Academy of Science and continue to meet at least annually. In 1948, URSI established a new commission named "Extra-terrestrial Radio Noise," that was renamed in 1950 as the commission on "Radio Astronomy." The acronym URSI is derived from the French name, *Union Radio-Scientifique Internationale*.
14 Jansky (1932).
15 KGJ Work Report for September 1932.
16 Southworth (1956).
17 KGJ to parents, 21 December 1932.
18 KGJ to family, 15 February 1933. Like many people do early in the calendar year, Jansky apparently miswrote the date of his letter as 1932.
19 KGJ to parents, 10 June 1933.
20 KGJ to parents, 5 May 1933.
21 WJZ was then part of the RCA's NBC Blue Network, that later became ABC.
22 C. M. Jansky Jr. to C. M. Jansky, 16 May 1933; KGJ to parents, 25 May 1933.
23 KGJ to his parents, 10 May 1933.
24 Jansky (1933a).
25 Jansky (1932).
26 Jansky (1933b).
27 Jansky (1933c).
28 Kraus (1984).
29 Jansky (1935).
30 Sullivan (1978).
31 G. Reber to J. Pfeiffer, 31 January 1955. No record of such a proposal has been found in the Bell Labs Archives or among any of Karl Jansky's existing personal papers and letters.
32 Friis and Feldman (1937). MUSA, consisted of six phased rhombic antennas located near Manahawkin, New Jersey, was used during WWII by President Roosevelt to communicate with British Prime Minister Winston Churchill.
33 Jansky (1937, 1939).
34 Karl's brother, Moreau, was perhaps the first person to document the existence of shortwave "skip zones" between nearby "ground wave" and long distance "sky wave" regions of enhanced radio reception, which was later understood to be the result of refraction in the ionosphere. DeSoto (1936), p. 103.
35 Pfeiffer (1956), pp. 16–17.
36 Edmondson (1956).
37 C. M. Jansky Jr. (1958).
38 Friis (1965).
39 Friis (1965, 1971). In a 1965 interview with Ray Kestenbaum of Bell Labs Public Relations Department, Friis firmly and repeatedly claimed that Jansky never asked to continue his star noise work. NAA-Jansky Papers, Oral Histories.
40 Alice Jansky Knopp to C. M. Jansky, Jr., 11 February 1958.

41 L. Espenschied to G. Southworth, 21 February 1965. NAA-Kellermann Papers, Publications, Open Skies, chapter 1.
42 G. Southworth to L. Espenschied, 23 February 1965. NAA-Kellermann Papers, Publications, Open Skies, chapter 1.
43 Russell Ohl, oral history conducted in 1965 by F. Polkinghorn, IEEE History Center, Hoboken, NJ; R. Ohl interview by L. Hoddeson, 20 August 1976, American Institute of Physics.
44 KGJ to family, 15 February 1933 (erroneously dated 1932).
45 www.nobelprize.org/nomination/archive/show_people.php?id=4554 (last accessed 21 November 2022).
46 Contopoulos and Jappel (1974), p. 166.
47 In 1994, Grote Reber donated most of his papers, notes, books, chart recordings, and other records, along with his Wheaton receivers and other instrumentation from the 1930s and 1940s, to NRAO/AUI. The receivers and other equipment as well as many personal artifacts are housed in the Green Bank Observatory Science Center. His papers, interviews, and some artifacts are kept at the NRAO/AUI Archives and may be searched at www.nrao.edu/archives/reber-finding-aid (last accessed 21 November 2022).
48 Reber's amateur radio call sign, W9GFZ, is now assigned to the National Radio Astronomy Observatory Amateur Radio Club and is used from time to time for amateur radio communications.
49 Reber (1958, 1984).
50 KIK interview with Grote Reber, 13 June 1994. NAA-Kellermann Papers, Oral Histories.
51 Alberta Adamson interview with Grote Reber at Wheaton History Center, 19 October 1985, transcribed at NRAO. NAA-Reber Papers, Oral History Interviews.
52 Reber (1958).
53 Reber (1958, 1984).
54 Reber's paper can be found at the NAA at www.nrao.edu/archives/items/show/13224 (last accessed 21 November 2022).
55 Reber to KGJ, 26 April 1937, NAA-Reber Papers, General Correspondence I, www.nrao.edu/archives/items/show/18458 (last accessed 21 November 2022).
56 Grote's father had died in 1933, shortly after Grote graduated from college. The original site of Reber's dish is now a parking lot for the local telephone company.
57 Reber built the square backup structure out of 20 foot long standard 2 by 4s. With a 28 foot diagonal plus a 1.5 foot overhang his circular disk had a final diameter of 31 feet 5 inches.
58 The dependence of intensity on frequency and temperature is governed by Planck's black-body radiation law. At radio frequencies where $h\nu \ll kT$, the intensity is given by the Raleigh-Jeans radiation law, $I(T) = 2\nu^2kT/c^2$, where ν is the frequency in Hertz, $k = 1.38 \times 10^{-23}$ Joules/K is the Boltzmann constant, and the speed of light, $c = 300{,}000$ km/s.
59 The angular resolution of a radio telescope is proportional to frequency and inversely proportional to the diameter of the telescope. The resolution of Reber's 31 foot diameter dish was 0.5, 4.5, 13 degrees at 3.3 GHz, 480 MHz, and 160 MHz, respectively. In his various publications, e.g., Reber (1940a, 1940b) incorrectly claimed that his resolution or "acceptance cone" at 160 MHz was "about 3 degrees," which, as pointed out by Sullivan (1982), p. 41 and others, is far smaller than theoretically possible. It is hard to understand how Reber could have made such an elementary error.
60 Reber (1958).
61 In 2021, it still takes 56 minutes each way to travel between Wheaton and Chicago.
62 Scuyler Reber to G. Reber, 16 December 1938, NAA-Reber Papers, General Correspondence I, www.nrao.edu/archives/items/show/18467 (last accessed 21 November 2022).
63 F. Whipple to Reber, 22 December 1938, NAA-Reber Papers, General Correspondence I, www.nrao.edu/archives/items/show/18468 (last accessed 21 November 2022).
64 Sullivan interview with Grote Reber, 25 October 1975, NAA-Sullivan Papers, Working Files, Individuals, www.nrao.edu/archives/items/show/15140 (last accessed 21 November 2022).
65 JLG, "I Was There in the Early Years of Radio Astronomy," unpublished manuscript. NAA-Kellermann Papers, Collected Publications.

66 Alberta Adamson interview with Grote Reber at Wheaton History Center, 19 October 1985, transcribed at NRAO. NAA-Reber Papers Oral History Interviews.
67 Levy (1993).
68 Reber (1940a).
69 Reber suggested that the galactic radio emission that he observed could be explained as the result of the interaction of electrons and protons in the ionized interstellar medium, assuming that the interstellar plasma was at a temperature of 10,000 to 15,000 consistent with what was known at the time. This is commonly called magneto thermal or bremsstrahlung radiation, as the electron velocities are determined by the temperature of the interstellar plasma.
70 Reber (1940b).
71 Reber (1982).
72 JLG, "I Was There in the Early Years of Radio Astronomy," unpublished manuscript. NAA-Kellermann Papers, Collected Publications.
73 Reber and Greenstein (1947).
74 Reber (1942).
75 Reber (1944). This paper was also delayed by Struve due to Reber's rather cavalier attitude toward preparation of illustrations for the paper.
76 Reber (1944, 1948a).
77 Reber (1948a). See Chapters 3 and 4 for the subsequent identification of these sources of radio emission with their optical counterparts.
78 Reber (1948b).
79 Reber (1948b, 1949a, 1949b, 1950).
80 Reber and Ellis (1956).
81 See George, Orchiston, and Wielebinski (2017) for a more detailed discussion of Grote Reber's time in Tasmania.
82 Reber (1984).
83 Jeff Reber to KIK, 24 June 2003, NAA-Kellermann Papers, Reber.
84 www.nobelprize.org/nomination/archive/show_people.php?id=7597 (last accessed 21 November 2022).
85 Reber (1966).
86 Reber (1964) claimed that bean pods that grew on vines that he forced to wind in the opposite direction from their natural right-hand screw pattern had a larger ratio of beans to shucks than a control sample that grew on vines that were allowed to wind naturally.
87 Cyril M. Jansky Sr. to KGJ, 18 May 1933, NAA-Kellermann Papers, Jansky.
88 Reber (1984).
89 Zwicky (1969), p. 90; Reber (1984); Greenstein (1984). Millikan had won fame for his oil drop experiment to measure the charge on the electron, and he ruled Caltech with an iron hand. Zwicky later claimed that they could have done their experiment for only $200, although Greenstein and Potapenko later estimated that it would have cost closer to $1,000.
90 Sullivan (2009), p. 44.
91 Sullivan (2009), p. 113; NAA-Sullivan Papers, Working Files, Individuals, DeWitt; Fränz (1942), Sander (1946).
92 Reber (1940a, 1940b) reported his results in terms of watts per square centimeter per circular degree per kilocycle bandwidth, a forerunner of the modern Jansky, that is defined as 10^{-26} Watts per square meter per Hertz.
93 KGJ to his parents, 22 January 1934.
94 Jansky (1935).
95 Henyey and Keenan (1940).
96 Langer (1936).
97 Whipple and Greenstein (1937).
98 Townes (1946).
99 KIK interview with Grote Reber, 25 May 1991, NAA-Kellermann Papers, Oral Interviews.
100 van de Hulst (1945).
101 Elder, Langmuir, and Pollock (1948).
102 Schott (1912), Vladimirsky (1948), Schwinger (1949).

103 With $E \sim 1$ GeV and $B \sim 10^{-4}$ Gauss, $\nu_{max} \sim 1$ GHz.
104 Alfvén and Herlofson (1950), Kiepenheuer (1950).
105 Ginzburg (1951), Shklovsky (1952).
106 Getmantsev (1952).

Chapter 2

1 Hey (1973), p. 16.
2 See, e.g., Sullivan (1982), p. 145; Sullivan (2009), p. 83.
3 Wilsing and Scheiner (1896).
4 Nordmann (1902). See also Débarbat, Lequeux, and Orchiston (2007) and Sullivan (2009), p. 23.
5 Deslandres and Décombre (1902). In those early years, the terms "Hertzian waves" or "electromagnetic waves" were used to describe what were later called "radio waves."
6 Kraus (1984); Kraus (1995), p. 30.
7 Sullivan (2009), p. 113.
8 Heightman (1946); Ham (1975).
9 Appleton and Hey (1946).
10 Nakagami and Miya (1939). This story was related by Tanaka (1984), who heard it from Miya after Nakagami had died in an accident.
11 Orchiston (2005a); Sullivan (2009), p. 83.
12 The development of radar and its impact on the Second World War and radio astronomy is told by Buderi (1996).
13 Schott (1947). We are indebted to Woody Sullivan for providing an English translation of this paper.
14 Hey (1973), p. 15.
15 Churchill (1950), p. 113.
16 *New York Times*, 13 February 1942; Churchill (1950), p. 114.
17 Churchill (1950), p. 115.
18 Hey (1973), pp. 14–16.
19 Hey's March 1942 secret report, "Notes on G. L.[Gun Laying] interference on 27 and 28th February," AVIA 7/3544, is filed at the UK Public Record Office in Kew, England. Copies are found at the Jodrell Bank Library in Manchester, at the CSIRO ATNF library in Sydney, Australia, and the NRAO-AUI Archives (including a copy of Hey's handwritten draft), NAA-Sullivan Papers, Sullivan Publications, Classics in Radio Astronomy. See also Sullivan (2009), p. 81.
20 Hey (1946).
21 Appleton (1945).
22 Appleton and Hey (1946).
23 Hey (1973), p. 18.
24 In Hey's 1945 report, Appleton's name is listed at the end of the report among those sent copies. NAA-Sullivan Papers, Sullivan Publications, Classics in Radio Astronomy.
25 Sullivan (2009), pp. 90–91.
26 Heightman (1946).
27 Edlén (1946); Stratton (1946).
28 Elizabeth Alexander's unique career is described by her daughter, Mary Harris (2019). See also Orchiston and Slee (2002); Orchiston (2005b); and Sullivan (2009), pp. 85, 128.
29 Alexander (1945, 1946).
30 Orchiston and Slee (2002); Orchiston (2004).
31 Orchiston (2004) gives a detailed account of Slee's extensive contributions to radio astronomy.
32 Southworth (1945). See also Southworth (1962), p. 247 for a personal account of Southworth's pioneering investigation.
33 Southworth (1956).
34 Sullivan (2009), pp. 97–99.
35 Dicke and Beringer (1946).
36 Reber (1944).

37 Here, Reber (1944) is using traditional optical magnitudes where each five magnitudes is a factor of 100.
38 Reber (1946); Reber and Greenstein (1947).
39 Reber, private communication to Kellermann, 1965.
40 Reber (1946); See also Reber (1958).
41 Reber (1946).
42 Reber (1948c).
43 Sullivan (2009), pp. 84, 228.
44 Charles Seeger, the older brother of the well-known folk singer, Pete Seeger, began radio observations of the Sun at Cornell University in 1946. Later at Cornell, Marshall Cohen developed an innovative polarimeter that he used to study the polarization of solar radio bursts at 201 MHz (1.5 m), Cohen (1958a, 1958b).
45 See, e.g., Frater, Goss, and Wendt (2017), p. 139.
46 Covington (1984a, 1984b, 1988) has described the early development of Canadian solar radio astronomy.
47 Denisse (1984); Orchiston et al. (2009).
48 Tanaka (1984); Orchiston, Nakamura, and Ishiguro (2016).
49 Lovell (1977); Bowen (1987).
50 Joe Pawsey's life and career are described by Lovell (1964) and by Goss, Hooker, and Ekers (2023).
51 Martyn (1946); Pawsey (1946). Goss, Hooker, and Ekers (2023), chapter 14, discuss the Australian solar observations in more detail, concentrating on the controversy between Martyn and Pawsey for priority. Although both Pawsey and Martyn seemingly ignored the earlier report of Reber's (1944) detection of the radio corona, Pawsey, at least, was aware of Reber's paper, as evidenced by his incorrect criticism of Reber's conclusion that the observed galactic radio emission cannot be the sum of the radio emission from many stars like the Sun.
52 Pawsey, Payne-Scott, and McCready (1946).
53 McCready, Pawsey, and Payne-Scott (1947). Apparently, this now classic paper was originally submitted with Joe Pawsey as the first author but, following journal policy, the editors listed the authors alphabetically.
54 Payne-Scott, Yabsley, and Bolton (1947).
55 Wild (1950); Wild and McCready (1950); Boischot (1957); Boischot and Denisse (1957). See also Frater and Ekers (2012) and Frater, Goss, and Wendt (2017), p. 91.
56 *New York Times*, 29 August 2018. See also Goss and McGee (2010) and Goss (2013) for a more detailed account of the life and career of Ruby Payne-Scott.
57 See Bowen (1987) for an autobiographical account of Bowen's life and career.
58 See Frater and Ekers (2012) and Frater, Goss, and Wendt (2017), p. 91 for further details of Paul Wild's career.
59 Martin Ryle's life and career are described by his brother-in-law Sir Francis Graham-Smith (1986).
60 Michelson (1920), Michelson and Pease (1921).
61 Ryle and Vonberg (1946).
62 Ginzburg (1946); Shklovsky (1946).
63 Khaikin and Chikhachev (1948). Further background on the Russian eclipse expedition is given by Dogel et al. (2012). See also Sullivan (2009), p. 217.
64 Hagen et al. (1948) was based on a previous classified report. See also Sullivan (2009), p. 207.
65 Hagen (1951) and Reber and Beck (1951) presented the results of their Attu eclipse expedition at a meeting of the American Astronomical Society at Haverford, Pennsylvania. Hagen, Haddock, and Reber (1951) later reported on the weather difficulties encountered.
66 Orchiston and Steinberg (2007).
67 Pick et al. (2011).
68 Covington (1984a).
69 Covington (1947).
70 Covington (1948).
71 The history of the 2,800 MHz Canadian solar intensity monitoring is told by Covington (1984a). At the 1984 SDRA Workshop in Green Bank, after Covington's talk, Hanbury Brown pointed out

that the magnetron looks like it was bored on the same machine used to make Colt revolvers and suggested that the 10 cm wavelength was determined by the size of the Colt cartridge. Alan Moffet added (p. 113) that he had heard that the wavelength (size) of the magnetron was set by the size of the British 1 pence piece that was used as a copper plate.

72 Wild, Smerd, and Weiss (1963) and Bastian, Benz, and Gary (1998) discuss the types of observed solar radio emission and their implication for understanding the complex physical processes in the Sun.
73 Christiansen and Warburton (1953, 1955); Christiansen, Mathewson, and Pawsey (1957); Christiansen (1984).
74 Wild (1967). See also Frater, Goss, and Wendt (2017), p. 108.
75 Bracewell and Swarup (1961); Bracewell (2005).
76 Maxwell, Swarup, and Thompson (1958).
77 In the twenty-first century, the National Science Foundation funded a long needed major upgrade to the Very Large Array in New Mexico and contributed to the construction of the $1.2 billion dollar Atacama Large Millimeter Array that was built in Chile. At the same time, the NSF funded built the D. K. Inoue Solar Telescope (DKIST) on Haleakala on the island of Maui in Hawaii, and NASA has funded the Solar Dynamics Observatory (SDO), the Solar and Heliospheric Observatory (SOHO), and the Solar Terrestrial Relations Observatory (STEREO), covering a wide range of non-radio wavelengths.
78 Lovell (1969).
79 Davis et al. (1978).
80 Kellermann and Pauliny-Toth (1966b); Seaquist (1967). Seaquist also reported a marginal detection of radio emission from the red supergiant π Aurigae.
81 Hjellming (1988).
82 Hjellming and Wade (1970).
83 Wade (1984).
84 Wade and Hjellming (1970).
85 Gower et al. (1967).
86 Clark and Murdin (1978).
87 Stephenson and Sanduleak (1977).
88 Fabian and Rees (1979); Margon et al. (1979); Hjellming and Johnston (1981); Jeffrey et al. (2016).
89 Mirabel and Rodriguez (1994).
90 See Güdel (2002) for a review of stellar radio emission.

Chapter 3

1 Bolton, Stanley, and Slee (1949).
2 Hey (1973).
3 Grote Reber's 161 MHz map of the Cygnus region is confused by the emission from the extended galactic source Cygnus X (Reber 1944). However, his 408 MHz map, made with better angular resolution, shows the two peaks corresponding to Cygnus X and what was later recognized as the radio galaxy Cygnus A (Reber 1948a).
4 Hey, Parsons, and Phillips (1946); Hey, Phillips, and Parsons (1946); Sullivan (2009), p. 103. Hey inferred that the intensity of a source could not vary significantly on times shorter than the light (or radio) travel time across the source. Otherwise signals arriving at the same time would be coming from different parts of the source where the intensity of the emission would be different and the variations would appear smeared out in time.
5 John Bolton's life and radio astronomy career are described by Kellermann (1996) and Robertson (2017).
6 Gordon Stanley's career is summarized in Kellermann, Orchiston, and Slee (2005).
7 Bolton (1982).
8 Bolton (1982).
9 Bolton and Stanley (1948a, 1948b); Bolton (1982); Stanley (1994); Slee (1994); Robertson, Orchiston, and Slee (2014); Robertson (2017), p. 72. In fact, the Bolton and Slee position was in error by about one degree. See Sullivan (2009), p. 138 and p. 317.

10 Bolton and Stanley (1948b).
11 Bolton (1948, 1982). Three of the claimed six new sources were never confirmed and were probably blends of weaker sources.
12 Bolton (1982).
13 Bruce Slee's long career in radio astronomy is documented by Orchiston (2005b).
14 Bolton and Stanley (1948b); Bolton (1982). These results were later confirmed by simultaneous observations made at Cambridge and Jodrell Bank with antenna separations of about 200 miles Graham-Smith (1950); Little and Lovell (1950).
15 For extended objects like planets, the path through the atmosphere is different for light coming from different parts of the planet, and so the twinkling is different, and tends to average out when averaged across the planet's disk. The same is true for extended radio sources, so only small diameter radio sources are observed to fluctuate or scintillate.
16 Bolton (1973). The Crab Nebula was well known to astronomers as the remains of a supernova that exploded in the year 1052 and was so bright that it was visible to Chinese astronomers in the daytime.
17 Bolton and Stanley (1949).
18 Bolton, Stanley, and Slee (1949).
19 Evans (1949). Bolton to Oort, 5 November 1949, Papers of Jan Henrick Oort, Department of Western Manuscripts, University Library, Leiden, File 102a. We are grateful to Miller Goss for bringing this letter to our attention.
20 Bolton, Stanley, and Slee (1954); Bolton (1955).
21 November 1959 Bolton comment to KIK. At the time KIK was Bolton's student at Caltech.
22 KIK interview with J. G. Bolton, August 1989. It was unlikely that this or other Radiophysics papers of that era were sent for peer review, although there was a rigorous internal review process before any paper was sent out for publication. It is possible that Bolton was concerned about this internal review, probably by Pawsey.
23 Bolton (1990).
24 Stanley and Slee (1950); Bolton, Stanley, and Slee (1954).
25 Martin Ryle held the British amateur radio license G3CY.
26 Ryle (1949).
27 Ryle, Smith, and Elsmore (1950).
28 Using a 218 foot parabolic dish, Hanbury Brown and Hazard (1950) barely detected the radio emission from the Andromeda Nebula (M31), the closest galaxy to the Milky Way in the northern hemisphere, or from another nearby spiral galaxy, M81. Hanbury Brown and Hazard (1953a).
29 Gold (1951); Edge and Mulkay (1976), p. 97.
30 Bolton (1990).
31 Mills and Thomas (1951). David Dewhurst (1951) independently suggested an identification with the same galaxy.
32 Ryle and Smith (1948).
33 Graham-Smith (1951, 1952).
34 Mills (1952a).
35 Spitzer and Baade (1951).
36 Baade and Minkowski (1954).
37 Hanbury Brown and Hazard (1952).
38 Minkowski (1959).
39 Shklovsky (1960a).
40 Högbom and Shakeshaft (1961); Findlay, Hvatum, and Waltman (1965).
41 Edge et al. (1959).
42 Elsmore, Ryle, and Leslie (1959); Bennett (1962).
43 For a history of the Owens Valley Radio Observatory see Cohen (1994, 2007) and Kellermann, Bouton, and Brandt (2020), pp. 295–300.
44 Morris, Palmer, and Thompson (1957); Allen et al. (1962a, 1962b).
45 Minkowski (1960a, 1960b).
46 Anderson, Palmer, and Rowson (1962).
47 The Caltech OVRO interferometer observations, e.g., Maltby, Matthews, and Moffet (1963); Matthews, Morgan, and Schmidt (1964); Moffet (1966) had a resolution of about 1 arcminute.

The Jodrell Bank work used much longer baselines with resolutions as good as 10 arcseconds, but gave incomplete structural information, e.g., note 38.
48 Burbidge (1959a, 1959b). See also Moffet (1966) for energy estimates based on the newer OVRO interferometry measurements.
49 Hargrave and Ryle (1974).
50 Perley, Dreher, and Cowan (1984).
51 Shklovsky (1953a); Mayer, McCullough, and Sloanaker (1957), and references therein.
52 Mayer, McCullough, and Sloanaker (1962a).
53 Pawsey and Bracewell (1955); Bracewell (2000). See also Thompson and Frater (2010) for a discussion of Bracewell's life and career.
54 Little and Bracewell (1961).
55 Whiteoak (1994).
56 Bracewell (2002) gives his account of the Parkes polarization observations. Somewhat conflicting stories of who did what, when they did it, and why they did it, are given by Price (1984); Robertson (1992), p. 223; Whiteoak (1994), Haynes et al. (1996), p. 251; Cooper (1998); and Robertson (2017), p. 235.
57 Bracewell, Cooper, and Cousins (1962).
58 Robertson (2017), p. 237.
59 Price (1984).
60 Shklovsky (1960b), p. 199.
61 Cooper and Price (1962); Price (1984).
62 Gardner and Whiteoak (1962).
63 Bracewell, Cooper, and Cousins (1962), Bracewell (2002).
64 Mayer, McCullough, and Sloanaker (1962b) for an abstract of the 1962 presentation. The full paper was not published until much later by Mayer, McCullough, and Sloanaker (1964).
65 Haynes et al. (1996), p. 251; Robertson (2017), p. 238. Frater, Goss, and Wendt (2017), p. 143 claim to have seen evidence that Bowen indeed, "persuaded the editors of *Nature* to delay publication of the Bracewell, Cooper, and Cousins letter."
66 See, e.g., Wielebinski (2012).

Chapter 4

1 See note 15.
2 Seyfert (1943).
3 Shklovsky (1962).
4 Jeans (1929).
5 Ambartsumian (1958).
6 Although Allen et al. (1962a, 1962b) and Rowson (1962) were not published until later, Henry Palmer had shared the Jodrell Bank measurements with Alan Moffet at the IAU General Assembly held in Berkeley, CA in August 1961.
7 Bolton to Pawsey, 16 November 1960. National Archives of Australia, C3830 z3/1/X. We are grateful to Miller Goss for making these letters available.
8 Bolton (1990).
9 Bolton to Pawsey, 19 December 1960. National Archives of Australia, C3830 z3/1/X.
10 Matthews et al. (1960). At the time the AAS did not publish abstracts of late papers, and the AAS records of the 107th AAS meeting have been mysteriously lost.
11 Un-authored article in *Sky and Telescope*, vol. 21, p. 148.
12 Greenstein (1961). The only other contemporaneous published report of 3C 48 as a star appears to be in the 1960–1961 Annual Report of the Director, Mount Wilson and Palomar Observatories (Washington: Carnegie Institution of Washington), p. 80.
13 Matthews and Sandage (1963) was not published until July 1963, with a section discussing "3C 48 as a Galaxy" added in proof after its redshift had been determined by Greenstein and Schmidt (1963).
14 Greenstein, J. L., The Radio Star 3C 48. Unpublished 1962 preprint. NRAO/AUI Archives, papers of K. I. Kellermann, Star Noise. See also note 37.

15 Greenstein, J. L., private communication to the author, January 1995. NRAO/AUI Archives, papers of K. I. Kellermann, Star Noise.
16 Allen et al. (1962a, 1962b).
17 Lequeux (1962).
18 Read (1962, 1963); Fomalont et al. (1964). Although not published until 1964, the OVRO interferometer positions were available much earlier to the Caltech and Mount Wilson/Palomar astronomers.
19 Read (1963).
20 Schmidt to Kellermann, 22 November 1993, and a copy of the Palomar 200 inch observing logs kindly made available by Maarten Schmidt. NRAO/AUI Archives, Papers of K. I. Kellermann, Star Noise.
21 Hazard (1962).
22 The date is erroneously given as 15 April in the publication by Hazard, Mackey, and Shimmins (1963).
23 Both the Radiophysics Chief, Taffy Bowen, and the University Head of Physics, Harry Messel, had strong egos. They competed for funds and recognition and greatly disliked each other (Robertson 2017), p. 253.
24 Bolton (1990). Hazard et al. (2018) have also discussed the three occultations of 3C 273 and the subsequent events that led to its optical identification.
25 Data from the May occultation were recorded at 410 MHz (73 cm) only.
26 The predicted time of emersion was in error, as it was based on an incorrect position of the Parkes telescope. See Hazard et al. (2018).
27 An additional occultation was observed on 26 October at 410 MHz and 1,420 MHz (21 cm) when only the immersion was visible from Parkes.
28 Hazard, Mackey, and Shimmins (1963).
29 The Polaroid print that Minkowski had with him in Australia was most likely a copy of a 200 inch plate taken by Schmidt in March 1962.
30 Letter from John Bolton to Maarten Schmidt, 20 August 1962. NRAO/AUI Archives, Papers of K. I. Kellermann, Star Noise. We are grateful to Maarten Schmidt for making this and other relevant correspondence available to us.
31 Schmidt (1984, 2011).
32 Schmidt to Bolton, 8 January 1963. NRAO/AUI Archives, Papers of K. I. Kellermann, Star Noise.
33 In various places Schmidt reports that this occurred on 5 or 6 February, and later he acknowledged that he was unclear on the precise date.
34 The new position was determined by W. Nicholson at the UK Nautical Almanac Office using data from the Parkes August and October occultations. Bolton to Schmidt, 26 January 1963. NRAO/AUI Archives, Papers of K. I. Kellermann, Star Noise. Bolton suggested that "maybe it's moving" to explain a small sub-arcsecond discrepancy between the positions derived from the different occultations and between the best radio and optical positions, and concluded with, "I can then get back to my cyclotron stars." A day earlier, Bolton gave a lecture to a group of undergraduate students on "Observing Radio Sources," and stated that, "4 genuine radio stars have been identified." At that time only three "radio stars" were known, 3C 48, 3C 196, and 3C 286, so Bolton apparently still considered 3C 273 to be among the class of galactic radio stars. Ronald Ekers, who was an undergraduate student at the time, attended the lecture, Ekers to Kellermann, March 2013. On 14 February 1963, Bolton wrote again to Schmidt, "I believe the star is the right choice." Hazard also stated that, at the time, Bolton told him that he believed that many radio sources were associated with galactic stars. Kellermann interview with C. Hazard, 22 March 2013. NRAO/AUI Archives, Papers of K. I. Kellermann. Oral Histories.
35 Like M 87, the radio and optical morphologies of 3C 273 are nearly identical.
36 Schmidt (1984, 2011).
37 Earlier, Greenstein had circulated a preprint of his paper among the Caltech faculty, but when he realized that his two year effort to understand the spectrum of 3C 48 was all wrong, he meticulously went around and collected all of the preprints he had given out. However, one of the present authors, KIK, was a graduate student at the time and he had made and carefully retained an undocumented photocopy of Tom Matthew's preprint copy. NRAO/AUI Archives Papers of K. I. Kellermann, Star Noise.

38 Schmidt to Hazard, 8 February 1963. NRAO/AUI Archives, Papers of K. I. Kellermann, Star Noise.
39 Hazard, Mackey, and Shimmins (1963).
40 Schmidt (1963).
41 Oke (1963). Oke did not report when his 100 inch observations were made. Sandage (1999), p. 477, has claimed that, "The breaking of the redshift code in the spectrum of 3C 273 was done by Oke and Schmidt together when they combined their wavelength data and realized that they were seeing the hydrogen Balmer series."
42 Greenstein and Matthews (1963).
43 Copy of original manuscript submitted to *Nature* by the Radiophysics Lab publication office. NRAO/AUI Archives, Papers of K. I. Kellermann, Star Noise. We thank Miller Goss for making this paper available to us. In their paper, Haynes and Haynes (1993) interpreted the word "delete" to mean that it was redundant, since Hazard acknowledges in the text of the paper that he was a guest observer from the University of Sydney. *Nature* (1963, vol. 198, p. 19) later issued an erratum clarifying Hazard's correct affiliation.
44 Greenstein and Schmidt (1963).
45 Chiu (1964).
46 Schmidt (1970).
47 Smith and Hoffleit (1963).
48 Schmidt (2011).
49 Letter from Greenstein to Kellermann, 30 March 1996. NRAO/AUI Archives, Papers of K. I. Kellermann, Star Noise.
50 Matthews interview with Kellermann on 28 April 2012. NRAO/AUI Archives, Papers of K. I. Kellermann, Oral Histories.
51 Bolton (1990).
52 Bolton to Sullivan, 28 February 1990, NRAO/AUI Archives, Papers of W. T. Sullivan III, Working Files, Individuals.
53 Oke to Alar Toomre, 1975, conveyed to Kellermann, 3 January 2021, NRAO/AUI Archives, Papers of K. I. Kellermann, Star Noise.
54 E.g., Waluska (2007). The Mount Wilson and Palomar Observatories were later renamed the Hale Observatories.
55 Sandage (1965).
56 Zwicky (1963). Paper presented at the 113th meeting of the AAS in Tucson, AZ. Zwicky had a long history of claiming prior discovery of many phenomena, including dark matter and the polarization of distant galaxies. Although largely rejected or ignored by the astronomy community, including the Caltech and Mount Wilson and Palomar astronomers, history has shown many of Zwicky's claims to have been valid.
57 Zwicky (1965).
58 Zwicky and Zwicky (1971), p. xix.
59 Kinman (1965); Lynds and Villere (1965).
60 Kellermann et al. (1989, 2016).
61 Stebbins and Whitford (1947); Reber (1948a).
62 Piddington and Minnett (1951); McGee and Bolton (1954). Goss and McGee (1996) and Goss, Brown, and Lo (2003) have given detailed historical accounts of the discovery of Sgr A and Sgr A*, respectively.
63 McGee and Bolton (1954).
64 See Chapter 7.
65 Blaauw et al. (1960).
66 Balick and Brown (1974). The infrared source had been discovered the previous year by George Rieke and Frank Low (Rieke and Low 1973).
67 Goss, Brown, and Lo (2003).
68 Balick interview with KIK, 15 October 2020.
69 Kellermann et al. (1977).
70 Lynden-Bell and Rees (1971).
71 Andrea Ghez and Reinhard Genzel shared half of the 2020 Nobel Prize in Physics for their leadership and meticulous observations made with the Keck 10 m and ESO 8 m telescopes,

respectively. Ghez and Genzel, along with their teams, independently traced the orbits of stars within a few hundred Astronomical Units from the Galactic Center and showed that the orbits enclosed a mass about four million times the mass of the Sun.

72 Kendra Shifflett (neé Brown) to K. Kellermann, 7 October 2020.
73 Akiyama et al. (2022).
74 Balick (2005).
75 From the 85 MHz surveys of Mills, Slee, and Hill (1958, 1960, 1961). Prepublication copies of these Mills Cross surveys were made available to the Caltech group.
76 Harris and Roberts (1960).
77 Kellermann et al. (1962).
78 Bolton, Gardner, and Mackey (1963, 1964).
79 Price (1969).
80 See Chapter 6 for the discovery of the Cosmic Microwave Background by Penzias and Wilson (1965).
81 Kardashev (1962); Conway, Kellermann, and Long (1963); Kellermann, Pauliny-Toth, and Williams (1969).
82 Shimmins, Bolton, and Wall (1968).
83 Pauliny-Toth et al. (1972).
84 Allen et al. (1962a, 1962b).
85 Kellermann et al. (1962).
86 Slysh (1963).
87 Shklovsky (1965a, 1965b).
88 Sholomitsky (1965a, 1965b, 1966).
89 Maltby and Moffet (1965).
90 Inverse Compton radiation becomes important when the brightness temperature of the sources approaches 10^{12} K. At this level, the inverse Compton X-ray emission due to the interaction between the relativistic electrons and the radiation field exceeds the radio synchrotron emission, further increasing the radiation field, and the runaway effect depletes the relativistic electron reservoir. Kellermann and Pauliny-Toth (1969).
91 We are grateful to Leonid Gurvits who has provided us with the details of the Soviet ADU-1000 space tracking antenna in Crimea.
92 Dent (1965a, 1965b, 1966).
93 Pauliny-Toth and Kellermann (1966); Kellermann and Pauliny-Toth (1968).
94 Knight et al. (1971).
95 Cohen et al. (1971).
96 Whitney et al. (1971).
97 Rogers and Morrison (1972).
98 For an object moving at a velocity $v = \beta c$ at an angle θ to the line-of-sight, the apparent transverse motion during a time, τ, is $v\tau \sin\theta$. During this time, the radiating source has moved a distance $v\tau \cos\theta$ toward the observer so the apparent elapsed time interval is only $\tau(1-\beta\cos\theta)$. The apparent transverse velocity, in terms of c, is then $\beta\,\mathrm{app} = \beta\sin\theta/(1-\beta\cos\theta)$, which has a maximum value, γc that occurs at an angle $\theta = 1/\gamma$, where the Lorentz factor $\gamma = \left(1-\beta^2\right)^{-1/2}$. The boost in luminosity depends somewhat on the source geometry and spectrum, but when $\theta = 1/\gamma$, the apparent luminosity is enhanced by a factor of about γ^3. For a typical observed value of $\gamma = 10, \theta \sim 6$ deg, and the luminosity is enhanced by a factor of about one thousand. See Cohen et al. (2007).
99 Rees (1966, 1967).
100 Ozernoy and Sazonov (1969).
101 E.g., Cohen et al. (1977); Kellermann et al. (2004).
102 Ginzburg (1961); Hoyle and Fowler (1963).
103 Lynden-Bell (1969).

Chapter 5

1 Ryle (1955).
2 The expansion of the Universe at a speed proportional to distance is described by the Hubble law, $v = H_0 D$, where v is the expansion velocity in km/sec determined by the redshift $z = \Delta\lambda/\lambda$, D is the distance in Megaparsecs (Mpc) and where $\Delta\lambda$ is the difference between the observed and rest

wavelength. Early estimates of the Hubble constant, H_0, by Hubble and others were near $H_0 = 500$ km/sec/Mpc that led to an age of the Universe of less than 2 billion years, in apparent contradiction to the age of the Earth, estimated from radioactive decay, of about 3 to 5 billion years. A revision of the Hubble constant by Walter Baade in 1952 doubled the size of the Universe and correspondingly the Hubble time scale. Later investigations became very controversial, with claimed values ranging from H_0 near 50 km/sec/Mpc to 100 km/sec/Mpc. More recent work indicates H_0 is near 72 km/sec/Mpc, but determinations made using different techniques still appear to differ by more than the claimed uncertainties. In recognition of his early contributions, in 2018, the International Astronomical Union renamed the velocity–distance relation as the "Hubble-Lemaitre Law."

3 Sandage (1970).
4 Formally, the big-bang was the beginning of space as well as time.
5 Bondi and Gold (1948); Hoyle (1948).
6 Hoyle (1982).
7 Ryle (1950); Bolton and Westfold (1951); Westerhout and Oort (1951); Mills (1952b); Hanbury Brown and Hazard (1953b); Bolton, Stanley, and Slee (1954).
8 Hanbury Brown and Hazard (1950, 1953a); Mills (1952c).
9 Kragh (1996), p. 309.
10 Ryle (1968).
11 For example, von Hoerner (1964).
12 See for example Hubble's *Realm of the Nebulae* (Hubble 1936a), pp. 70–72, 182–189.
13 As described by Hubble (1936a), pp. 193–196, the brightness of galaxies is diminished with increasing redshift faster than the square of the distance because there are fewer photons reaching the observer by an amount $1 + z$ (the Number–Effect) and because the energy of each photon is reduced by the same factor (the Energy–Effect) ($z = \Delta\lambda/\lambda$). Also, due to the redshift of the spectrum, the observer sees the emission that originated at a frequency higher by a factor of $1 + z$. To evaluate this so-called K correction one needs to know the intrinsic spectrum of the source. The situation at radio wavelengths is different, since radio astronomers observe with a fixed bandwidth and the redshift compresses the observed bandwidth and so increases the observed flux density by a factor of $1 + z$. This nearly cancels the K correction in radio astronomy since the spectral dependence of flux density is typically close to $(\text{frequency})^{-0.8}$.
14 Bolton, Stanley, and Slee (1954).
15 "Confusion" is the term used by radio astronomers to describe the effect of having multiple sources in the antenna beam. Because of the random distribution of sources on the sky, at least an average of 20 beam widths per source are considered the minimum necessary to mitigate the effect of confusion.
16 Shakeshaft et al. (1955). The full survey included 1,936 sources but 30 were of large angular size and located near the plane of the Milky Way so were assumed to be of galactic origin.
17 Scheuer (1957) derived a source count analysis that was based on the probability distribution of interferometer deflections on the telescope chart recorder output and was independent of counting individual sources. This became known as the P(D) analysis and has subsequently been used to investigate the radio source distribution from other more recent confusion limited surveys.
18 Ryle (1955). Halley Lecture, delivered in Oxford, 6 May 1955.
19 Ryle and Scheuer (1955).
20 Edge and Mulkay (1976), p. 162.
21 The Mills Cross consisted of two orthogonal linear arrays each 45 m (148 feet) long. When the signals from the two arrays were combined, the resultant nearly circular antenna beam was somewhat smaller than 1 degree, but the technique resulted in high side–lobes resulting in spurious sources. In the Mills Cross type of telescope, this generated spurious "extended sources" that were, in fact, blends of multiple independent sources.
22 Mills and Slee (1957).
23 Mills and Slee (1957). See also Mills (1984).
24 Ryle (1957).
25 Pawsey (1957).
26 Kragh (1996), p. 315.

27 Ryle (1956a).
28 Mills (1956, 1984).
29 Ryle (1956b).
30 Mills (1984).
31 Scheuer (1957).
32 Hazard and Walsh (1959).
33 Edge et al. (1959).
34 The blending of sources in the Cambridge interferometers was more subtle than in the Mills Cross, as the different sources combined in random phase which could either add or subtract depending on their separation. The 2C Catalogue contained about one source per beam width in the sky, and so was horribly confused. Hewish (1961) used Scheuer's method to calculate a slope of −1.8 for the new 4C survey, in good agreement with the value obtained from counting individual strong sources.
35 Edge, Scheuer, and Shakeshaft (1958).
36 Ryle (1958).
37 Mills, Slee, and Hill (1958). It should be pointed out that the quoted error in the cumulative source count slope here and elsewhere is statistically incorrect. See note 49.
38 Mills, Slee, and Hill (1960).
39 Bennett (1962).
40 Mills (1959).
41 Ryle (1959).
42 Scott, Ryle, and Hewish (1961); Pilkington and Scott (1965); Gower, Scott, and Wills (1967). Earlier, in an elegant implementation of aperture (Earth rotation) synthesis using non-tracking antenna elements, Ryle and Neville (1962) surveyed a small region near the north celestial pole reaching a source density of more than 10,000 sources per steradian and showed the convergence of the source count for the weaker sources. See Chapter 10 for a discussion of synthesis imaging.
43 Scott and Ryle (1961); Bolton, Gardner, and Mackey (1964); Gower (1966). As was customary in both Sydney and Cambridge, the scientific analysis was often published well ahead of the detailed catalogs and data used to derive the earlier published result.
44 Ryle and Clark (1961). See also Hewish (1961); Ryle (1961, 1968).
45 Hoyle and Narlikar (1961, 1962).
46 Ryle (1962) and discussion following Ryle's presentation.
47 Ryle (1963).
48 Eddington (1913).
49 Jauncey (1967, 1975). In a 1967 interview with KIK, Ryle demonstrated how he determined the slope of the count and its error by taking a straight-edge to a plot of the integral source count and estimating the range of reasonable fits to the data. Crawford, Jauncey, and Murdoch (1970) later showed how to derive an unbiased estimate of the source count and its error from ungrouped data.
50 E.g., Bolton (1960).
51 Mills (1984).
52 Kellermann (1972); Kellermann and Wall (1987).
53 Hubble (1936b) and later Kellermann (1972) speculated on the implications of this apparent coincidence of the optical and radio source counts.
54 Ryle and Neville (1962).
55 Pearson and Kus (1978).
56 Condon and Mitchell (1984).
57 Condon et al. (2012).
58 Shimmins, Bolton, and Wall (1968). John Shimmins was an engineer turned scientist who oversaw the operation of the Parkes antenna. Jasper Wall was a Canadian graduate student from the Australian National University who later went on to become Director of the Royal Greenwich Observatory in the United Kingdom. Wall was present in 1969 when the Parkes antenna was used to communicate with the Apollo 11 lunar landing.
59 Pooley (1968); Kellermann, Davis, and Pauliny-Toth (1971); Kellermann (1972); Jauncey (1975).
60 21 cm neutral hydrogen observations of nearly 3,000 galaxies (Haynes and Giovanelli 1986) and optical spectroscopic observations show that, contrary to Ryle's claim, there are structures in the local Universe extending out to hundreds of million light years (Geller and Huchra 1989).

61 As shown by Yahil (1972); Pearson (1974); Pauliny-Toth et al. (1978); and Shaver and Pierre (1989), the distribution of strong radio sources is not isotropic.
62 Hoyle and Narlikar (1961, 1962); Hoyle (1968); Kellermann (1972); Kellermann and Wall (1987).
63 Kellermann (1972).
64 Edge and Mulkay (1976), pp. 152–184; Kragh (1996), pp. 305, 327.
65 Penzias and Wilson (1965).
66 Burbidge (1971); Burbidge and Narlikar (1976).
67 Longair (1966) used information about the radio luminosity derived from optically measured redshifts (distances) and the integrated sky background to derive the radio luminosity function to show that the Cambridge 178 MHz source count required the steep evolution of only the most powerful radio sources up a redshift of about 4, corresponding to a cosmic age of only 500 million years. At higher redshifts (earlier epochs) the density or luminosity of radio sources rapidly decreased.
68 Hoyle (1959).
69 Sandage (1961).
70 Miley (1971). Earlier, a more limited study by Thomas Legg (1970) in Canada had hinted at a similar angular size–redshift relation, while a later more extensive analysis based on 511 radio galaxies and quasars by Vijay Kapahi (1987) confirmed Miley's conclusions of a simple linear dependence on the redshift.
71 Kellermann (1972).
72 E.g., Rees and Setti (1968).
73 Kellermann (1993). An extension of this study to include a sample of 330 compact sources by Gurvits, Kellermann, and Frey (1999), including the dependence of linear size on luminosity, gave similar results.
74 Schmidt (1968). The change in evolution with radio luminosity was first noted by Longair (1966).
75 Schmidt (1970); Schmidt and Green (1983).
76 Schmidt, Schneider, and Gunn (1995).
77 Shaver et al. (1996).
78 E.g., Hoyle and Burbidge (1966); Field, Arp, and Bahcall (1973); Hoyle (1981); Arp (1987, 1998).
79 Terrell (1964).
80 Burbidge and Burbidge (1967); Burbidge (1968).
81 Field, Arp, and Bahcall (1973); Arp (1987, 1998).
82 E.g., Lopez-Corredoira (2011).

Chapter 6

1 Penzias and Wilson (1965).
2 The maser (**m**icrowave **a**mplification by **s**imulated **e**mission of **r**adiation) was invented in 1953 by Charles Townes at Columbia University. Townes had previously worked at Bell Labs during the Second World War developing centimeter wavelength radar systems. Due to its low noise properties, the maser was used as a very sensitive amplifier for experimental communications systems as well as for radio astronomy systems. However, as maser amplifiers needed to be cooled to only 4 K (above absolute zero), with the subsequent development of transistor amplifiers that needed less cooling, masers ultimately fell out of favor with radio astronomers.
3 The book by Peebles, Page, and Partridge (2009) includes first person reports by most of the people involved in the complex story of the experimental and theoretical history of the cosmic microwave background.
4 Wilkinson and Peebles (1984). See also Peebles, Page, and Partridge (2009).
5 Short autobiographies of Penzias and Wilson are given on the Nobel Prize Web pages: www.nobelprize.org/prizes/physics/1978/penzias/biographical/, and www.nobelprize.org/prizes/physics/1978/wilson/biographical/ (both last accessed 23 November 2022)
6 Crawford, Hogg, and Hunt (1961).
7 Wilson (1980, 1984, 1992, 2009); Penzias (1992, 2009).

8 Ohm (1961).
9 Crawford, Hogg, and Hunt (1961).
10 Doroshkevich and Novikov (1964).
11 The liquid helium cold load was designed and built by Penzias (1965).
12 Dicke (1946).
13 Dicke et al. (1946).
14 Van Allen (1966).
15 Wilkinson and Peebles (1984).
16 The versions told by Penzias (2009), Wilson (1984, 2009) and by Burke (2009), as to why Burke called Penzias, differ in detail and also in regard to who called who. We are indebted to Bob Wilson, who shared with us his recollections of the events of February 1965.
17 Wilkinson (2009).
18 Penzias and Wilson (1965). It should be noted that the ± 1.0 degree uncertainty quoted by Penzias and Wilson was meant to reflect the absolute uncertainty in the calibration and not the significance of the detection. One of the present authors (KIK) first learned of this remarkable result when he picked up a copy of the *International Herald Tribune* at the Paris Orly Airport, arriving in transit to the United States after two weeks in the USSR, isolated from all news of the world.
19 Dicke et al. (1965).
20 Dicke et al. (1965).
21 Dicke et al. (1946).
22 Wilkinson and Peebles (1984).
23 Roll and Wilkinson (1966).
24 Penzias and Wilson (1967).
25 Howell and Shakeshaft (1966, 1967).
26 Gush, Halpern, and Wisnow (1990); Mather (2006). See also, e.g., Smoot (2006) for a compilation of CMB temperature measurements between 0.408 and 266 GHz (1.1 mm to 73.5 cm).
27 E.g., Kaufman (1965); Wolfe and Burbidge (1969); Hoyle, Burbidge, and Narlikar (1994).
28 Penzias, Schraml, and Wilson (1969).
29 Adams (1941).
30 McKellar (1941). See also McKellar (1940).
31 Field (2009).
32 Field and Hitchcock (1966); Field, Herbig, and Hitchock (1966).
33 Thaddeus and Clauser (1966). See also Thaddeus (1972).
34 Shklovsky (1966).
35 Woolf (2009).
36 Alpher, Bethe, and Gamow (1948); Gamow (1948, 1949); and Alpher and Herman (1988) have reviewed their work over a period spanning nearly three decades on the synthesis of the elements during the first few seconds of the Universe and their calculations of the current cosmic background temperature.
37 Alpher and Herman (1948).
38 Alpher and Herman (1949, 1951); Alpher, Follin, and Herman (1953).
39 Gamow (1950). Alpher and Herman (1990) suggest that Gamow took this number from their own work.
40 Peebles (2014) has discussed the complex set of eleven 1948 publications by Gamow, Alpher, and Herman that he suggests were perhaps published faster than the authors could reflect on what they were doing.
41 Alpher and Herman (1988).
42 Ohm (1961).
43 Doroshkevich and Novikov (1964); Novikov (2009).
44 Shmaonov's (1957) work is discussed by Kragh (1996), p. 343; Novikov (2009); and Kaidanovsky (2012).
45 Kragh (1996), p. 343.
46 Jakes (1963).
47 Wilkinson and Peebles (1984).

48 Alpher and Herman (1988, 1990); Alpher (2014).
49 Wilkinson and Peebles (1984).
50 Wilson (1980).
51 E.g., Silk (1967); Rees and Sciama (1968).
52 E.g., Sachs and Wolfe (1967); Boynton and Partridge (1973), see also the Nobel Lecture by George Smoot (2006).
53 Conklin (1967, 1969a, 1969b, 1972).
54 Smoot, Gorenstein, and Muller (1977); Gorenstein and Smoot (1981); Smoot (2006); Jarosik et al. (2007) give a more precise value of the large scale anisotropy of 3.358 $\pm$ 0.017 mK corresponding to a solar motion of 368 $\pm$ 2 km/sec relative to the Universe.
55 E.g., Geller and Huchra (1989).
56 Following the pioneering observations by Conklin and Bracewell (1967) observers in many countries tried, unsuccessfully, to measure the small scale fluctuations in the CMB. See Partridge (1980) for a summary of observations up to 1979 and Smoot (2006) for his remarks about the theoretical predictions.
57 Smoot (2006). See also Smoot and Davidson (1993) for a highly personal account of the history of the CMB.
58 Fixsen et al. (2011) determined that the nonthermal cosmic background has a spectrum of the form $T = (2.41 \pm 2.1) \times (\text{frequency}/310\ \text{MHz})^{-2.6}$ K.
59 Singal et al. (2018).

Chapter 7

1 Bell (1968), p. 231.
2 Clarke (1964a, 1964b, 1964c).
3 Clarke (1964a).
4 Clarke (1964a), appendix II.
5 Hewish and Wyndham (1963).
6 Hewish (1974).
7 Hewish, Scott, and Wills (1964).
8 Bolton, Clarke, and Ekers (1965); Bolton et al. (1965); Clarke, Bolton, and Shimmins (1966); Shimmins, Clarke, and Ekers (1966).
9 Longair (2009).
10 This section on the discovery of pulsars is based in part on the first-hand accounts of Bell (1968); Hewish (1974, 2008); Bell Burnell (1977, 1984, 2015); and Longair (2009).
11 A balun is a device to transform the signal from the **bal**anced dipoles to the **un**balanced coaxial cables connecting the array.
12 Bell Burnell (1977).
13 All quotations in this and the remainder of this section are taken from Bell Burnell (1984), unless otherwise noted.
14 Bell Burnell (1977).
15 Astronomical parallax measurements refer to the change in apparent direction from different positions in the Earth's orbit around the Sun. Hewish et al. did not measure any parallax greater than 2 arcminutes.
16 Hewish et al. (1968).
17 The pulse period was quoted as 1.3372795 $\pm$ 0.0000020 seconds.
18 Pilkington et al. (1968).
19 Longair (2009).
20 Mitton (2005), p. 288.
21 Gold (1968).
22 Pacini (1967).
23 Baade and Zwicky (1934).
24 Bell (1968), p. 231.
25 This section is based largely on the report of Schisler (2008).

26 Following the deactivation of the Clear radars, and the declassification of their records, Schisler established contact with radio astronomers. He was invited to an August 2007 symposium, held in Montan Areal, Canada, celebrating the 40th anniversary of the discovery of pulsars, where he disclosed, for the first time, the remarkable story of his independent discovery of pulsars.
27 Hewish and Okoye (1965); Bell and Hewish (1967).
28 Shklovsky (1953a).
29 Oort and Walraven (1956).
30 Minkowski (1942).
31 Woltjer (1968).
32 Staelin and Reifenstein (1968).
33 Comella et al. (1969).
34 American Institute of Physics, *Moments of Discovery, Unit 2: A Pulsar Discovery*. Available at: https://history.aip.org/exhibits/mod/pulsar/pulsar2/09.html (last accessed 23 November 2022).
35 Cocke, Disney, and Taylor (1969); https://history.aip.org/exhibits/mod/pulsar/pulsar1/05.html (last accessed 23 November 2022).
36 Bell Burnell (2015). See https://rahist.nrao.edu/1-Bell_Hawaii_PSR_Refln.pptx (last accessed 23 November 2022).
37 Pilkington and Scott (1965).
38 Rickard and Cronyn (1979).
39 Zwicky (1939).
40 Backer's paper was refereed by W. M. Goss, who recommended publication, but only after revisions to clarify the presentation. We are grateful to Miller Goss for bringing this correspondence to our attention and for clarifying several issues surrounding the discovery of millisecond pulsars. Backer's submitted manuscript, Goss's referee report, and Stewart Pottash's (*A&A* editor) letter to Backer may be found in NRAO/AUI Archives, Papers of Donald C. Backer, Millisecond Pulsars.
41 Backer et al. (1983). Interstellar scintillations, due to turbulence in the interstellar medium, cause fluctuations in intensity on time scales of hours and are observed only in very small diameter radio sources such as pulsars.
42 Djorgovski (1982).
43 Helfand (1982).
44 The pulsar PSR 1937+214 has been measured to have a stability at least as good as the best atomic frequency standards with which it has been compared.
45 E.g., Hyman et al. (2005); Mickaliger et al. (2012).
46 McLaughlin et al. (2006).
47 Lorimer et al. (2007).
48 Based on the observed dispersion measure of 375 cm/pc^3 and making some assumptions about the dispersion measure from the Milky Way and from the host galaxy, Lorimer et al. concluded that the source was at a redshift of about 0.12 or about 500 Mpc distant.
49 Burke-Spolaor et al. (2011).
50 Burke-Spolaor et al. (2011).
51 Bagchi, Nieves, and McLaughlin (2012); Kulkarni et al. (2014).
52 Thornton et al. (2013) found four FRBs with dispersion measures between 550 and 1,100 cm^{-3} pc.
53 Spitler et al. (2014); Masui (2015).
54 Petroff et al. (2015).
55 Spitler et al. (2016).
56 Chatterjee et al. (2017).
57 Bannister et al. (2019). ASKAP includes 36 12-m diameter antennas, of which only 24 were used in this investigation. Each of the ASKAP antennas forms 36 beams in the sky, so they were able to search a relatively large area for an FRB and measure its position.
58 Berger et al. (2001); Hyman et al. (2005).
59 Bochenek et al. (2020). See Kaspi and Beloborodov (2017) for a review of magnetars.
60 Colgate (1975); Rees (1977).
61 CHIME is composed of an array of parabolic cylinders whose origin can be traced to an early Australian design for the SKA that was abandoned in favor of the phased array feeds used for ASKAP.

Chapter 8

1 1944 comment from Jan Oort to H. C. van de Hulst that led to the latter's now famous, but misleading, paper on the prospects for detecting features in the radio spectrum from atomic hydrogen in the Milky Way. van de Hulst (1957).
2 This section is based in large part on the papers and recollections of Harold (Doc) Ewen held in the NRAO/AUI Archives, on the papers of Edward Purcell in the Harvard University Archives, on Ewen's Harvard doctoral dissertation, and on interviews with Ewen, Purcell, and van de Hulst by W. T. Sullivan III that are also held in the NRAO/AUI Archives. Other discussions of the discovery of the 21 cm line have been reported by Stephan (1990); Buderi (1996), p. 291; van Woerden and Strom (2006); Sullivan (2009), p. 394; and Kellermann, Bouton, and Brandt (2020), p. 48.
3 van de Hulst (1945).
4 The radio recombination line frequency corresponding to a transition from an energy level having a quantum number n to quantum number $n-1$ is given by $R\left[\frac{1}{[n-1]^2}-\frac{1}{n^2}\right]$, where the Rydberg constant, $R = 3.29 \times 10^9$ MHz.
5 Sullivan (1982), p. 299 reanalyzed van de Hulst's working notes and discovered that van de Hulst inverted an exponent of 5/3 in the wavelength dependence of line broadening, thus underestimating the strength of recombination lines at high frequencies.
6 The hand-written notes of van de Hulst's 15 April 1944 talk to the Dutch *Astronomenclub* were translated to English by Adriaan Blauuw who wrote "one may perhaps expect an observable intensity." NAA-Sullivan Papers, Working Papers, H. van de Hulst. In his 1944 talk, van de Hulst was apparently more positive about the possibility of detecting the 21 cm hydrogen line than he was in his 1945 publication.
7 Reber and Greenstein (1947).
8 J. L. Pawsey to E. G. Bowen, 23 January 1948, National Archives of Australia, file C4659/8. We are grateful to W. M. Goss for making a copy of this letter available to us.
9 Shklovsky (1949).
10 Shklovsky (1982), in Russian, was translated for the authors by one of Shklovsky's former students, Vladimir Slysh, who was himself an important contributor to radio astronomy in the USSR.
11 Kusch and Prodell (1950).
12 Kleppner and Horowitz (2016) have written a delightful short paper describing Purcell's AAA&S proposal.
13 Due to the Doppler shift, a motion of 1 km/sec along the line of sight corresponds to a shift of 5 kHz in frequency.
14 Dicke (1946).
15 NAA-Ewen Papers. Available at: www.nrao.edu/archives/static/Ewen/ewen_HI_slide19.shtml (last accessed 30 November 2022).
16 Ewen to Purcell, 19 February 1978, Harvard University Archives, Papers of E. M. Purcell.
17 Some reports have suggested that van de Hulst was the first to point out to Ewen that his negative result might be because his frequency separation was too small and that the hydrogen line might appear in both channels and, therefore, be subtracted out. However, Ewen apparently did not meet van de Hulst until mid-April, after he had detected the line.
18 Ewen to E. Purcell, 19 February 1978.
19 Ewen comment at a May 2001 meeting in Green Bank, West Virginia, NAA-Ewen Papers.
20 Ewen and Purcell (1951a, 1951b).
21 Ewen and Purcell (1951c); Muller and Oort (1951).
22 Christiansen and Hindman (1952a, 1952b); van de Hulst, Muller, and Oort (1954).
23 Oort, Kerr, and Westerhout (1958).
24 www.nrao.edu/archives/static/Ewen/ewen_harvard24and60.shtml (last accessed 30 November 2022).
25 Shklovsky (1982).
26 Wild's two 1949 reports *The Radio-Frequency Line-Spectrum of Atomic Hydrogen, I. The Calculation of Frequencies of Possible Transitions*, and 2. *The Calculation of Transition Probabilities* were distributed as Radiophysics Laboratory bound reports RPL 33 and 34, in

February and May 1949, respectively. See Goss, Hooker, and Ekers (2023), chapter 20, for a more detailed discussion of the early Australian 21 cm investigations.
27 Wild (1952). Wild's concern about line broadening may have been the result of his interest in detecting RRLs in the Sun, where, of course, the density is much greater than in the interstellar medium.
28 Bolton (1990).
29 Saha (1946).
30 Olof Rydbeck interview with W. T. Sullivan, 15 September 1978, NAA-Sullivan Papers, Working Files, Interviewees. Available at: www.nrao.edu/archives/items/show/15155 (last accessed 30 November 2022).
31 See Kellermann, Bouton, and Brandt (2020), p. 53 for a more complete discussion of the early years of the Harvard radio astronomy program following the detection of the 21 cm hydrogen line.
32 Lilley and McClain (1956).
33 Davies and Fennison (1964).
34 Field (1962a); Roberts (1975a). Lilley and McClain incorrectly assumed that the fractional change in frequency is identical to the fractional change in wavelength (the redshift) determined by the optical astronomers. Radio astronomers now can and do measure spectral line emission from distant quasars and galaxies.
35 van de Hulst (1945, original emphasis).
36 Kardashev (1959).
37 Yuri Parijsky presented the paper by Drovskikh, Dravskikh, and Kolbasov on the "Detection of Excited H-Line Radiation at Pulkovo." Victor Vitkevitch presented the paper on behalf of Sorochenko and Borodzich on the "Detection of 3 cm Radio Line in the Omega Nebula." In *Transactions of the International Astronomical Union*, **XIIB**, ed. J.-C. Pecker (London: Academic Press), 360.
38 Dravskikh et al. (1966); Sorochenko and Borodzich (1966).
39 Gordon and Sorochenko (2002), p. 17.
40 We are indebted to Jim Moran who told us of the interaction between Menzel and Lilley, as told to him by Lilley.
41 Hoglund and Mezger (1965a, 1965b).
42 Lilley et al. (1966); Dieter (1967); Gardner and McGee (1967); Churchwell and Mezger (1970); Reifenstein et al. (1970). See Dupree and Goldberg (1970); Brown et al. (1978); Gordon and Sorchenko (2002); and Mezger and Palmer (1968) for good reviews of radio recombination lines.
43 This section is based in part on the paper by Barrett (1984).
44 Shklovsky (1949, 1953b).
45 C. Townes to W. Kornberg, 28 March 1979. NAA-NRAO Records, Director's Office, Topical Correspondence, Molecular Astronomy. Available at: www.nrao.edu/archives/items/show/35854 (last accessed 30 November 2022).
46 Townes (1957).
47 Ehrenstein, Townes, and Stevenson (1959).
48 Barrett (1984).
49 Barrett and Lilley (1957).
50 Weinreb et al. (1963). Weinreb's novel 20 channel digital spectrometer is described in Weinreb (1961).
51 Bolton et al. (1964a); Gardner et al. (1964).
52 Bolton et al. (1964b); Goldstein et al. (1964); Robinson et al. (1964).
53 Cheung et al. (1968).
54 Townes (1999), p. 173.
55 Snyder and Buhl (1969a). See also Snyder and Buhl (1969b).
56 Townes (1999), p. 175.
57 NRAO Assistant to the Director, W. E. Howard III, in *SDRA*, p. 288.
58 Snyder et al. (1969).
59 Palmer et al. (1967).
60 A. A. Penzias, R. W. Wilson, and K. B. Jefferts to W. E. Howard III, 27 February 1969, NA-NRAO, Tucson Operations, 36 Foot Telescope, Box 4.
61 Wilson, Jefferts, and Penzias (1970).

62 Gundermann (1965). Apparently after being scooped by Berkeley, Gundermann never published what seemingly was the first discovery of OH emission. We are grateful to Jim Moran for providing us with a copy of Gundermann's Harvard thesis.
63 Weaver et al. (1965); McGee et al. (1965).
64 Weinreb et al. (1965); Zuckerman, Lilley, and Penfield (1965); Dieter, Weaver, and Williams (1966).
65 Weinreb et al. (1965).
66 Davies, De Jager, and Verschuur (1966).
67 Barrett and Rogers (1966).
68 Barrett and Rogers (1966). Weinreb et al. (1965) also mentioned "amplification of a maser-type population inversion" as one of four possibilities to explain the anomalous polarization of the OH emission. No doubt, Alan Barrett, who had been a student of Townes and so was familiar with the maser concept, originated the maser idea discussed in both of these papers.
69 Cudaback, Read, and Rougoor (1966); Rogers et al. (1966); Moran (1967a, 1967b); Moran et al. (1968). At Cornell, Perkins, Gold, and Salpeter (1966) and at MIT, Litvak et al. (1966) put the maser interpretation on a sound theoretical footing. Later interferometric observations showed that there were multiple co-located features in the maser source known as W3 (OH) that exhibited both left- and right-hand circular polarization, thus providing "the most convincing evidence yet obtained for the Zeeman effect in cosmic masers" (Moran et al. 1978).
70 Menzel (1937). We are indebted to Jim Moran for bringing this early paper to our attention.
71 Buhl et al. (1969); Knowles et al. (1969a, 1969b).
72 Burke et al. (1970, 1972); Moran et al. (1973).
73 Cheung et al. (1969).
74 Barrett, Schwartz, and Waters (1971); Snyder and Buhl (1974).
75 Claussen and Lo (1986).
76 Nakai, Inoue, and Miyoshi (1993).
77 Miyoshi et al. (1995); Hernstein et al. (1999); Humphreys et al. (2013); Reid, Pesce, and Riess (2019).
78 Caputo, Marconi, and Musella (2002); Humphreys et al. (2013).
79 Pesce et al. (2020).
80 Reid, Pesce, and Riess (2019); Pesce et al. (2020).
81 Verde, Treu, and Riess (2019).

Chapter 9

1 Payne-Gaposhkin (1954) on the expected temperature of the dark side of Mercury.
2 An Astronomical Unit (AU) is the mean distance from the Earth to the Sun and is 149,597,871 km or 92,955,807 miles.
3 In terms of degrees Celsius (Centigrade) the numbers need to be replaced by 119, 70, and 4, respectively, while for Fahrenheit degrees they become 246, 150, and 39, respectively. See Kellermann (1966) for further background on radio measurements of planetary temperatures.
4 Dicke and Beringer (1946). See also Sullivan (2009), p. 200.
5 Piddington and Minnett (1949). See also Sullivan (2009), p. 277.
6 Schiaperelli (1890); *PASP*, **2**, 79 (1890). Actually, because of Mercury's very elliptical orbit and corresponding variation in orbital velocity, there would be a slight variation in the apparent position of the Sun as seen from Mercury producing small zones where the Sun rises and sets.
7 Sander (1963) and Columbo and Shapiro (1966) give authoritative accounts of the history of visual measurements of the rotation period.
8 Dolfuss (1953).
9 Payne-Gaposchkin (1954), p. 181.
10 Howard, Barrett, and Haddock (1962).
11 Pettit and Nicholson (1936); Kuiper (1952).
12 Kellermann (1965a, 1966), *New York Times*, 29 August 1964, "Mars Gives a Hint of Radiation Belt," p. 46.
13 Field (1962b); Kozyrev (1964); Kellermann (1965a).
14 Pettengill and Dyce (1965); Pettengill (1984).

15 Columbo (1965); Peale and Gold (1965); Columbo and Shapiro (1966). More precisely, the orbital period of Mercury is measured to be 87.97 days and the rotation period 58.7 days.
16 Pettengill (1984).
17 McGovern, Gross, and Rasool (1965).
18 Lissauer and de Pater (2013), p. 234.
19 See, e.g., Morrison and Sagan (1967); Morrison and Klein (1970); Mitchell and de Pater (1994). This was later confirmed by the NASA Messenger spacecraft that orbited Mercury for four years.
20 Kraus (1956).
21 Mayer, McCullough, and Sloanaker (1958a); Mayer (1959, 1984).
22 Wildt, (1940); Sagan (1960).
23 Mayer (1984).
24 Sagan (1960); Walker and Sagan (1960); Jones (1961).
25 See Barrett (1965); Kellermann (1966); Pollack and Sagan (1965), and references therein for a summary of the radio observations and interpretation.
26 Drake (1962, 1964).
27 Mayer, McCullough, and Sloanaker (1958b, 1958c).
28 Barrett (1965); Kellermann (1965b); Briggs and Drake (1972); Rudy et al. (1987).
29 Morrison, Sagan, and Pollack (1969).
30 Sullivan (1964).
31 Kellermann (1965b) measured Martian temperatures of 192 ± 26, 162 ± 18, and 190 ± 41 K at 6, 11.3, and 21.3 cm, respectively.
32 Clancey, Grossman, and Muhleman (1992).
33 Drake and Ewen (1958); Mayer, McCullough, and Sloanaker (1958b, 1958c); Cook et al. (1960); Barrett (1965); Berge (1966); Kellermann (1966); Kellermann and Pauliny-Toth (1966b).
34 Low (1964, 1966).
35 Mayer, McCullough, and Sloanaker (1958b); Mayer (1959). Using the same antenna, but with a sensitive maser radiometer, Alsop et al. (1958) later measured a temperature of 165 ± 17 K at 3.15 cm (9.15 GHz). The infrared temperature was measured by Murray and Wildey (1963).
36 Berge (1968).
37 de Pater et al. (2019).
38 Drake and Ewen (1958).
39 Drake (1962); Seling (1970).
40 Low (1964); Kellermann (1966).
41 Kellermann and Pauliny-Toth (1966a); Klein and Seling (1966); Low (1966); Berge (1968).
42 Kellermann and Pauliny-Toth (1966a); Berge (1968).
43 See, e.g., de Pater (1990) for an extensive review of the observations and interpretation of radio observations of the giant planets.
44 Öpik (1962); Trafton (1964).
45 Altenhoff et al. (1988).
46 KIK interview with Edward Fomalont, 22 August 2021.
47 Kellermann, Bouton, and Brandt (2020).
48 B. F. Burke interview with W. T. Sullivan III, 21 January 1972, NAA-Sullivan Papers, Working Files, Individuals Unit, Burke. Available at: www.nrao.edu/archives/items/show/913 (last accessed 30 November 2022).
49 The sequence of events behind the discovery of the intense radio bursts from Jupiter is based primarily on the accounts of Franklin (1959, 1984).
50 Franklin (1959).
51 B. F. Burke interview with W. T. Sullivan III, 21 January 1972, NAA-Sullivan Papers, Working Files, Individuals Unit, Burke. Available at: www.nrao.edu/archives/items/show/913 (last accessed 30 November 2022).
52 *New York Times*, "Sound on Jupiter Is Picked Up in U.S." 6 April 1955, p. 31.
53 Burke and Franklin (1955a, 1955b). See also Franklin (1959, 1984).
54 Franklin (1959).
55 Burke and Franklin (1955b); Franklin and Burke (1956).
56 Franklin and Burke (1956); Burke (2006).

57 Shain (1955, 1956); Carr et al. (1958).
58 Burke (2006). The Jupiter radio emission can be as strong as 10,000 megawatts.
59 Bigg (1964).
60 Goldreich and Lynden-Bell (1969).
61 Franklin (1959).
62 http://radiojove.gsfc.nasa.gov (last accessed 30 November 2022).
63 Van Allen et al. (1958).
64 McClain and Sloanaker (1959). Subsequently, additional measurements by Sloanaker suggested a value closer to 700 K (Roberts and Stanley 1959).
65 The Green Bank 85 foot radio telescope was purchased from the Blaw-Knox corporation and was named after Howard Tatel who had earlier played a key role in the discovery of the dekametric radio burst on Jupiter. Tatel had designed the 85 foot telescope but died before it was completed. See Kellermann, Bouton, and Brandt (2020), p. 166.
66 Drake and Hvatum (1959). See also Drake (1984), p. 258 for a first-hand account of the discovery of Jupiter's radiation belts.
67 Drake (1984); Bolton (1990). John Bolton had left Australia in 1955 to build the Caltech Owens Valley Radio Observatory (OVRO) two element interferometer near Big Pine, California.
68 Roberts and Stanley (1959).
69 Radhakrishnan and Roberts (1960); Morris and Berge (1962).
70 de Pater, Schultz, and Brecht (1997).
71 Pettengill (1984).
72 See Buderi (1996) for an authoritative account of the development of radar and the application of radar to investigate the Solar System.
73 DeWitt and Stodola (1949); Clark (1980). See also Sullivan (2009), p. 264.
74 *New York Times*, 25 January 1946, "Contact with Moon Achieved by Radar in Test by the Army," p. 1; 26 January1946, "Radar Code to the Planets Envisioned for Future," p. 1; *Time*, 4 February 1946, p. 11; *Newsweek*, February 1946, p. 64. On 27 January 1946, a recording of the lunar radar echoes was broadcast by the Mutual Broadcasting System to listeners around the country.
75 Webb (1946). See also Kauffman (1946) for a personal account by the person who heard the first pulse echo.
76 Bay (1946). See also Sullivan (2009), p. 271.
77 Butrica (1996) and Buderi (1996) give detailed accounts of the early planetary radar experiments.
78 Price (1959); *New York Times*, 20 March 1959, "Venus is Reached by Radar Signals," p. 1.
79 Evans and Taylor (1959).
80 Green and Pettengill (1960).
81 Victor and Stevens (1961).
82 Pettengill and Price (1961); Staff of the Millstone Radar Observatory (1961); Pettengill et al. (1962).
83 *Pravda*, 12 May 1961, *Izvestia*, 12 May 1961; Kotelnikov (1961); Butrica (1996), p. 45.
84 Kotelnikov (1963).
85 Smith (1963).
86 Pettengill et al. (1962).
87 Goldstein and Carpenter (1963). Uranus is the only other planet in the Solar System that rotates in a counter-clockwise or retrograde direction.
88 Drake (1964).
89 Dick (1996), pp. 183–188.
90 Bailes, Lyne, and Shemar (1991a).
91 Bailes, Lyne, and Shemar (1991b).
92 Wolszczan and Frail (1992).
93 Wolszczan (1994).
94 Lyne and Bailes (1992).
95 Mayor and Queloz (1995).
96 Marcy and Butler (1998).
97 We thank Dale Frail and Alex Wolszczan for sharing their memories on the discovery of the first exoplanets with the authors.

98 Zhu and Dong (2021) and references therein.
99 Winn and Fabrycky (2015) and references therein.

Chapter 10

1 Einstein (1936) on the practical possibility of observing the effects of gravitational lensing.
2 Einstein (1916).
3 Pound and Rebka (1960).
4 Vessot et al. (1980).
5 Einstein (1911).
6 Dyson, Eddington, and Davidson (1920).
7 See, e.g., Will (1983, 2015); Kennefick (2009).
8 von Klüber (1960).
9 Seielstad, Sramek, and Weiler (1970).
10 Muhleman, Ekers, and Fomalont (1970).
11 Sramek (1970).
12 Counselman et al. (1974); Weiler et al. (1974, 1975).
13 Fomalont and Sramek (1975, 1976).
14 Brans and Dicke (1961); Dicke (1974).
15 Fomalont et al. (2009).
16 Robertson, Carter, and Dillinger (1991); Lebach et al. (1995); Shapiro et al. (2004).
17 Shapiro (1964).
18 Shapiro (1968, 1971).
19 Reasenberg et al. (1979).
20 Demorest et al. (2010); Antoniadis et al. (2013); Cromartie et al. (2020).
21 Einstein (1936).
22 In 1920, Eddington, and in 1924, Orest Khvolson also discussed gravitational lensing but Einstein was apparently unaware of this work. See Will (2015).
23 Zwicky (1937a, 1937b, 1937c).
24 Barnothy (1965, 1966a, 1966b); Barnothy and Barnothy (1966, 1967, 1968, 1969a, 1969b, 1971, 1972).
25 Barnothy (1963) named his cosmology “FIB,” meaning not truthful, since messages carried by signals can get distorted during their propagation.
26 This discussion is based on the colorful story of the discovery of the first gravitation lens by Dennis Walsh (1989).
27 Porcas et al. (1980).
28 Walsh, Carswell, and Weymann (1979).
29 Porcas et al. (1981).
30 Greenfield, Roberts, and Burke (1985).
31 Refsdal and Surdej (1994).
32 Hewitt et al. (1988).
33 Refsdal (1964).
34 In practice, it has been difficult to accurately determine the value of H_0 using gravitational lensing due to uncertainties in the mass distribution in the lensing galaxy or cluster of galaxies.
35 Refsdal and Surdej (1994); Will (2015).
36 Weber (1960, 1968, 1969, 1970).
37 Garwin (1974); Tyson and Giffard (1978); Yodh and Wallis (2001); Levine (2004).
38 Yodh and Wallis (2001).
39 The following paragraphs are based in part on the first-hand accounts of the discovery of the binary pulsar 1913+16 given in their Nobel Prize lectures and at the 184th meeting of the AAS by Hulse (1994a, 1994b) and Taylor (1994a, 1994b).
40 Taylor (1974).
41 Hulse and Taylor (1974a).
42 We are grateful to Joe Taylor for sharing with us his memory of events surrounding the discovery of the binary pulsar.

43 Hulse and Taylor (1974b); Taylor and Hulse (1974).
44 Hulse and Taylor (1975).
45 Taylor and McCulloch (1980).
46 Taylor and Weisberg (1982); Weisberg and Huang (2016). See also Taylor, Fowler, and McCulloch (1979) for an earlier preliminary report of gravitational radiation from PSR 1913+16.
47 LIGO, the Laser Interferometer Gravitational-Wave Observatory, uses optical interferometry to detect the minute changes in the lengths of 4-km (2.5 mile) long interferometer arms. See Abbott et al. (2016).
48 Lyne et al. (2004); Kramer et al. (2021).
49 Antoniadis et al. (2022).
50 Zwicky (1933).
51 Slipher (1914).
52 Babcock (1939).
53 See the reviews by Faber and Gallagher (1979); Rubin (2000); Sofue and Rubin (2001); Cohen (2005); and Roberts (2008).
54 E.g., Rubin and Ford (1970).
55 Gottesman, Davies, and Reddish (1966); Roberts (1966); Shobbrook and Robinson (1967).
56 See Cohen (2005) and references therein.
57 Burbidge (1975).
58 Roberts and Whitehurst (1975).
59 Roberts and Rots (1973); Roberts (1975b); Bosma (1978); Bosma and van der Kruit (1979); and Cohen (2005).
60 Ostriker, Peebles, and Yahil (1974); Einasto, Kaasik, and Saar (1974); Einasto et al. (1974).
61 Roberts (2008).
62 Milgrom (1983). See also Milgrom (2002) for a less technical account.

Chapter 11

1 Adaptation from the film *Field of Dreams* where the hero, an Iowa farmer, hears a voice telling him that if he builds a baseball field the players and fans will come.
2 Stephen Chu, 2015, *Bulletin of the American Academy of Arts and Science*, Summer Issue. Stephen Chu won the 1997 Nobel Prize in Physics for his “development of methods to cool and trap atoms with laser light,” and later was the US Secretary of Energy from 2009 to 2013. As an undergraduate, he spent the summer of 1972 working with one of the authors (KIK) and other radio astronomers at the National Radio Astronomy Observatory in Green Bank, West Virginia and in Charlottesville, Virginia.
3 Wilkinson et al. (2004).
4 Reber has variously referred to the diameter of his dish as 31 feet 5 inches and 31.5 feet.
5 It is the custom in the US lumber industry to refer to the cross-section of a standard piece of lumber as 2 by 4, although the actual dimensions are 1.5 by 3.5 inches (3.8 $\times$ 8.9 cm).
6 Kellermann, Bouton, and Brandt (2020), p. 161 give more details on the NRL 50 foot antenna.
7 Haddock (1984).
8 This section is based in part on the review articles by Altschuler (2002) and Cohen (2009).
9 Gordon (1958).
10 Bowles (1958).
11 Drake (1984).
12 Gordon and LaLonde (1961).
13 Marshall Cohen (2009) was then a professor in the Cornell Department of Electrical Engineering.
14 Pettengill (1984), p. 275.
15 Further details on the history of the Haystack Observatory can be found in Weiss (1965) and Kellermann, Bouton, and Brandt (2020), p. 292.
16 Herb Weiss to Irwin Shapiro e-mail, 10 April 2022. It is not clear to what extent the reporter was also referring to the controversial Westford Needles Project that orbited nearly 500 million thin copper dipoles, referred to as “needles,” that were intended to create an artificial reflecting medium for global military radio communications. After a successful 1963 launch, the Needles

Project was abandoned and replaced by the development of more reliable communication satellites. As expected, the orbiting needles decayed after only a few years, much to the relief of radio astronomers who were concerned about the possible reflection of interference from terrestrial radio transmitters.

17 Kellermann, Bouton, and Brandt (2020), p. 170 discuss the problems surrounding the construction of the NRAO 140 foot radio telescope.

18 McClain (1960); *New York Times*, 19 June 1959.

19 D. R. Lord to G. Kistiakowski, 8 September 1960, Dwight D. Eisenhower Presidential Library, Radar and Radio Astronomy, Box 5, Records of the US President's Science Advisory Committee.

20 Lovell (1984).

21 Blackett and Lovell (1941).

22 Lovell (1984).

23 Lovell (1987, 1990).

24 Lovell (1987), p. 239.

25 As there were no antennas in the United States capable of tracking the Pioneer rocket at such a great distance, the US Air Force, in a secret meeting, appealed to Lovell to use the Jodrell Bank antenna to track it. Unfortunately, the rocket failed to reach the Moon and, after reaching a distance of only 79,000 miles (127,000 km) the rocket fell back to Earth and burned up in the atmosphere over the Pacific Ocean (Lovell 1987), p. 215.

26 A more detailed account of the events surrounding the funding and construction of the GBT are given in Kellermann, Bouton, and Brandt (2020), p. 487.

27 Condon et al. (1994).

28 *New York Times*, 17 November 1988.

29 NRAO Director Paul Vanden Bout and Green Bank Director George Seielstad were present at these discussions and later shared their notes with the authors.

30 In conventional radio telescopes, the feed structure is placed on the electrical axis and is supported by three or four symmetrically placed "feed legs." The focal structure and the feed legs partially block and scatter the incoming celestial radio waves, resulting in a decrease in effective sensitivity and increased sidelobes from the scattered radiation. Moreover, reflections between the dish surface and feed structure introduce instabilities in spectroscopic observations. All of these problems are avoided by placing the feed structure off the electrical axis so it does not block the dish. The typical home satellite TV dish has an unblocked aperture but, prior to the 300 foot replacement project, the largest asymmetric off-axis structure was a 7.5 m radio telescope at Bell Laboratories.

31 At the end of each fiscal year, with little or no debate, the US Congress passes an emergency funding bill to pay for the cost of natural tragedies such as fires, earthquakes, and floods that occurred during the previous year, but it is also used as a convenient place to fund projects of interest to individual Members of Congress.

32 To placate the NSF and the astronomical community, and counteract the perception of apparent "pork," 1989 became the first year of funding for a new NSF Major Research Equipment (MRE), now called Major Research Equipment and Facilities Construction (MREFC), standing budget line that later supported a variety of new telescopes as well as other large NSF facilities.

33 Although Congress had made $75 million available for the project, NRAO reserved $20 million for administrative costs of project management and for instrumentation.

34 While the true cost of construction was not an issue, the litigation was focused on who was responsible for the increased costs.

35 This distinction was later surpassed by the enclosure built to shield the Chernobyl nuclear power plant.

36 Condon (2008).

37 The dramatic development of radio astronomy imaging using interferometer and aperture synthesis is described in more detail by Kellermann and Moran (2001), by Goss, Hooker, and Ekers (2023), and by Thompson, Moran, and Swenson (2017).

38 van de Hulst (1945).

39 Struve (1949).

40 Michelson (1890).

41 McCready, Pawsey, and Payne-Scott (1947). Fourier synthesis is a well-known mathematical technique by which a complex function is represented by a series of sine and cosine functions. Each interferometer spacing measures one Fourier component of the function or, in the case of astronomical imaging, the celestial brightness distribution. Although Michelson had derived all the relevant equations half a century earlier, McCready et al. appear to be the first to point out the Fourier transform relation between interferometer observations and the celestial image.
42 Ryle and Vonberg (1946).
43 The funding and events leading up to the construction of the WSRT are discussed by Elbers (2017), p. 169.
44 The complex and lengthy process leading up to the funding and construction of the VLA is given by Kellermann, Bouton, and Brandt (2020), p. 319. Napier, Thompson, and Ekers (1983) give a detailed report on the technical specifications and performance of the VLA, while Trimble and Zaich (2006) have reported on the huge impact of the VLA on astronomy.
45 The resolution of HST is about 0.04 arcsec. JWST is about 2.5 times larger than HST but operates at longer infrared wavelengths, with a resolution of about 0.1 arcsec.
46 Readhead and Wilkinson (1978); Cotton (1979).
47 Hanbury Brown (1984) gives a first person account of the development of long baseline interferometry at Jodrell Bank.
48 In a conventional Michelson interferometry it is necessary to retain both the amplitude and phase of the incoming celestial radio signals. Traditionally, this is accomplished by sending a common reference local oscillator signal to each of the interferometer antenna elements and returning the intermediate frequency signals to a common location for correlation. The connections between the elements are normally made using cable, waveguide, or radio links.
49 Hanbury Brown, Jennison, and Das Gupta (1952); Jennison and Das Gupta (1953).
50 Baade and Minkowski (1954).
51 Hanbury Brown and Twiss (1956).
52 In 1965, Carr et al. also used intensity interferometry to measure the structure of Jupiter bursts.
53 Allen et al. (1962a).
54 Palmer et al. (1967).
55 See Broten (1988); Kellermann and Cohen (1988); Kellermann and Moran (2001).
56 Levy et al. (1986).
57 See Hirabayashi, Edwards, and Murphy (2000) for a technical description and the scientific results from the HALCA satellite.
58 Kardashev et al. (2013).
59 The image of the black hole in M87 was obtained by the Event Horizon Telescope and a six station global array operating at 1.3 mm wavelength or 230 GHz (Akiyama et al. 2019).
60 Reid and Honma (2014).
61 Matveyenko, Kardashev, and Sholomitsky (1965); Matveyenko (2013).
62 A summary of the activities connected with the planning for the Square Kilometre Array is given by Ekers (2013) and Schilizzi et al. (2023).
63 E.g., Swarup (1981); Parijisky (1992); Noordam (2013).
64 By agreement, the British spelling of "kilometre" was universally used.
65 Taylor and Braun (1999); Carilli and Rawlings (2004); Bourke et al. (2015).
66 Smolders and van Haarlem (1999); Hall (2005).
67 Hotan (2021).
68 www.skatelescope.org/wp-content/uploads/2012/06/117_SSAC.Report.pdf (last accessed 30 November 2022).
69 www.skatelescope.org/wp-content/uploads/2012/06/120_ANZSCC.response.to_.SSAC_.report-covering.letter.pdf (last accessed 30 November 2022).
70 SKAO-Prospectus, January 2022. Available at: www.skatelescope.org/ska-prospectus/ (last accessed 30 November 2022).
71 Blandford (2010).
72 Amiri et al. (2022).

Chapter 12

1 Charles Townes, in a letter to Warren Kronberg, 28 March 1979.
2 Harwit (1981, 2013, 2021).
3 Verschuur (1969).
4 Livingston (1954); Ekers (2014).
5 Frail to Kellermann, 23 February 2022, email. NAA-Kellermann Papers, Star Noise.
6 Bell (1968).
7 Bolton to Bowen, 1 December 1964. NAA-Kellermann Papers, Star Noise.
8 Remark made by Wilson at the 1983 conference on Serendipitous Discoveries in Radio Astronomy, Available at: http://library.nrao.edu/public/collection/02000000000280.pdf, p. 244 (last accessed 30 November 2022).
9 E.g., Harwit (1984).
10 Pettengill (1984).
11 Harwit (2021), p. 289.
12 We are grateful to Duncan Lorimer for sharing with us a copy of the editor's report.
13 Broderick (1984); Gold (1984).
14 Gold (1984).
15 Note by J. R. Pierce on the occasion of the 8 June 1998 Dedication of the Jansky Monument in Holmdel, New Jersey.
16 Kragh (2017) has drawn attention to remarks made by Hans Bethe (2003), who, at the time of his 1967 Prize ceremony, was told the story about Nobel's wife by a member of the Nobel Foundation.
17 Interestingly, Genzel started his career as a radio astronomer with a PhD thesis on water masers in star-forming regions.
18 Mitton (2005), p. 328.
19 Hulse and Taylor (1975).
20 Other winners of the Nobel Prize in Physics who were women are Marie Curie (1903), Maria Goeppert-Mayer (1963), and Donna Strickland (2018).
21 Bell Burnell (1977).
22 *Washington Post*, Obituary, Antony Hewish, 97: Astronomer won the Nobel for role in Discovery of Pulsars, 27 September 2021, p. B6. See also *New York Times* documentary, available at: www.youtube.com/watch?v=NDW9zKqvPJI (last accessed 30 November 2022).
23 Lyne and Bailes (1992); Wolszczan and Frail (1992); Mayor and Queloz (1995).
24 https://archive.ph/20180320091111/http://archive.defense.gov/Transcripts/Transcript.aspx?TranscriptID=2636 (last accessed 30 November 2022).
25 Reiss et al. (1998); Perlmutter et al. (1999).
26 Jaffe et al. (2001); Chae et al. (2002).
27 Burke-Spolaor et al. (2019); Antoniadis et al. (2022).
28 Particularly in the critical 87 to 107 MHz FM broadcast band, meteor scatter makes this experiment difficult from almost everywhere in the world.
29 Bowman et al. (2018).
30 Hills et al. (2018).
31 These include LOFAR in the Netherlands, the MWA in Australia, the LWA in the United States, and SKA Low, discussed in Chapter 11, and HERA – the Hydrogen Epoch of Reionization Array being built in South Africa.
32 Fixsen et al. (2011).
33 Singal et al. (2018).
34 Drake and Sobel (1992); Kellermann, Bouton, and Brandt (2020), p. 229.
35 Cocconi and Morrison (1959).
36 Townes (1983).
37 Dyson (1960) suggested that advanced civilizations might build a sphere around their star in order to harness all of its radiated energy.
38 See Webb (2015) for an extensive discussion of the Fermi Paradox. Drake and Sobel (1992) and Tarter (2001) give good summaries of SETI research.

39 Drake (1986).
40 "C.T.A.-102." *Younger than Yesterday*. Available at: www.youtube.com/watch?v=1q5PQJFsDyM (last accessed 30 November 2022).
41 Ehman (1998). Gray (2012) gives a detailed account of the circumstances surrounding the WOW report and other SETI programs.
42 Gray and Marvel (2001) and Harp et al. (2020) have reported on the negative results of sensitive searches using the VLA and Allen Telescope Array, respectively.
43 Horowitz and Sagan (1993).
44 Bracewell (1960).
45 Harwit (2021) indicates that, for astronomy as a whole, the rate of increase has been maintained through the year 2010.

Glossary: Abbreviations and Acronyms

AAA&S American Academy of Arts and Sciences
AAS American Astronomical Society
AGN Active galactic nuclei
ALMA Atacama Large Millimeter/submillimeter Array
AORG Army Operational Research Group [UK]
ARCADE Absolute Radiometer for Cosmology, Astrophysics and Diffuse Emission
ARPA Advanced Research Projects Agency
ASG Agencies SKA Group
ASKAP Australian SKA Pathfinder
AU Astronomical Units
BAL Broad absorption line quasar
BMEWS Ballistic Missile Early Warning System
BSO Blue stellar object
BTL Bell Telephone Laboratories
BuDocks Navy Bureau of Yards and Docks [US]
Cas A Cassiopeia A
CHIME Canadian Hydrogen Intensity Mapping Experiment
CMB Cosmic microwave background
COBE Cosmic Microwave Background Explorer
COBRA COsmic Background RAdiation [rocket experiment]
COMSAT Communications Satellite Corporation
CRT Cathode ray tube
CSIR Council for Scientific and Industrial Research [Australia]
CSIRO Commonwealth Scientific and Industrial Research Organization [Australia]
DEW line Distant early warning line
DoD Department of Defense [US]

DTM	Department of Terrestrial Magnetism [Carnegie Institution]
EHT	Event Horizon Telescope
EMP	Electromagnetic pulse
EoR	Epoch of Reionization
FASR	Frequency Agile Solar Radiotelescope
FAST	Five-hundred-meter Aperture Spherical Telescope
FIRAS	Far Infrared Absolute Spectrometer
FRB	Fast radio burst
GBI	Green Bank Interferometer
GBT	Green Bank Telescope
GR	General Relativity
HST	Hubble Space Telescope
IAU	International Astronomical Union
IFAG	Informal Funding Agencies Group [SKA]
IPS	Interplanetary Scintillations
IRE	Institute of Radio Engineers
ISPO	International SKA Project Office
ISSC	International SKA Steering Committee
JLG	Jesse L. Greenstein
JPL	Jet Propulsion Laboratory
JWST	James Webb Space Telescope
KGJ	Karl Guthe Jansky
KIK	Kenneth I. Kellermann
LGM	Little Green Men
LIGO	Laser Interferometer Gravitational-Wave Observatory
LINER	Low-ionization nuclear emission-line region
LNSD	Large number of small dishes [US SKA concept]
LOFAR	Low Frequency Array
LWA	Long Wavelength Array
Maser	Microwave amplification by simulated emission of radiation
MERLIN	Multi-Element Radio Linked Interferometer Network
MOND	MOdified Newtonian Dynamics
MTRLI	Multi-Telescope Radio Linked Interferometer
MUSA	Multiple Unit Steerable Antenna
MWA	Murchison Widefield Array
MWPO	Mount Wilson and Palomar Observatories
NASA	National Aeronautics and Space Administration
NBS	National Bureau of Standards
NEROC	North East Radio Observatory Corporation

ngVLA	next generation VLA
NRAO	National Radio Astronomy Observatory
NRL	Naval Research Laboratory [US]
ONR	Office of Naval Research [US]
OVRO	Owens Valley Radio Observatory [Caltech]
PAF	Phased array feed
QSG	Quasi-stellar galaxy
QSO	Quasi-stellar object
RDL	Radio Development Laboratory [NZ]
RRATS	Rotating Radio Transits
RRLs	Radio Recombination Lines
SKA	Square Kilometre Array
SKAO	SKA Organization
SMBH	Supermassive black hole
SMC	Small Magellanic Cloud
SPDO	SKA Project Development Office
SSEC	SKA Science and Engineering Committee
TDRSS	Tracking and Data Relay Satellite System (NASA)
TESS	Transiting Exoplanet Survey Satellite
URSI	International Union of Radio Science
USSKAC	US SKA Consortium
VLA	Very Large Array
VLBA	Very Long Baseline Array
VLBI	Very Long Baseline Interferometry
WARC	World Administrative Radio Conference
WSRT	Westerbork Synthesis Radio Telescope

Citation Abbreviations for NRAO/AUI Archives Materials

NAA	NRAO/AUI Archives
SDRA	*Serendipitous Discoveries in Radio Astronomy.* Ed. K. I. Kellermann and B. Sheets. Green Bank: NRAO/AUI, 1984

Journal Abbreviations Used

A&A	Astronomy and Astrophysics
A&A Rev.	Astronomy and Astrophysics Reviews
AJ	Astronomical Journal
Ann. Astrophys.	Annales d'Astrophysique
Ann. NY Acad. Sci.	Annals of the New York Academy of Sciences
ApJ	Astrophysical Journal
ApJS	Astrophysical Journal Supplement
ARAA	Annual Review of Astronomy and Astrophysics
Astron. Astrophys. Trans.	Astronomy and Astrophysics Transactions
Astron. Circ.	Astronomical Circular [USSR Academy of Sciences]
Astron. Nach.	Astronomische Nachrichten
Astron. Rpts.	Astronomy Reports
Astron. Zh.	Astronomicheskii Zhurnal
Astrophys. Let.	Astrophysics Letters
Astrophys. Space Sci.	Astrophysics and Space Science
Austrl. J. Phys. A	Australian Journal of Physics, Series A
Austrl. J. Sci. Res. A	Australian Journal of Scientific Research, Series A
BAN	Bulletin of the Astronomical Institutes of the Netherlands
Bell Syst. Tech. J.	Bell System Technical Journal
Bull. Astron. Soc. India	Bulletin of the Astronomical Society of India
Comptes Rendus	Comptes rendus de l'Académie des Sciences [France]
Dokl. Akad. Nauk SSSR	Doklady Akademiia Nauk SSSR
IAU Inf. Bull. Var. Stars	IAU Information Bulletin on Variable Stars

IEEE Trans. Ant. Prop.	IEEE Transactions on Antennas and Propagation
IEEE Trans. Mil. Elec.	IEEE Transactions on Military Electronics
Int. J. Astron. Astrophys.	International Journal of Astronomy and Astrophysics
Int. J. Mod. Phys. D	International Journal of Modern Physics D
Izvestiya Akad. Nauk	Izvestiya Akademiia Nauk SSSR
J. Astron. Hist. Heritage	Journal of Astronomical History and Heritage
J. Brit. Astron. Assoc.	Journal of the British Astronomical Association
J. Brit. Inst. Radio Eng.	Journal of the British Institute of Radio Engineers
J. Geophys. Res.	Journal of Geophysical Research
J. Hist. Astron.	Journal for the History of Astronomy
J. Inst. Elec. Eng.	Journal of the Institute of Electrical Engineers
J. Inst. Elec. Eng. Japan	Journal of the Institute of Electrical Engineers, Japan
JRASC	Journal of the Royal Astronomical Society of Canada
Math. Proc. Cambridge Phil. Soc.	Mathematical Proceedings of the Cambridge Philosophical Society
Mem. RAS	Memoirs of the Royal Astronomical Society
Mem. Roy. Soc.	Biographical Memoirs of the Fellows of the Royal Astronomical Society
MNRAS	Monthly Notices of the Royal Astronomical Society
New Astron. Rev.	New Astronomy Reviews
Obs.	The Observatory
PASA	Publications of the Astronomical Society of Australia
PASP	Publications of the Astronomical Society of the Pacific
Phil. Mag. Ser. 5	Philosophical Magazine, Series 5
Phil. Trans. Roy. Soc. A	Philosophical Transactions of the Royal Society A
Phys. Lett. A	Physics Letters A
Phys. Rev.	Physical Review
Phys. Rev. Let.	Physical Review Letters
Phys. Today	Physics Today
Plan. Space Sci.	Planetary and Space Science

PNAS	Proceedings of the National Academy of Sciences
Proc. Am. Phil. Soc.	Proceedings of the American Philosophical Society
Proc. IRE	Proceedings of the Institute of Radio Engineers
Proc. Nat. Electronic Conf.	Proceedings of the National Electronic Conference
Proc. Phys. Soc. A	Proceedings of the Physical Society, Section A
Proc. Roy. Soc. London	Proceedings of the Royal Society of London
Pub. Astron. Soc. Austrl.	Publications of the Astronomical Society of Australia
Rev. Mod. Phys.	Reviews of Modern Physics
Rev. Sci. Instr.	Review of Scientific Instruments
Rpts. Prog. Phys.	Reports on Progress in Physics
SciAm	Scientific American
Sky Tel.	Sky and Telescope
Sov. Astron.	Soviet Astronomy
Sov. Phys. Doklady	Soviet Physics Doklady
Sov. Phys. Usp.	Soviet Physics Uspekhi
Vistas Astron.	Vistas in Astronomy

Bibliography

Abbott, B. P. et al. 2016, Observation of Gravitational Waves from a Binary Black Hole Merger, *Phys. Rev. Let.*, **116**, 061102.

Adams, W. S. 1941, Some Results with the COUDÉ Spectrograph of the Mount Wilson Observatory, *ApJ*, **93**, 11.

Akiyama, K. et al. 2019, First M87 Event Horizon Telescope Results. II. Array and Instrumentation, *ApJ*, **875**, L2.

Akiyama, K. et al. 2022, First Sagittarius A* Event Horizon Telescope Results. I. The Shadow of the Supermassive Black Hole in the Center of the Milky Way, *ApJ*, **930**, L12.

Alexander, F. E. S. 1945, Report on the Investigation of the "Norfolk Island Effect," Radio Development Laboratory, Department of Scientific & Industrial Research.

Alexander, F. E. S. 1946, The Sun's Radio Energy, *Radio Electron.*, **1**(1), 16.

Alfvén, H. and Herlofson, N. 1950, Cosmic Radiation and Radio Stars, *Phys. Rev.*, **78**, 616.

Allen, L. R. et al. 1962a, Observations of 384 Radio Sources at a Frequency of 158 Mc/s with a Long Baseline Interferometer, *MNRAS*, **124**, 477.

Allen, L. R. et al. 1962b, An Analysis of the Angular Sizes of Radio Sources, *MNRAS*, **125**, 57.

Alpher, R. A. and Herman, R. C. 1948, Evolution of the Universe, *Nature*, **162**, 774.

Alpher, R. A. and Herman, R. C. 1949, Remarks on the Evolution of the Expanding Universe, *Phys. Rev.*, **75**, 1089.

Alpher, R. A. and Herman, R. C. 1951, Neutron-Capture Theory of Element Formation in an Expanding Universe, *Phys. Rev.*, **84**, 60.

Alpher, R. A. and Herman, R. C. 1988, Reflections on Early Work on "Big Bang" Cosmology, *Phys. Today*, **41**(8), 24.

Alpher, R. A. and Herman, R. C. 1990, Early Work on "Big-Bang" Cosmology and the Cosmic Blackbody Radiation. In *Modern Cosmology in Retrospect*, ed. B. Bertotti, R. Balbinot, and S. Bergia (Cambridge: Cambridge University Press), 129.

Alpher, R. A., Bethe, H. A., and Gamow, G. 1948, The Origin of Chemical Elements, *Phys. Rev.*, **73**, 803.

Alpher, R. A., Follin, J. W., and Herman, R. C. 1953, Physical Conditions in the Initial Stages of the Expanding Universe, *Phys. Rev.*, **92**, 1347.

Alpher, V. 2014, Ralph A. Alpher, George Antonovich Gamow, and the Prediction of the Cosmic Microwave Background Radiation, *Asian J. Phys.*, **23**, 17.

Alsop, L. E. et al. 1958, Observations Using a Maser Radiometer at 3-cm Wave Length, *AJ*, **63**, 30.

Altenhoff, W. et al. 1988, First Radio Astronomical Estimate of the Temperature of Pluto, *A&A*, **190**, L15.

Altschuler, D. R. 2002, The National Astronomy and Ionospheric Center's (NAIC) Arecibo Observatory in Puerto Rico. In *ASPC* **278**, *Single–Dish Radio Astronomy: Techniques and Applications*, ed. S. Stanimirovic et al. (San Francisco: Astronomical Society of the Pacific), 1.

Ambartsumian, V. 1958, On the Evolution of Galaxies. In *La Structure et l'Evolution de l'Univers*, ed. R. Stoops (Brussels: Coudenberg), 266.

Amiri, M. et al. 2022, An Overview of CHIME, the Canadian Hydrogen Intensity Mapping Experiment, *ApJS*, **261**, 29.

Anderson, B., Palmer, H. P., and Rowson, B. 1962, Brightness Distribution of the Radio Source 14N5A, *Nature*, **195**, 165.

Antoniadis, J. et al. 2013, A Massive Pulsar in a Compact Relativistic Binary, *Science*, **340**, 448.

Antoniadis, J. et al. 2022, The International Pulsar Timing Array Second Data Release: Search for an Isotropic Gravitational Wave Background, *MNRAS*, **510**, 4873.

Appleton, E. 1945, Departure of Long-Wave Solar Radiation from Black-Body Intensity, *Nature*, **156**, 534.

Appleton, E. and Hey, J. S. 1946, Solar Radio Noise, *Phil. Mag*., **7**, 73.

Arp, H. C. 1987, *Quasars, Redshifts, and Controversies* (Berkeley: Interstellar Media).

Arp, H. C. 1998, *Seeing Red: Redshifts, Cosmology, and Academic Science* (Montreal: Apeiron).

Baade, W. and Minkowski, R. 1954, Identification of the Radio Sources in Cassiopeia, Cygnus A, and Puppis A, *ApJ*, **119**, 206.

Baade, W. and Zwicky, F. 1934, Remarks on Super-Novae and Cosmic Rays, *Phys. Rev*., **46**, 76.

Babcock, H. W. 1939, The Rotation of the Andromeda Nebula, *Lick Observatory Bulletin*, No. 498.

Backer, D. C. et al. 1983, A Millisecond Pulsar, *Nature*, **300**, 615.

Bagchi, M., Nieves, A. C., and McLaughlin, M. 2012, A Search for Dispersed Radio Bursts in Archival Parkes Multibeam Pulsar Survey Data, *MNRAS*, **425**, 250.

Bailes, M., Lyne, A., and Shemar, S. I. 1991a, A Planet Orbiting the Neutron Star PSR1829–10, *Nature*, **352**, 311.

Bailes, M., Lyne, A., and Shemar, S. 1991b, Radio Astronomers Claim Discovery of First Planet Outside Solar System, *J. Brit. Astron. Assoc*., **101**, 256.

Balick, B. 2005, The Discovery of Sgr A*. In *The New Astronomy: Opening the Electromagnetic Window and Expanding Our View of Planet Earth*, ed. W. Orchiston (Dordrecht: Springer), 183.

Balick, B. and Brown, R. L. 1974, Intense Sub-Arcsecond Structure in the Galactic Center, *ApJ*, **194**, 265.

Bannister, K. W. et al. 2019, A Single Fast Radio Burst Localized to a Massive Galaxy at Cosmological Distance, *Science*, **365**, 565.

Barnothy, J. M. 1963, Astronomical Consequences of the FIB Theory, *PASP*, **75**, 430.

Barnothy, J. M. 1965, Quasars and the Gravitational Image Intensifier, *AJ*, **70**, 666.

Barnothy, J. M. 1966a, Two Observational Tests: Are Quasars Super-Luminous Objects or Optical Effects?, *AJ*, **71**, 154.

Barnothy, J. M. 1966b, An Observational Test of the Hypothesis that Quasars Are Produced by Gravitational Lenses, *Obs*., **86**, 115.

Barnothy, J. M. and Barnothy, M. F. 1966, Apparent Brightness of the Deflector Galaxy of Quasars Produced through Gravitational Lenses, *AJ*, **71**, 378.

Barnothy, J. M. and Barnothy, M. F. 1967, Observations Supporting the Gravitational Lens Explanation of Quasars, *AJ*, **72**, 291.

Barnothy, J. M. and Barnothy, M. F. 1968, Galaxies as Gravitational Lenses, *Science*, **162**, 348.

Barnothy, J. M. and Barnothy, M. F. 1969a, Anomalous Hubble Plot of Quasi-stellar Objects, *Nature*, **222**, 759.

Barnothy, J. M. and Barnothy, M. F. 1969b, Concentration of QSO's around z=2 Redshift, *BAAS*, **1**, 181.

Barnothy, J. M. and Barnothy, M. F. 1971, Rapid Differential Proper Motion in the Radio Fine Structure of 3C 279, *BAAS*, **3**, 472.

Barnothy, J. M. and Barnothy, M. F. 1972, Expected Density of Gravitational-Lens Quasars, *ApJ*, **174**, 477.

Barrett, A. H. 1965, Passive Radio Observations of Mercury, Venus, Mars, Saturn, and Uranus, *Radio Science*, **69D**, 1565.

Barrett, A. H. 1984, The Beginnings of Molecular Radio Astronomy. In *Serendipitous Discoveries in Radio Astronomy*, ed. K. I. Kellermann and B. Sheets (Green Bank: NRAO/AUI), 280.

Barrett, A. H. and Lilley, A. E. 1957, A Search for the 18-cm Line of OH in the Interstellar Medium, *AJ*, **62**, 5.

Barrett, A. H. and Rogers, A. E. E. 1966, Observations of Circularly Polarized OH Emission and Narrow Spectral Features, *Nature*, **210**, 188.

Barrett, A. H., Schwartz, P. R., and Waters, J. W. 1971, Detection of Methyl Alcohol in Orion at a Wavelength of ~1 Centimeter, *ApJ*, **168**, L102.

Bastian, T. S., Benz, A. O., and Gary, D. E. 1998, Radio Emission from Solar Flares, *ARAA*, **36**, 131.

Bay, Z. 1946, Reflection of Microwaves from the Moon, *Hungarian Physics Acta*, **1**, 1.

Beck, A. 1984, Personal Recollections of Karl Jansky. In *Serendipitous Discoveries in Radio Astronomy*, ed. K. I. Kellermann and B. Sheets (Green Bank: NRAO/AUI), 32.

Bell, S. J. 1968, The Measurement of Radio Source Diameters Using a Diffraction Method, PhD Dissertation, Cambridge University.

Bell, S. J. and Hewish, A. 1967, Angular Size and Flux Density of the Small Source in the Crab Nebula at 81.5 Mc/s, *Nature*, **213**, 1214.

Bell Burnell, S. J. 1977, Petit Four. In *Proceedings of the 8th Texas Symposium on Relativistic Astrophysics*, ed. M. D. Papagiannis. *Annals of the New York Academy of Science*, **302**, 685, reprinted under the title "Little Green Men, White Dwarfs or Pulsars," *Cosmic Search*, **1**, 16.

Bell Burnell, S. J. 1984, The Discovery of Pulsars. In *Serendipitous Discoveries in Radio Astronomy*, ed. K. I. Kellermann and B. Sheets (Green Bank: NRAO/AUI), 160.

Bell Burnell, S. J. 2015, Reflections on the Discovery of Pulsars. Paper presented at IAU General Assembly, Honolulu, HI, available at: https://rahist.nrao.edu/1-Bell_Hawaii_PSR_Refln.pptx (last accessed 23 November 2022).

Bennett, A. S. 1962, The Revised 3C Catalogue of Radio Sources, *MNRAS*, **68**, 162.

Berge, G. L. 1966, An Interferometric Study of Jupiter's Decimeter Radio Emission, *ApJ*, **146**, 767.

Berge, G. L. 1968, Recent Observations of Saturn, Uranus, and Neptune at 3.12 cm, *Astrophys. Lett.*, **2**, 127.

Berger, E. et al., 2001, Discovery of Radio Emission from the Brown Dwarf LP944–20, *Nature*, **410**, 338.

Bethe, H. 2003, My Life in Astrophysics, *ARAA*, **41**, 1.

Bigg, E. K. 1964, Influence of the Satellite Io on Jupiter's Decametric Emission, *Nature*, **203**, 1008.

Blaauw, A. et al. 1960, The New I.A.U. System of Galactic Coordinates (1958 Revision), *MNRAS*, **121**, 123.

Blackett, P. M. S. and Lovell, A. C. B. 1941, Radio Echoes and Cosmic Ray Showers, *Proc. R. Soc. A*, **177**, 183.

Blandford, R. D. (ed.) 2010, *New Worlds, New Horizons* (Washington: NAS Press).

Bochenek, C. D. et al. 2020, A Fast Radio Burst Associated with a Galactic Magnetar, *Nature*, **587**, 59.

Boischot, A. 1957, Caractères d'un Type d'Émission Herzienne Associé à Certaines Éruptions Chromosphériques, *Comptes Rendus*, **244**, 1326 [Characteristics of a Herzian Emission Type Associated with Certain Chromospheric Eruptions].

Boischot, A. and Denisse, J.-F. 1957, Les Émissions de Type IV et l'Origine des Rayons Cosmiques Associés aux Éruptions Chromosphériques, *Comptes Rendus*, **245**, 2194 [Type IV Emissions and the Origin of Cosmic Rays Associated with Chromospheric Eruptions].

Bolton, J. G. 1948, Discrete Sources of Galactic Radio Frequency Noise, *Nature*, **162**, 141.

Bolton, J. G. 1955, Australian Work on Radio Stars, *Vistas Astron.*, **1**, 568.

Bolton, J. G. 1960, The Discrete Sources of Cosmic Radiation, Introductory Talk at the URSI General Assembly, London, reprinted in *Observations of the California Institute of Technology Radio Observatory*, no. 5.

Bolton, J. G. 1973, Prospects of Astronomy in Australia, *Nature*, **246**, 282.

Bolton, J. G. 1982, Radio Astronomy at Dover Heights, *Proc. Astron. Soc. Austrl.*, **4**, 349. Reprinted in Kellermann and Sheets 1984, p. 312.

Bolton, J. G. 1990, The Fortieth Anniversary of Extragalactic Radio Astronomy: Radiophysics in Exile, *Pub. Astron. Soc. Austrl.*, **8**, 381.

Bolton, J. G. and Stanley, G. J. 1948a, Variable Source of Radio Frequency Radiation in the Constellation of Cygnus, *Nature*, **161**, 312.

Bolton, J. G. and Stanley, G. J. 1948b, Observations on the Variable Source of Cosmic Radio Frequency Radiation in the Constellation of Cygnus, *Austrl. J. Sci. Res. A*, **1**, 58.

Bolton, J. G. and Stanley, G. J. 1949, The Position and Probable Identification of the Source of Galactic Radio-Frequency Radiation Taurus-A, *Austrl. J. Sci. Res. A*, **2**, 139.

Bolton, J. G. and Westfold, K. C. 1951, Galactic Radiation at Radio Frequencies: IV. The Distribution of Radio Stars in the Galaxy, *Austrl. J. Sci. Res.*, **4**, 476.

Bolton, J. G., Clarke, M. E., and Ekers, R. D. 1965, Identification of Extragalactic Radio Sources Between Declinations –20° and –44°, *Austrl. J. Phys.*, **18**, 627.

Bolton, J. G., Gardner, F. F., and Mackey, M. B. 1963, A Radio Source with a Very Unusual Spectrum, *Nature*, **199**, 682.

Bolton, J. G., Gardner, F. F., and Mackey, M. B. 1964, The Parkes Catalogue of Radio Sources, Declination Zone –20° to –60°, *Austrl. J. Phys.*, **17**, 340.

Bolton, J. G., Stanley, G. J., and Slee, O. B. 1949, Positions of Three Discrete Sources of Galactic Radio-Frequency Radiation, *Nature*, **164**, 101.

Bolton, J. G., Stanley, G. J., and Slee, O. B. 1954, Galactic Radiation at Radio Frequencies. VIII. Discrete Sources at 100 Mc/s between Declinations +50° and –50°, *Austrl. J. Phys.*, **7**, 110.

Bolton, J. G. et al. 1964a, Observations of OH Absorption Lines in the Radio Spectrum of the Galactic Centre, *Nature*, **201**, 279.

Bolton, J. G. et al. 1964b, Distribution and Motions of OH Near the Galactic Centre, *Nature*, **204**, 30.

Bolton, J. G. et al. 1965, Identifications of Six Faint Radio Sources with Quasi-Stellar Objects, *ApJ*, **142**, 1289.

Bondi, H. and Gold, T. 1948, *MNRAS*, **108**, 252.

Bosma, A. 1978, The Distribution and Kinematics of Neutral Hydrogen in Spiral Galaxies of Various Morphological Types, PhD Thesis, Groningen University.

Bosma, A. and van der Kruit, P. C. 1979, The Local Mass-to-Light Ratio in Spiral Galaxies, *A&A*, **79**, 281.
Bourke, T. et al. (eds.) 2015, *Advancing Astrophysics with the Square Kilometre Array*, Proceedings of Science (Manchester: SKA Organization), available at: www .skatelescope.org/books/ (last accessed 30 November 2022).
Bowen, E. G. 1987, *Radar Days* (Bristol: A. Hilger).
Bowles, K. L. 1958, Observation of Vertical-Incidence Scatter from the Ionosphere at 41 Mc/sec., *Phys. Rev. Let.*, **1**, 454.
Bowman, J. D. et al. 2018, An Absorption Profile Centred at 78 Megahertz in the Sky-Averaged Spectrum, *Nature*, **555**, 67.
Bown, R. 1927, Transatlantic Radio Telephony, *Bell Syst. Tech. J.*, **6**, 248.
Boynton, P. E. and Partridge, R. B. 1973, Fine-Scale Anisotropy of the Microwave Background: An Upper Limit at $\lambda = 3.5$ Millimeters, *ApJ*, **181**, 243.
Bracewell, R. N. 1960, Communication from Superior Galactic Communities, *Nature*, **186**, 670.
Bracewell, R. N. 2000, *The Fourier Transform and its Applications*, 3rd ed. (New York: McGraw Hill).
Bracewell, R. N. 2002, The Discovery of Strong Extragalactic Polarization Using the Parkes Radio Telescope, *J. Astron. Hist. Heritage*, **5**, 107.
Bracewell, R. N. 2005, Radio Astronomy at Stanford, *J. Astron. Hist. Heritage*, **8**, 75.
Bracewell, R. N. and Swarup, G. 1961, The Stanford Microwave Spectro Heliograph Antenna, A Microsteradian Pencil Beam Antenna, *IRE Tr. Ant. Prop.*, **AP-9**, 22.
Bracewell, R. N., Cooper, B. F. C., and Cousins, T. E. 1962, Polarization in the Central Component of Centaurus A, *Nature*, **195**, 1289.
Brans, C. and Dicke, R. H. 1961, Mach's Principle and a Relativistic Theory of Gravitation, *Phys. Rev.*, **124**, 924.
Briggs, F. H. and Drake, F. D. 1972, Interferometric Observations of Mars at 21-cm Wavelength, *Icarus*, **17**, 543.
Broderick, J. 1984, The Buffalo Syndrome. In *Serendipitous Discoveries in Radio Astronomy*, ed. K. I. Kellermann and B. Sheets (Green Bank: NRAO/AUI), 221.
Broten, N. W. 1988, Early Days of Canadian Long-Baseline Interferometry: Reflections and Reminiscences, *JRASC*, **82**, 233.
Brown, R. L. et al. 1978, Radio Recombination Lines, *ARAA*, **16**, 445.
Buderi, C. 1996, *The Invention that Changed the World* (New York: Simon & Schuster).
Buhl, D. et al. 1969, An Investigation of the Spectra and Time Variations of Galactic Water-Vapor Sources, *ApJ*, **158**, 97.
Burbidge, G. R. 1959a, The Theoretical Explanation of Radio Emission. In IAU Symposium No. **9**, *Paris Symposium on Radio Astronomy*, ed. R. N. Bracewell (Stanford: Stanford University Press), 541.
Burbidge, G. R. 1959b, Estimates of the Total Energy in Particles and Magnetic Field in the Non-Thermal Radio Sources, *ApJ*, **129**, 849.
Burbidge, G. R. 1968, The Distribution of Redshifts in Quasi-Stellar Objects, N-Systems and Some Radio and Compact Galaxies, *ApJ*, **7**, 41.
Burbidge, G. R. 1971, Was There Really a Big Bang?, *Nature*, **233**, 36.
Burbidge, G. R. 1975, On the Masses and Relative Motions of Galaxies, *ApJ*, **196**, L7.
Burbidge, G. R. and Burbidge, E. M. 1967, *Quasi-stellar Objects* (San Francisco: Freeman).
Burbidge, G. R. and Narlikar, J. V. 1976, The Log N-Log S Curve for 3CR Radio Galaxies and the Problem of Identifying Faint Radio Galaxies, *ApJ*, **205**, 329.
Burke, B. F. 2006, Planetary Radio Astronomy, Fifty Years Ago and Fifty Years Hence. In *Planetary Radio Emissions VI, Proceedings of the 6th International Workshop*, ed. H. O. Rucker, W. S. Kurth, and G. Mann (Wein: Austrian Academy of Sciences Press), 1.

Burke, B. F. 2009, Radio Astronomy from First Contacts to the CMBR. In *Finding the Big Bang*, ed. J. E. Peebles, L. A. Page, and R. B. Partridge (Cambridge: Cambridge University Press), 176.
Burke, B. F. and Franklin, K. L. 1955a, High Resolution Radio Astronomy at 13.5 m Wavelength, *AJ*, **60**, 155.
Burke, B. F. and Franklin, K. L. 1955b, Observations of a Variable Radio Source Associated with the Planet Jupiter, *J. Geophys. Res.*, **60**, 213.
Burke, B. F. et al. 1970, Studies of H_2O Sources by Means of a Very-Long-Baseline Interferometer, *ApJ*, **160**, L63.
Burke, B. F. et al. 1972, Observations of Maser Radio Sources with an Angular Resolution of $0''.0002$, *Sov. Astron.*, **16**, 379; Russian original, *Astron Zh.*, **49**, 465.
Burke-Spolaor, S. et al. 2011, Radio Bursts with Extragalactic Spectral Characteristics Show Terrestrial Origins, *AJ*, **727**, 18.
Burke-Spolaor, S. et al. 2019, The Astrophysics of Nanohertz Gravitational Waves, *A&A Rev.*, **27**, 5.
Butrica, A. J. 1996, *To See the Unseen: A History of Planetary Radar Astronomy* (Washington: NASA).
Caputo, F., Marconi, M., and Musella, I. 2002, The Cepheid Period-Luminosity Relation and the Maser Distance to NGC 4258, *ApJ*, **566**, 833.
Carilli, C. C. and Rawlings, S. (eds.) 2004, *Science with the Square Kilometre Array* (Amsterdam: Elsevier).
Carr, T. D. et al. 1958, 18-Megacycle Observations of Jupiter in 1957, *ApJ*, **127**, 274.
Carr, T. D. et al. 1965, Post-Detector Correlation Interferometry of Jupiter at 18 Mc/s, *IEEE NEREM Record*, **7**, 222.
Chae, K.-H. et al. 2002, Constraints on Cosmological Parameters from the Analysis of the Cosmic Lens All Sky Survey, Radio-Selected Gravitational Lens Statistics, *Phys. Rev. Let.*, **89**, 151301.
Chatterjee, S. et al. 2017, A Direct Localization of a Fast Radio Burst and Its Host, *Nature*, **541**, 5.
Cheung, A. et al. 1968, Detection of NH3 Molecules in the Interstellar Medium by Their Microwave Emission, *Phys. Rev. Let.*, **21**, 170.
Cheung, A. et al. 1969, Detection of Water in Interstellar Regions by Its Microwave Radiation, *Nature*, **221**, 626.
Chiu, H.-Y. 1964, Gravitational Collapse, *Phys. Today*, **17**, 21.
Christiansen, W. N. 1984, The First Decade of Solar Radio Astronomy in Australia. In *The Early Years of Radio Astronomy*, ed. W. T. Sullivan III (Cambridge: Cambridge University Press), 113.
Christiansen, W. N. and Hindman, J. V. 1952a, 21 cm Line Radiation from Galactic Hydrogen, *Obs.*, **72**, 149.
Christiansen, W. N. and Hindman, J. V. 1952b, A Preliminary Survey of 1420 Mc/s. Line Emission from Galactic Hydrogen, *Austrl. J. Sci. Res.*, **A5**, 437.
Christiansen, W. N. and Warburton, J. A. 1953, The Distribution of Radio Brightness over the Solar Disk at a Wavelength of 21cm: I. A New Highly-Directional Aerial System, *Austrl. J. Phys.*, **6**, 190.
Christiansen, W. N. and Warburton, J. A. 1955, The Distribution of Radio Brightness over the Solar Disk at a Wavelength of 21cm: III. The Quiet Sun: Two Dimensional Observations, *Austrl. J. Phys.*, **8**, 474.
Christiansen, W. N., Mathewson, D. S., and Pawsey, J. L. 1957, Radio Pictures of the Sun, *Nature*, **180**, 944.
Churchill, W. S., 1950, *The Second World War, Vol. 4, The Hinge of Fate* (Boston: Houghton Mifflin Company).

Churchwell, E. and Mezger, P. G. 1970, On the Determination of Helium Abundance from Radio Recombination Lines, *Astrophys. Lett.*, **5**, 227.
Clancy, R. T., Grossman, A. W., and Muhleman, D. O. 1992, Mapping Mars Water Vapor with the Very Large Array, *Icarus*, **100**, 48.
Clark, D. 1980, How Diana Touched the Moon, *IEEE Spectrum*, **17**, 44.
Clark, D. H. and Murdin, P. 1978, An Unusual Emission-Line Star/X-ray Source Radio Star, Possibly Associated with an SNR, *Nature*, **276**, 44.
Clarke, M. E. 1964a, Two Topics in Radiophysics, PhD Dissertation, Cambridge University.
Clarke, M. E. 1964b, The Determination of the Positions of 88 Radio Sources, *MNRAS*, **127**, 405.
Clarke, M. E. 1964c, Some Radio Source Flux Density Measurements at 178 MHz, *Obs.*, **85**, 67.
Clarke, M. E., Bolton, J. G., and Shimmins, A. J. 1966, Identification of Extragalactic Radio Sources between Declinations 0° and +20°, *Austrl. J. Phys.*, **19**, 375.
Claussen, M. J. and Lo, K. Y. 1986, Circumnuclear Water Vapor Masers in Active Galaxies, *ApJ*, **308**, 592.
Cocconi, G. and Morrison, P. 1959, Searching for Interstellar Communications, *Nature*, **184**, 844.
Cocke, W. J., Disney, M., and Taylor, D. J. 1969, Discovery of Optical Signals from Pulsar NP 0532, *Nature*, **221**, 525.
Cohen, M. H. 1958a, The Cornell Radio Polarimeter, *Proc. IRE*, **46**, 183.
Cohen, M. H. 1958b, Radio Astronomy Polarization Measurements, *Proc. IRE*, **46**, 172.
Cohen, M. H. 1994, The Owens Valley Radio Observatory: Early Years, *Eng. Sci.*, **57**, 8.
Cohen, M. H. 2005, Dark Matter and the Owens Valley Radio Observatory. In *The New Astronomy: Opening the Electromagnetic Window and Expanding Our View of the Planet Earth*, ed. W. Orchiston (Dordrecht: Springer), 169.
Cohen, M. H. 2007, A History of OVRO Part II, *Eng. Sci.*, **70**, 33.
Cohen, M. H. 2009, Genesis of the 1000-Foot Arecibo Dish, *J. Astron. Hist. Heritage*, **12**, 141.
Cohen, M. H. et al. 1971, The Small-Scale Structure of Radio Galaxies and Quasi-Stellar Sources at 3.8 Centimeters, *ApJ*, **170**, 207.
Cohen, M. H. et al. 1977, Radio Sources with Superluminal Velocities, *Nature*, **268**, 405.
Cohen, M. H. et al. 2007, Relativistic Beaming and the Intrinsic Properties of Extragalactic Radio Jets, *ApJ*, **658**, 232.
Colgate, S. A. 1975, Electromagnetic Pulse from Supernovae, *ApJ*, **198**, 439.
Colombo, G. 1965, Rotational Period of the Planet Mercury, *Nature*, **208**, 575.
Colombo, G. and Shapiro, I. I. 1966, The Rotation of the Planet Mercury, *ApJ*, **145**, 296.
Comella, J. M. et al. 1969, Crab Nebula Pulsar NP 0532, *Nature*, **221**, 453.
Condon, J. J. 2008, ZAPPED! ... by Hostile Space Aliens! In ASPC **398**, *Frontiers of Astrophysics: A Celebration of NRAO 50th Anniversary*, ed. A. H. Bridle, J. J. Condon, and G. C. Hunt (San Francisco: Astronomical Society of the Pacific), 323.
Condon, J. J. and Mitchell, K. J. 1984, A Deeper VLA Survey of the α = 08h 52m 15s, δ = +17° 16' Field, *AJ*, **89**, 610.
Condon, J. J. et al. 1994, A 4.85 GHz Sky Survey. III. Epoch 1986 and Combined (1986–1987) Maps Covering $0° < \delta < +75°$, *AJ*, **107**, 1829.
Condon, J. J. et al. 2012, Resolving the Radio Source Background: Deeper Understanding through Confusion, *ApJ*, **768**, 37.
Conklin, E. K. 1967, Isotropy of the Cosmic Background Radiation at 10 690 MHz, *Phys. Rev. Let.*, **18**, 614.

Conklin, E. K. 1969a, Anisotropy and Inhomogeneity in the Cosmic Background Radiation, PhD Dissertation, Stanford University.

Conklin, E. K. 1969b, Velocity of the Earth with Respect to the Cosmic Background Radiation, *Nature*, **222**, 97.

Conklin, E. K. 1972, Observations of Large-Scale Anisotropy in the 3 K Background Radiation. In IAU Symposium no. **44**, *External Galaxies and Quasi-Stellar Objects*, ed. D. E. Evans (Dordrecht: Reidel), 518.

Conklin, E. K. and Bracewell, R. N. 1967, Limits on Small Scale Variations in the Cosmic Background Radiation, *Nature*, **216**, 777.

Contopoulos, G. and Jappel, A. (eds.) 1974, *Transactions of the IAU XV: B* (Dordrecht: Reidel).

Conway, R. G., Kellermann, K. I., and Long, R. J. 1963, The Radio Frequency Spectra of Discrete Radio Sources, *MNRAS*, **125**, 261.

Cook, J. J. et al. 1960, Radio Detection of the Planet Saturn, *Nature*, **188**, 393.

Cooper, B. F. C. 1998, *Parkes, Centaurus A, and All That*, unpublished memo, NAA, Bracewell Papers, Stanford University, Centaurus A Research.

Cooper, B. F. C. and Price, R. M. 1962, Faraday Rotation Effects Associated with the Radio Source Centaurus A, *Nature*, **195**, 1064.

Cotton, W. D. 1979, A Method of Mapping Compact Structure in Radio Sources Using VLBI Observations, *AJ*, **84**, 1122.

Counselman III, C. C. et al. 1974, Solar Gravitational Deflection of Radio Waves Measured by Very-Long-Baseline Interferometry, *Phys. Rev. Let.*, **33**, 162.

Covington, A. E. 1947, Micro-Wave Solar Noise Observations during the Partial Eclipse of November 23, 1946, *Nature*, **159**, 404.

Covington, A. E. 1948, Solar Noise Observations at 10.7 Centimeters, *Proc. IRE*, **36**, 454.

Covington, A. E. 1984a, Early Radar Research and a Beginning in Radio Astronomy. In *Serendipitous Discoveries in Radio Astronomy*, ed. K. I. Kellermann and B. Sheets (Green Bank: NRAO/AUI), 105.

Covington, A. E. 1984b, Beginnings of Solar Radio Astronomy in Canada. In *The Early Years of Radio Astronomy*, ed. W. T. Sullivan III (Cambridge: Cambridge University Press), 317.

Covington, A. E. 1988, Origins of Canadian Radio Astronomy, *JRASC*, **82**, 165.

Crawford, A. B., Hogg, D. C., and Hunt, L. E. 1961, *Bell Syst. Tech. J.*, **40**, 1095.

Crawford, D. F., Jauncey, D. L., and Murdoch, H. S. 1970, Maximum-Likelihood Estimation of the Slope from Number-Flux Counts of Radio Sources, *ApJ*, **162**, 405.

Cromartie, H. T. et al. 2020, Relativistic Shapiro Delay Measurements of an Extremely Massive Millisecond Pulsar, *Nat. Astron.*, **4**, 72.

Cudaback, D. D., Read, R. B., and Rougoor, G. W. 1966, Diameters and Positions of Three Sources of 18-cm OH Emission, *Phys. Rev. Let.*, **17**, 452.

Davies, R. D. and Fennison, R. C. 1964, A Search for Intergalactic Neutral Hydrogen, I. The Observations, *MNRAS*, **128**, 123.

Davies, R. D., De Jager, G., and Verschuur, G. L. 1966, Detection of Circular and Linear Polarization in the OH Emission Sources near W3 and W49, *Nature*, **209**, 974.

Davis, R. J. et al. 1978, Interferometric Observations of Weak Radio Flares from a Red Dwarf Star, *Nature*, **273**, 644.

de Pater, I. 1990, Radio Images of the Planets, *ARAA*, **28**, 347.

de Pater, I., Schultz, M., and Brecht, H. 1997, Synchrotron Evidence for Amalthea's Influence on Jupiter's Electron Radiation Belt, *J. Geophys. Res.*, **102**, 22043.

de Pater, I. et al. 2019, First ALMA Millimeter-Wavelength Maps of Jupiter, with a Multiwavelength Study of Convection, *AJ*, **158**, 139.

Débarbat, S., Lequeux, J., and Orchiston, W., 2007, Highlighting the History of French Radio Astronomy. 1. Nordmann's Attempt to Observe Solar Radio Emission in 1901, *J. Astron. Hist. Heritage*, **19**, 3.

Demorest, P. B. et al. 2010, A Two-Solar-Mass Neutron Star Measured Using Shapiro Delay, *Nature*, **467**, 1081.

Denisse, J. G. 1984, The Early Years of Radio Astronomy in France. In *The Early Years of Radio Astronomy*, ed. W. T. Sullivan III (Cambridge: Cambridge University Press), 303.

Dent, W. A. 1965a, Variation in the Radio Emission of 3C273 and Other Quasi-Stellar Sources, *AJ*, **70**, 672.

Dent, W. A. 1965b, Quasi-Stellar Sources: Variation in the Radio Emission of 3C 273, *Science*, **148**, 1458.

Dent, W. A. 1966, Variation in the Radio Emission from the Seyfert Galaxy NGC 1275, *ApJ*, **144**, 843.

Deslandres, H. and Décombre, L. 1902, On the Search for Hertzian Radiation Emanating from the Sun, *Comptes Rendus*, **134**, 527 (In French). English translation 1982 in *Classics in Radio Astronomy*, ed. W. T. Sullivan III (Dordrecht: Reidel), 161.

DeSoto, C. B. 1936, *200 Meters and Down* (West Hartford: American Radio Relay League).

Dewhurst, D. 1951, 12 October 1951 Meeting of the Royal Astronomical Society, *Obs.*, **71**, 209.

DeWitt, J. M. and Stodola, E. K. 1949, Detection of Radio Signals Reflected from the Moon, *Proc. IRE*, **37**, 229.

Dick, S. J. 1996, *The Biological Universe* (Cambridge: Cambridge University Press).

Dicke, R. H. 1946, The Measurement of Thermal Radiation at Microwave Frequencies, *Rev. Sci. Instr.*, **17**, 7.

Dicke, R. H. 1974, The Oblateness of the Sun and Relativity, *Science*, **184**, 419.

Dicke, R. H. and Beringer, R. 1946, Microwave Radiation from the Sun and Moon, *ApJ*, **103**, 375.

Dicke, R. H. et al. 1946, Atmospheric Absorption Measurements with a Microwave Radiometer, *Phys. Rev.*, **70**, 340.

Dicke, R. H. et al. 1965, Cosmic Black-Body Radiation, *ApJ*, **142**, 414.

Dieter, N. H. 1967, Observations of the Hydrogen Recombination Line 158α in Galactic H II Regions, *ApJ*, **150**, 435.

Dieter, N. H., Weaver, H., and Williams, D. R. W. 1966, Secular Variations in the Radio-Frequency Emission of OH, *AJ*, **71**, 160.

Djorgovski, G. 1982, Optical Identification of the Millisecond Pulsar 1937+214, *Nature*, **300**, 618.

Dogel, B. A. et al. 2012, Radio Astronomy Studies at the Lebedev Physical Institute. In *A Brief History of Radio Astronomy in the USSR*, ed. S. Braude and K. I. Kellermann (Dordrecht: Springer), 1; English translation of 1985 Russian edition.

Dolfuss, A. 1953, Observation Visuelle et Photographique des Planètes Mercure et Vénus a L'observatoire du Pic du Midi, *L'Astronomie*, **67**, 61.

Doroshkevich, A. G. and Novikov, I. D. 1964, Mean Density of Radiation in the Metagalaxy and Certain Problems in Relativistic Cosmology, *Sov. Phys. Doklady*, **9**, 11.

Drake, F. 1962, 10 cm Observations of Venus Near Superior Conjunction, *Nature*, **195**, 894.

Drake, F. 1964, Microwave Observations of Venus, 1962–1963, *AJ*, **69**, 62.

Drake, F. 1984, Discovery of Jupiter Radiation Belts. In *Serendipitous Discoveries in Radio Astronomy*, ed. K. I. Kellermann and B. Sheets (Green Bank: NRAO/AUI), 258.

Drake, F. 1986, The Search for Extraterrestrial Intelligence. In *Proceedings of the NRAO Workshop on the Search for Extraterrestrial Intelligence*, ed. K. I. Kellermann and G. A. Seielstad (Green Bank: NRAO/AUI), 17.

Drake, F. and Ewen, H. I. 1958, A Broad-Band Microwave Source Comparison Radiometer for Advanced Research in Radio Astronomy, *Proc. IRE*, **46**, 53.
Drake, F. and Hvatum, H. 1959, Non-thermal Microwave Radiation from Jupiter, *AJ*, **645**, 329.
Drake, F. D. and Sobel, D. 1992, *Is Anyone Out There? The Scientific Search for Extraterrestrial Intelligence* (New York: Delacorte Press).
Dravskikh, A. F. et al. 1966, Investigation of the Radio Line of Excited Hydrogen at a Wavelength of 5 cm Using a Quantum Paramagnetic Amplifier, *Sov. Phys. Dokl.*, **10**, 627; Russian original: *Dokl. Akad. Nauk SSSR*, **163**, 332, 1960.
Dupree, A. K. and Goldberg, L. 1970, Radio Frequency Recombination Lines, *ARAA*, **8**, 231.
Dyson, F. W., Eddington, A. S., and Davidson, C. 1920, A Determination of the Deflection of Light by the Sun's Gravitational Field, from Observations Made at the Total Eclipse of May 29, 1919, *Phil. Trans. Roy. Soc. A*, **220**, 291.
Dyson, R. 1960, Search for Artificial Stellar Sources of Infrared Radiation, *Science*, **131**, 1667.
Eddington, A. S. 1913, On a Formula for Correcting Statistics for the Effects of a Known Error of Observation, *MNRAS*, **73**, 359.
Edge, D. O. and Mulkay, J. M. 1976, *Astronomy Transformed* (New York: Wiley)
Edge, D. O., Scheuer, P. A. G., and Shakeshaft, J. R. 1958, Evidence on the Spatial Distribution of Radio Sources Derived from a Survey at a Frequency of 159 Mcs^{-1}, *MNRAS*, **118**, 183.
Edge, D. O. et al. 1959, A Survey of Radio Sources at a Frequency of 159 Mc/s, *Mem. RAS*, **68**, 37.
Edlén, B. 1946, Untitled Letter, *Nature*, **157**, 297.
Edmondson, F. 1956, Review of "The Changing Universe. The Story of the New Astronomy," *Science*, **124**, 541.
Ehman, J. R. 1998, The Big Ear Wow! Signal, available at: www.bigear.org/wow20th.htm (last accessed 1 December 2022).
Ehrenstein, G., Townes, C. H., and Stevenson, M. J. 1959, Ground State Λ-Doubling Transitions of OH Radical, *Phys. Rev Let.*, **3**, 40.
Einasto, J., Kaasik, A., and Saar, E. 1974, Dynamic Evidence on Massive Coronas of Galaxies, *Nature*, **250**, 309.
Einasto, J. et al. 1974, Missing Mass around Galaxies: Morphological Evidence, *Nature*, **252**, 111.
Einstein, A. 1911, Über den Einfluß der Schwerkraft auf die Ausbreitung des Lichtes, *Annalen der Physik*, **35**, 898; English translation, On the Influence of Gravitation on the Propagation of Light, in 1952, *The Principle of Relativity* (New York: Dover) 97.
Einstein, A. 1916, The Foundation of the General Theory of Relativity, *Annalen der Physik*, **49**, 769 (in German).
Einstein, A. 1936, Lens-Like Action of a Star by the Deviation of Light in the Gravitational Field, *Science*, **84**, 506.
Ekers, R. D. 2013, The History of the Square Kilometer Array (SKA) Born Global. In *Resolving the Sky: Radio Interferometry: Past, Present, and Future*, ed. M. A. Garrett and J. C. Greenwood (Manchester: SKA Organization), 68.
Ekers, R. D. 2014, Non-Thermal Radio Astronomy, *Astropar. Phys.*, **53**, 152.
Elbers, A. 2017, *The Rise of Radio Astronomy in the Netherlands* (Cham: Switzerland).
Elder, F. R., Langmuir, R. V. and Pollock, H. C. 1948, Radiation from Electrons Accelerated in a Synchrotron, *Phys. Rev.*, **74**, 52.
Elsmore, B., Ryle, M., and Leslie, P. R. R. 1959, The Positions, Flux Densities and Angular Diameters of 64 Radio Sources Observed at a Frequency of 178 Mc/s, *Mem. RAS*, **68**, 61.

Evans, D. S. 1949, Photometry of NGC 5128, *MNRAS*, **109**, 94.
Evans, J. V. and Taylor, G. N. 1959, Radio Echo Observations of Venus, *Nature*, **184**, 1358.
Ewen, H. I. and Purcell, E. M. 1951a, Radiation from Hyperfine Levels of Interstellar Hydrogen, *Phys. Rev.*, **83**, 881.
Ewen, H. I. and Purcell, E. M. 1951b, Radiation from Hyperfine Levels of Interstellar Hydrogen, *AJ*, **56**, 125.
Ewen, H. I. and Purcell, E. M. 1951c, Observation of a Line in the Galactic Radio Spectrum, *Nature*, **168**, 356.
Faber, S. M. and Gallagher, J. S. 1979, Masses and Mass-to-Light Ratios of Galaxies, *ARAA*, **17**, 135.
Fabian, A. C. and Rees, M. J. 1979, SS 433: A Double Jet in Action?, *MNRAS*, **187**, 13.
Feain, I. J. et al. 2011, The Radio Continuum Structure of Centaurus A at 1.4 GHz, *ApJ*, **740**, 17.
Field, G. 1962a, Absorption by Intergalactic Hydrogen, *ApJ*, **135**, 684.
Field, G. 1962b, Atmosphere of Mercury, *AJ*, **67**, 575.
Field, G. 2009, Cyanogen and the CMBR. In *Finding the Big Bang*, ed. P. J. E. Peebles, L. A. Page, and R. B. Partridge (Cambridge: Cambridge University Press), 75.
Field, G. and Hitchcock, J. L. 1966, Cosmic Black-Body Radiation at λ=2.6 mm, *Phys. Rev. Let.*, **16**, 817.
Field, G., Arp, H., and Bahcall, J. N. 1973, *The Redshift Controversy* (Reading: Benjamin).
Field, G., Herbig, G. H., and Hitchcock, J. 1966, Radiation Temperature of Space at λ2.6 mm, *AJ*, **71**, 161.
Findlay, J. W., Hvatum, H., and Waltman, W. B. 1965, An Absolute Flux-Density Measurement of Cassiopeia A, *ApJ*, **141**, 873.
Fixsen, D. J. et al. 2011, ARCADE 2 Measurement of the Absolute Sky Brightness at 3–90 GHz, *ApJ*, **734**, 5.
Fomalont, E. B. and Sramek, R. A. 1975, A Confirmation of Einstein's General Theory of Relativity by Measuring the Bending of Microwave Radiation in the Gravitational Field of the Sun, *ApJ*, **199**, 749.
Fomalont, E. B. and Sramek, R. A. 1976, Measurements of the Solar Gravitational Deflection of Radio Waves in Agreement with General Relativity, *Phys. Rev. Let.*, **36**, 147.
Fomalont, E. B. et al. 1964, Accurate Right Ascensions for 226 Radio Sources, *AJ*, **69**, 772.
Fomalont, E. B. et al. 2009, Progress in Measurements of the Gravitational Bending of Radio Waves Using the VLBA, *ApJ*, **699**, 1395.
Franklin, K. L. 1959, An Account of the Discovery of Jupiter as a Radio Source, *AJ*, **64**, 37.
Franklin, K. L. 1984, The Discovery of Jupiter Bursts. In *Serendipitous Discoveries in Radio Astronomy*, ed. K. I. Kellermann and B. Sheets (Green Bank: NRAO/AUI), 252.
Franklin, K. L. and Burke, B. F. 1956, Radio Observations of Jupiter, *AJ*, **61**, 177.
Fränz, K. 1942, Measurement of the Sensitivity of Short Wave Receivers, *Hochfrequenztechnik. und Elektroakustik*, **59**, 105 (in German).
Frater, R. H. and Ekers, R. D. 2012, John Paul Wild 1923–2008, *Mem. Roy. Soc.*, **58**, 327.
Frater, R. H., Goss, W. M., and Wendt, H. W. 2017, *Four Pillars of Radio Astronomy: Mills, Christiansen, Wild, Bracewell* (Cham: Springer).
Friis, H. T. 1965, Karl Jansky: His Career at Bell Telephone Laboratories, *Science*, **149**, 841.
Friis, H. T. 1971, *Seventy-Five Years in an Exciting World* (San Francisco: San Francisco Press).
Friis, H. T. and Feldman, C. B. 1937, A Multiple Unit Steerable Antenna for Short-Wave Radio Reception, *Proc. IRE*, **25**, 841.

Gamow, G. 1948, The Evolution of the Universe, *Nature*, **162**, 680.
Gamow, G. 1949, On Relativistic Cosmology, *Rev. Mod. Phys.*, **21**, 367.
Gamow, G. 1950, Half an Hour of Creation, *Phys. Today*, **3**, 16.
Gardner, F. F. and McGee, R. X. 1967, Detection of β-Transitions in the Recombination Spectrum of Hydrogen Near 9 cm Wavelength, *Nature*, **213**, 480.
Gardner, F. F. and Whiteoak, J. B. 1962, Polarization of 20-cm Wavelength Radiation from Radio Sources, *Phys. Rev. Let.*, **9**, 197.
Gardner, F. F. et al. 1964, Detection of the Interstellar OH Lines at 1612 and 1720 Mc/sec, *Phys. Rev. Let.*, **13**, 3.
Garwin, R. L. 1974, Detection of Gravity Waves Challenged, *Phys. Today*, **27**, 9.
Geller, M. J. and Huchra, J. P. 1989, Mapping the Universe, *Science*, **246**, 897.
George, M., Orchiston, W., and Wielebinski, R. 2017, The History of Early Low Frequency Radio Astronomy in Australia. 8: Grote Reber and the 'Square Kilometre Array' near Bothwell, Tasmania, in the 1960s and 1970s, *J. Astron. Hist. Heritage*, **20**, 195.
Getmantsev, G. G. 1952, Cosmic Electrons as a Source of Radio Emission from the Galaxy, *Dokl. Akad. Nauk SSSR*, **83**, 557 (in Russian).
Ginzburg, V. L. 1946, On Solar Radiation in the Radio Spectrum, *Dokl. Akad. Nauk, SSSR*, **52**, 487 (in Russian).
Ginzburg, V. L. 1951, Cosmic Rays as a Source of Galactic Radio Emission, *Dokl. Akad. Nauk SSSR*, **76**, 377 (in Russian). English translation in *Classics in Radio Astronomy*, ed. W. T. Sullivan III (Dordrecht: Reidel), 93.
Ginzburg, V. L. 1961, On the Nature of the Radio Galaxies, *Sov. Astron.*, **5**, 282; English translation of *Astron. Zh.*, **38**, 380.
Gold, T. 1951, The Origin of Cosmic Radio Noise. In *Proceedings of the Conference on Dynamics of Ionized Media*, ed. R. I. Boyd (London: University College London), 105. Reprinted in 1979 in *A Source Book in Astronomy and Astrophysics, 1900–1975*, ed. K. R. Lang and O. Gingerich (Cambridge, MA: Harvard University Press), 783.
Gold, T. 1968, Rotating Neutron Stars as the Origin of the Pulsating Radio Sources, *Nature*, **218**, 731.
Gold, T. 1984, The Reception of New Ideas by the Scientific Community. Paper Presented at the *Third Annual Meeting of the Society for Scientific Exploration*, October 1984.
Goldreich, P. and Lynden-Bell, D. 1969, Io, a Jovian Unipolar Inductor, *ApJ*, **156**, 59.
Goldstein, R. M. and Carpenter, R. L. 1963, Rotation of Venus: Period Estimated from Radar Measurements, *Science*, **139**, 910.
Goldstein, S. et al. 1964, OH Absorption Spectra in Sagittarius, *Nature*, **203**, 65.
Gordon, M. A. and Sorochenko, R. L. 2002, *Radio Recombination Lines* (Dordrecht: Kluwer).
Gordon, W. E. 1958, Incoherent Scattering of Radio Waves by Free Electrons with Applications to Space Exploration by Radar, *Proc. IRE*, **46**, 1824.
Gordon, W. E. and LaLonde, L. 1961, The Design and Capabilities of an Ionospheric Radar Probe, *Trans. IRE*, **AP-9**, 17.
Gorenstein, M. V. and Smoot, G. 1981, Large-Angular-Scale Anisotropy in the Cosmic Background Radiation, *ApJ*, **244**, 361.
Goss, W. M. 2013, *Making Waves: The Story of Ruby Payne-Scott: Australian Pioneer Radio Astronomer* (Berlin: Springer).
Goss, W. M. and McGee, R. X. 1996, The Discovery of the Radio Source Sagittarius A (Sgr A). In ASPC **102**, *The Galactic Center*, ed. R. Gredel (San Francisco: Astronomical Society of the Pacific), 369.

Goss, W. M. and McGee, R. X. 2010. *Under the Radar: The First Woman in Radio Astronomy: Ruby Payne-Scott* (Berlin: Springer).
Goss, W. M., Brown, R. L., and Lo, K. Y. 2003, The Discovery of Sgr A*. In *Proceedings of the Galactic Center Workshop 2002: The Central 300 Parsecs of the Milky Way*, ed. A. Cotera et al. *Astron. Nachtr.*, 324, 497.
Goss, W. M., Hooker, C., and Ekers, R. D. 2023, *Joe Pawsey and the Founding of Australian Radio Astronomy: Early Discoveries, from the Sun to the Cosmos* (Cham: Springer).
Gottesman, S. T., Davies, R. D., and Reddish, V. C. 1966, A Neutral Hydrogen Survey of the Southern Regions of the Andromeda Nebula, *MNRAS*, **133**, 359.
Gower, J. F. R. 1966, The Source Counts from the 4C Survey, *MNRAS*, **133**, 151.
Gower, J. F. R., Scott, P. F., and Wills, D. 1967, A Survey of Radio Sources in the Declination Ranges –07° to 20° and 40° to 80°, *Mem. RAS*, **71**, 4.
Graham-Smith, F. 1950, Origin of the Fluctuations in the Intensity of Radio Waves from Galactic Sources: Cambridge Observations, *Nature*, **165**, 422.
Graham-Smith, F. 1951, An Accurate Determination of the Positions of Four Radio Stars, *Nature*, **168**, 555.
Graham-Smith, F. 1952, Apparent Angular Sizes of Discrete Radio Sources: Observations at Cambridge, *Nature*, **170**, 1065.
Graham-Smith, F. 1986, Martin Ryle 1918–1984, *Mem. Roy. Soc.*, **32**, 495.
Gray, R. H. 2012, *The Elusive WOW: Search for Extraterrestrial Intelligence* (Chicago: Palmer Square Press).
Gray, R. H. and Marvel, K. B. 2001, A VLA Search for the Ohio State "Wow," *ApJ*, **546**, 1171.
Green, P. E. and Pettengill, G. H. 1960, Exploring the Solar System by Radar, *Sky Tel.*, **20**, 9.
Greenfield, P. D., Roberts, D. H., and Burke, B. F. 1985, The Gravitationally Lensed Quasar 0957+561: VLA Observations and Mass Models, *ApJ*, **293**, 356.
Greenstein, J. L. 1961, The First True Radio Star. *Eng. Sci.*, **24**, 18.
Greenstein, J. L. 1984, Optical and Radio Astronomers in the Early Years. In *Serendipitous Discoveries in Radio Astronomy*, ed. K. I. Kellerman and B. Sheets (Green Bank: NRAO/AUI), 79.
Greenstein, J. L. and Matthews, T. A. 1963, Redshift of the Radio Source 3C 48, *Nature*, **197**, 1041.
Greenstein, J. L. and Schmidt, M. 1963, The Quasi-Stellar Radio Sources 3C 48 and 3C 273, *ApJ*, **140**, 1.
Güdel, M. 2002, Stellar Radio Astronomy: Probing Stellar Atmospheres from Protostars to Giants, *ARAA*, **40**, 217.
Gundermann, E. 1965, Observations of the Interstellar Hydroxyl Radical, PhD Dissertation, Harvard University.
Gurvits, L., Kellermann, K. I., and Frey, L. 1999, The "Angular Size–Redshift Relation" for Compact Radio Structures in Quasars and Radio Galaxies, *A&A*, **342**, 378.
Gush, H. P., Halpern, M., and Wisnow, E. H. 1990, Rocket Measurement of the Cosmic-Background Radiation mm Wave Spectrum, *Phys. Rev. Let.*, **65**, 537.
Haddock, F. 1984, U.S. Radio Astronomy Following World War II. In *Serendipitous Discoveries in Radio Astronomy*, ed. K. I. Kellermann and B. Sheets (Green Bank: NRAO/AUI), 115.
Hagen, J. P. 1951, Naval Research Laboratory Eclipse Expedition to Attu, Alaska, September 12, 1950, *AJ*, **56**, 39.
Hagen, J. P., Haddock, F. T., and Reber, G., 1951, NRL Aleutian Radio Eclipse Expedition, *Sky Tel.*, **10**, 111.
Hagen, J. P. et al., 1948, Observations on the May 20, 1947, Total Eclipse of the Sun, *NRL Report*, March 1948.

Hall, P. (ed.) 2005, *The SKA; An Engineering Perspective* (Dordrecht: Springer).
Ham, R. A. 1975, The Hissing Phenomena, *J. Br. Astron. Assoc.*, **85**, 317.
Hanbury Brown, R. 1984, The Development of Michelson and Intensity Long Baseline Interferometry. In *Serendipitous Discoveries in Radio Astronomy*, ed. K. I. Kellermann and B. Sheets (Green Bank: NRAO/AUI), 133.
Hanbury Brown, R. and Hazard, C. 1950, Radio-Frequency Radiation from the Great Nebula in Andromeda (M.31), *Nature*, **166**, 901.
Hanbury Brown, R. and Hazard, C. 1952, Radio-Frequency Radiation from Tycho Brahe's Supernova (A.D. 1572), *Nature*, **170**, 364.
Hanbury Brown, R. and Hazard, C. 1953a, Radio-Frequency Radiation from the Spiral Nebula Messier 81, *Nature*, **172**, 853.
Hanbury Brown, R. and Hazard, C. 1953b, A Survey of 23 Localized Radio Sources in the Northern Hemisphere, *MNRAS*, **113**, 123.
Hanbury Brown, R. and Twiss, R. Q. 1956, A Test of a New Type of Stellar Interferometer on Sirius, *Nature*, **178**, 1046.
Hanbury Brown, R., Jennison, R. C., and Das Gupta, M. K. 1952, Apparent Angular Sizes of Discrete Radio Source, *Nature*, **170**, 1061.
Hargrave, P. J. and Ryle, M. 1974, Observations of Cygnus A with the 5-km Radio Telescope, *MNRAS*, **166**, 305.
Harp, G. R. et al. 2020, An ATA Search for a Repetition of the Wow Signal, *AJ*, **160**, 162.
Harris, D. E. and Roberts, J. A. 1960, Radio Source Measurements at 960 Mc/s, *PASP*, **72**, 237.
Harris, M. 2019, Untitled, *Radio Science Bulletin*, **371**, 91.
Harwit, M. 1981, *Cosmic Discovery* (New York: Basic Books).
Harwit, M. 1984, Observational Discovery vs. Theoretical Discovery. In *Serendipitous Discoveries in Radio Astronomy*, ed. K. I. Kellermann and B. Sheets (Green Bank: NRAO/AUI), 197.
Harwit, M. 2013, *In Search of the True Universe* (New York: Cambridge University Press).
Harwit, M. 2019, *Cosmic Discovery*, revised issue (Cambridge: Cambridge University Press).
Harwit, M. 2021, *Cosmic Messengers* (Cambridge: Cambridge University Press).
Haynes, M. P. and Giovanelli, R. 1986, The Connection between Pisces–Perseus and the Local Supercluster *ApJ*, **306**, L55.
Haynes, R. F. and Haynes, D. H. 1993, 3C 273: The Hazards of Publication, *PASA*, **10** 355.
Haynes, R. et al. 1996, *Explorers of the Southern Sky: A History of Australian Radio Astronomy* (Cambridge: Cambridge University Press).
Hazard, C. 1962, The Method of Lunar Occultations and Its Application to a Survey of Radio Source 3C 212, *MNRAS*, **124**, 343.
Hazard, C. and Walsh, D. 1959, An Experimental Investigation of the Effects of Confusion in a Survey of Localized Radio Sources, *MNRAS*, **119**, 648.
Hazard, C., Mackey, M. B., and Shimmins, A. J. 1963, Investigation of the Radio Source 3C 273 by the Method of Lunar Occultations, *Nature*, **197**, 1037.
Hazard, C. et al. 2018, The Sequence of Events that Led to the 1963 Publication in *Nature* of 3C 273, the First Quasar and the First Extragalactic Radio Jet, *PASA*, **35**, 6.
Heightman, D. W. 1946, Signals from the Sun, *Wireless World*, March, 99.
Helfand, D. J. 1982, A Superfast New Pulsar, *Nature*, **300**, 573.
Henyey, L. G. and Keenan, P. C. 1940, Interstellar Radiation from Free Electrons and Hydrogen Atoms, *ApJ*, **91**, 625.
Hernstein, J. R. et al. 1999, A Geometric Distance to the Galaxy NGC 4258 from Orbital Motions in a Nuclear Gas Disk, *Nature*, **400**, 539.

Hewish, A. 1961, Extrapolation of the Number-Flux Density Relation of Radio Stars by Scheuer's Statistical Methods, *MNRAS*, **123**, 167.

Hewish, A. 1974, Pulsars and High Density Physics, 1974 Nobel Lecture. In *Nobel Lectures, Physics 1971–1980*, ed. S. Lundqvist (Singapore: World Scientific Publishing).

Hewish, A. 2008, Background to Discovery: Some Recollections. In *40 Years of Pulsars: Millisecond Pulsars, Magnetars, and More*, ed. C. G. Bassa et al. (New York: American Institute of Physics), 3.

Hewish, A. and Okoye, S. E. 1965, Evidence for an Unusual Source of High Radio Brightness Temperature in the Crab Nebula, *Nature*, **207**, 59.

Hewish, A. and Wyndham, J. D. 1963, The Solar Corona in Interplanetary Space, *MNRAS*, **126**, 469.

Hewish, A., Scott, P. F., and Wills, D. 1964, Interplanetary Scintillation of Small Diameter Radio Source, *Nature*, **203**, 1214.

Hewish, A. et al. 1968, Observations of a Rapidly Pulsating Radio Source, *Nature*, **217**, 709.

Hewitt, J. N. et al. 1988, Unusual Radio Source MG1131+0456: A Possible Einstein Ring, *Nature*, **333**, 537.

Hey, J. S. 1946, Solar Radiation in the 4–6 Metre Radio Wave-Length Band, *Nature*, **157**, 47.

Hey, J. S. 1973, *The Evolution of Radio Astronomy* (New York: Science History Publications).

Hey, J. S., Parsons, S. J., and Phillips, J. W. 1946, Fluctuations in Cosmic Radiation at Radio Frequencies, *Nature*, **158**, 234.

Hey, J. S., Phillips, J. W., and Parsons, S. J. 1946, Cosmic Radiations at 5 Metre Wavelength, *Nature*, **157**, 296.

Hills, R. et al. 2018, Concerns about Modelling of the EDGES Data, *Nature*, **564**, 32.

Hirabayashi, H., Edwards, P. G., and Murphy D. W. (eds.) 2000, *Astrophysical Phenomena Revealed by Space VLBI* (Sagamihara: Institute of Space and Astronautical Science).

Hjellming, R. M. 1988, Radio Stars. In *Galactic and Extragalactic Radio Astronomy*, ed. G. L. Verschuur and K. I. Kellermann (Berlin: Springer), 384.

Hjellming, R. M. and Johnston, K. J. 1981, An Analysis of the Proper Motions of SS 433 Radio Jets, *Nature*, **290**, 100.

Hjellming, R. M. and Wade, C. M. 1970, Radio Novae, *ApJ*, **162**, L1.

Högbom, J. A. and Shakeshaft, J. R. 1961, Secular Variations of the Flux Density of the Radio Source Cassiopeia A, *Nature*, **189**, 561.

Hoglund, B. and Mezger, P. G. 1965a, The Detection of the Hydrogen Emission Line N110–N109 at the Frequency 5009 Mc/Sec in Galactic H II Regions, *AJ*, **70**, 678.

Hoglund, B. and Mezger, P. G. 1965b, Hydrogen Emission Line $n_{110} \rightarrow n_{109}$: Detection at 5009 Megahertz in Galactic H II Regions, *Science*, **150**, 339.

Horowitz, P. and Sagan, C. 1993, Five Years of Project Meta: An All-Sky Narrow-Band Radio Search for Extraterrestrial Signals, *ApJ*, **415**, 218.

Hotan, A. W. 2021, Australian Square Kilometre Array Pathfinder: I. System Description, *PASA*, **38**, 9.

Howard, W. E., Barrett, A. H., and Haddock, F. T. 1962, Measurement of the Microwave Radiation from the Planet Mercury, *ApJ*, **136**, 995.

Howell, T. F. and Shakeshaft, J. R. 1966, Measurement of the Minimum Cosmic Background Radiation at 20.7-cm Wave-Length, *Nature*, **210**, 1318.

Howell, T. F. and Shakeshaft, J. R. 1967, Spectrum of the 3° K Cosmic Microwave Radiation, *Nature*, **216**, 753.

Hoyle, F. 1948, A New Model for the Expanding Universe, *MNRAS*, **108**, 372.

Hoyle, F. 1959, The Relation of Radio Astronomy to Cosmology. In IAU Symposium no. **9**, *Paris Symposium on Radio Astronomy*, ed. R. N. Bracewell (Stanford: Stanford University Press), 529.

Hoyle, F. 1968, The Bakerian Lecture: Review of Recent Developments in Cosmology, *Proc. Roy. Soc. London*, **308**, 1.

Hoyle, F. 1981, *The Quasar Controversy Resolved* (Cardiff: University College Cardiff Press).

Hoyle, F. 1982, The Universe Past and Present, *ARAA*, **220**, 1.

Hoyle, F. and Burbidge, G. 1966, On the Nature of the Quasi-Stellar Objects, *ApJ*, **144**, 534.

Hoyle, F. and Fowler, W. A. 1963, Nature of Strong Radio Sources, *Nature*, **197**, 533.

Hoyle, F. and Narlikar, J. 1961, On the Counting of Radio Sources in the Steady-State Cosmology, *MNRAS*, **123**, 133.

Hoyle, F. and Narlikar, J. 1962, On the Counting of Radio Sources in the Steady-State Cosmology, II, *MNRAS*, **125**, 13.

Hoyle, F., Burbidge, G., and Narlikar, J. V. 1994, Astrophysical Deductions from the Quasi-Steady-State Cosmology, *MNRAS*, **267**, 1007.

Hubble, E. 1936a, *The Realm of the Nebulae* (New Haven: Yale University Press), reprinted 1958 by Dover Publications Inc.

Hubble, E. 1936b, Effects of Red Shifts on the Distribution of Nebulae, *PNAS*, **22**, 621.

Hulse, R. A. 1994a, The Discovery of the Binary Pulsar, *Rev. Mod. Phys.*, **66**, 699.

Hulse, R. A. 1994b, The Discovery of the Binary Pulsar, *BAAS*, **26**, 971.

Hulse, R. A. and Taylor, J. H. 1974a, A High Sensitivity Pulsar Survey, *ApJ*, **191**, L59.

Hulse, R. A. and Taylor, J. H. 1974b, Discovery of a Pulsar in a Close Binary System, *BAAS*, **6**, 453.

Hulse, R. A. and Taylor, J. H. 1975, Discovery of a Pulsar in a Binary System, *ApJ*, **195**, L51.

Humphreys, E. M. L. et al. 2013, Toward a New Geometric Distance to the Active Galaxy NGC 4258. III. Final Results and the Hubble Constant, *ApJ*, **775**, 13.

Hyman, S. D. et al. 2005, A Powerful Bursting Radio Source toward the Galactic Center, *Nature*, **434**, 50.

Jaffe, A. H. et al. 2001, Cosmology from MAXIMA-1, BOOMERANG, and COBE DMR Cosmic Microwave Background Observations, *Phys. Rev. Let.*, **86**, 3475.

Jakes, W. C. 1963, Participation of the Holmdel Station in the Telstar Project, *Bell Syst. Tech. J.*, **42**, 1421.

Jansky, C. M, Jr. 1958, The Discovery and Identification by Karl Guthe Jansky of Electromagnetic Radiation of Extraterrestrial Origin in the Radio Spectrum, *Proc. IRE*, **46**, 13.

Jansky, K. G. 1932, Directional Studies of Atmospherics at High Frequencies, *Proc. IRE*, **20**, 1920.

Jansky, K. G. 1933a, Electrical Disturbances Apparently of Extraterrestrial Origin, *Proc. IRE*, **21**, 1387.

Jansky, K. G. 1933b, Radio Waves from Outside the Solar System, *Nature*, **132**, 66.

Jansky, K. G. 1933c, Electrical Phenomena that Apparently Are of Interstellar Origin, *Popular Astronomy*, **41**, 548.

Jansky, K. G. 1935, A Note on the Source of Interstellar Interference, *Proc. IRE*, **23**, 1158.

Jansky, K. G. 1937, Minimum Noise Levels Obtained on Short Wave Receiving Systems, *Proc. IRE*, **25**, 1517.

Jansky, K. G. 1939, An Experimental Investigation of the Characteristics of Certain Types of Noise, *Proc. IRE*, **27**, 763.

Jarosik, N. et al. 2007, Three-Year Wilkinson Microwave Anisotropy Probe (WMAP) Observations: Beam Profiles, Data Processing, Radiometer Characterization, and Systematic Error Limits, *ApJS*, **170**, 263.

Jauncey, D. 1967, Re-Examination of the Source Counts for the 3C Revised Catalogue, *Nature*, **216**, 877.
Jauncey, D. J. 1975, Radio surveys and Source Counts, *ARAA*, **13**, 23.
Jeans, J. 1929, *Astronomy and Cosmogony* (Cambridge: The University Press).
Jeffrey, R. M. et al. 2016, Fast Launch Speeds in Radio Flares, From a New Determination of the Intrinsic Motions of SS 433's Jet Bolides, *MNRAS*, **461**, 312.
Jennison, R. C. and Das Gupta, M. K. 1953, Fine Structure in the Extraterrestrial Radio Source Cygnus I, *Nature*, **172**, 996.
Jones, D. E. 1961, The Microwave Temperature of Venus, *Planet. Space Sci.*, **5**, 166.
Kaidanovsky, N. I. 2012, The Department of Radio Astronomy of the Main Astrophysical Observatory. In *A Brief History of Radio Astronomy in the USSR* (English ed.), ed. K. I. Kellermann (Dordrecht: Springer), 127.
Kapahi, V. K. 1987, The Angular Size-Redshift Relation as a Cosmological Tool. In IAU Symposium no. **124**, *Observational Cosmology*, ed. A. Hewitt, G. Burbidge, and L. Z. Fang (Dordrecht: Reidel), 251.
Kardashev, N. 1959, On the Possibility of Detection of Allowed Lines of Atomic Hydrogen in the Radio Frequency Spectrum, *Sov. Astron.*, **3**, 813; Russian original, *Astron. Zh.*, **36**, 838.
Kardashev, N. 1962, Nonstationarity of Spectra of Young Sources of Nonthermal Radio Emission, *Sov. Astron.*, **6**, 317; Russian original: *Astron. Zh.*, **39**, 393.
Kardashev, N. et al. 2013, RadioAstron-A Telescope with a Size of 300 000 km: Main Parameters and First Observational Results, *Astron. Rpts.*, **57**, 153. Russian original, *Astron. Zh.* 2013, **90**, 179.
Kaspi, V. M. and Beloborodov, A. M. 2017, Magnetars, *ARAA*, **55**, 261.
Kauffman, H. 1946, A DX Record: To the Moon and Back, *QST*, **30**, 65.
Kaufman, M. 1965, Limits on the Density of Intergalactic Ionized Hydrogen, *Nature*, **207**, 736.
Kellermann, K. I. 1964, The Spectra of Non-Thermal Radio Sources, *ApJ*, **140**, 969.
Kellermann, K. I. 1965a, 11 cm Observations of the Temperature of Mercury, *Nature*, **205**, 1091.
Kellermann, K. I. 1965b, Radio Observations of Mars, *Nature*, **206**, 1034.
Kellermann, K. I. 1966, The Thermal Radio Emission from Mercury, Venus, Mars, Saturn, and Uranus, *Icarus*, **5**, 478.
Kellermann, K. I. 1972, Radio Galaxies, Quasars, and Cosmology, *AJ*, **77**, 531.
Kellermann, K. I. 1993, The Cosmological Deceleration Parameter Estimated from the Angular-Size/Redshift Relation for Compact Radio Sources, *Nature*, **361**, 134.
Kellermann, K. I. 1996, John Gatenby Bolton (1922–1993), *PASP*, **108**, 729.
Kellermann, K. I. 1999, Grote Reber's Observations on Cosmic Static, *ApJ*, **525**, 37.
Kellermann, K. I. 2003, Grote Reber (1911–2002), *Nature*, **421**, 596.
Kellermann, K. I. 2004, Grote Reber (1911–2002), *PASP*, **116**, 703.
Kellermann, K. I. 2014, The Discovery of Quasars and Its Aftermath, *J. Astron. Hist. Heritage*, **17**, 267.
Kellermann, K. I., and Cohen, M. H. 1988, The Origin and Evolution of the NRAO-Cornell VLBI System, *JRASC*, **82**, 248.
Kellermann, K. I. and Moran, J. M. 2001, The Development of High-Resolution Imaging in Radio Astronomy, *ARAA*, **29**, 457.
Kellermann, K. I. and Pauliny-Toth, I. I. K. 1966a, Observations of the Radio Emission from Uranus, Neptune, and Other Planets at 1.9 cm, *ApJ*, **145**, L954.
Kellermann, K. I. and Pauliny-Toth, I. I. K. 1966b, A Search for Radio Emission from the Star Alpha Orionis, *ApJ*, **145**, 953.
Kellermann, K. I. and Pauliny-Toth, I. I. K. 1968, Variable Radio Sources, *ARAA*, **6**, 417.

Kellermann, K. I. and Pauliny-Toth, I. I. K. 1969, The Spectra of Opaque Radio Sources, *ApJ*, **155**, 71.
Kellermann, K. I. and Sheets, B. 1984, *Serendipitous Discoveries in Radio Astronomy* (Green Bank: NRAO).
Kellermann, K. I. and Wall, J. V. 1987, Radio Source Counts and Their Interpretation. In IAU Symposium **124**, *Observational Cosmology*, ed. A. Hewett et al. (Dordrecht: Reidel), 545.
Kellermann, K. I., Bouton, E. N., and Brandt, S. S. 2020, *Open Skies: The National Radio Astronomy Observatory and Its Impact on US Radio Astronomy* (Cham: Springer).
Kellermann, K. I., Davis, M. M., and Pauliny-Toth, I. I. K. 1971, Counts of Radio Sources at 6-Centimeter Wavelength, *ApJ*, **170**, L1.
Kellermann, K. I., Orchiston, W., and Slee, B. 2005, Gordon James Stanley and the Early Development of Radio Astronomy in Australia and the United States, *PASA*, **22**, 13.
Kellermann, K. I., Pauliny-Toth, I. I. K., and Williams, P. J. 1969, The Spectra of Radio Sources in the Revised 3C Catalogue, *ApJ*, 157, 1.
Kellermann, K. I. et al. 1962, A Correlation between the Spectra of Non-thermal Radio Sources and Their Brightness Temperatures, *Nature*, **195**, 692.
Kellermann, K. I. et al. 1977, The Small Radio Source at the Galactic Center, *ApJ*, **214**, L61.
Kellermann, K. I. et al. 1989, VLA Observations of Objects in the Palomar Bright Quasar Survey, *AJ*, **96**, 1195.
Kellermann, K. I. et al. 2004, Sub-Milliarcsecond Imaging of Quasars and Active Galactic Nuclei. III. Kinematics of Parsec-scale Radio Jets, *ApJ*, **609**, 539.
Kellermann, K. I. et al. 2016, Radio-Loud and Radio-Quiet QSOs, *ApJ*, **831**, 168.
Kennefick, D. 2009, Testing Relativity from the 1919 Eclipse: A Question of Bias, *Phys. Today*, **62**, 37.
Khaikin, S. E. and Chikhachev, B. M. 1948, Investigation of Radio Emission from the Sun during the Total Solar Eclipse of May 20, 1947, *Izvestiya Akad. Nauk SSSR*, **12**, 38 (in Russian).
Kiepenheuer, K. O. 1950, Cosmic Rays As the Source of General Galactic Radio Emission, *Phys. Rev.*, **79**, 738.
Kinman, T. D. 1965, The Nature of the Fainter Haro-Luyten Objects, *ApJ*, **142**, 1241.
Klein, M. J. and Seling, T. V. 1966, Radio Emission from Uranus at 8 Gc/s, *ApJ*, **146**, 599.
Kleppner, D. and Horowitz, P. 2016, A Perfect Proposal, *Phys. Today*, **69**, 48.
Knight, C. A. et al. 1971, Quasars: Millisecond-of-Arc Structure Revealed by Very-Long-Baseline Interferometry, *Science*, **172**, 52.
Knowles, S. G. et al. 1969a, Spectra, Variability, Size, Polarization of H_2O Microwave Emission in the Galaxy, *Science*, **163**, 1055.
Knowles, S. G. et al. 1969b, Galactic Water Vapor Emission: Further Observations of Variability, *Science*, **166**, 221.
Kotelnikov, V. 1961, Radar Contact with Venus, *J. Br. Inst. Radio Eng.*, **22**, 293.
Kotelnikov, V. 1963, Radar Observations of the Planet Venus in the Soviet Union in April, 1963, Science Report of the Institute of Radio Engineering and Electronics, Moscow. English translation by the Defense Documentation Center.
Kozyrev, N. A. 1964, The Atmosphere of Mercury, *Sky Tel.*, **27**, 339.
Kragh, H. 1996, *Cosmology and Controversy* (Princeton: Princeton University Press).
Kragh, H. 2017, The Nobel Prize System and the Astronomical Sciences, *J. Hist. Astron.*, **48**, 257.
Kramer, M. et al. 2021, Strong-Field Gravity Tests with the Double Pulsar, *Phys. Rev. X*, **11**, 041050.

Kraus, J. D. 1956, Impulsive Radio Signals from the Planet Venus, *Nature*, **178**, 33; Radio Observations of the Planet Venus at a Wave-length of 11 m., *Nature*, **178**, 103; Class II Radio Signals from Venus at a Wave-length of 11 Metres, *Nature*, **178**, 159.

Kraus, J. 1984, Karl Guthe Jansky's Serendipity, Its Impact on Astronomy and Its Lessons for the Future. In *Serendipitous Discoveries in Radio Astronomy*, ed. K. I. Kellermann and B. Sheets (Green Bank: NRAO/AUI), 57.

Kraus, J. 1995, *Big Ear Two Listening for Other Worlds* (Powell: Cygnus-Quasar Books).

Kuiper, G. P. 1952, *The Atmospheres of the Earth and Planets* (Chicago: University of Chicago Press).

Kulkarni, S. R. et al. 2014, Giant Sparks at Cosmological Distances?, *ApJ*, **797**, 70.

Kusch, P. and Prodell, A. G. 1950, On the Hyperfine Structure of Hydrogen and Deuterium, *Phys. Rev.*, **79**, 1009.

Langer, R. M. 1936, Radio Noise from the Galaxy, *Phys. Rev.*, **49**, 209.

Lebach, D. E. et al. 1995, Measurement of the Solar Gravitational Deflection of Radio Waves Using Very-Long-Baseline Interferometry, *Phys. Rev. Let.*, **75**, 1439.

Legg, T. H. 1970, Redshift and the Size of Double Radio Sources, *Nature*, **226**, 64.

Lequeux, J. 1962, Mesures Interférométriques a Haute Résolution du Diamétre et de la Structure des Principales Radiosources, *Ann. Astrophys.*, **25**, 221.

Levine, J. L. 2004, Early Gravity-Wave Detection Experiments, 1960–1975, *Phys. Pers.*, **6**, 42.

Levy, D. H. 1993, *The Man Who Sold the Milky Way: A Biography of Bart Bok* (Tucson: University of Arizona Press), 45.

Levy, G. S. et al. 1986, Very Long Baseline Interferometric Observations Made with an Orbiting Radio Telescope, *Science*, **234**, 187.

Lilley, A. E. and McClain, E. F. 1956, The Hydrogen-Line Red Shift of Radio Source Cygnus A, *ApJ*, **123**, 127.

Lilley, A. E. et al. 1966, Radio Astronomical Detection of Helium, *Nature*, **211**, 174.

Lissauer, J. J. and de Pater, I. 2013, *Fundamental Planetary Science: Physics, Chemistry, and Habitability* (New York: Cambridge).

Lister, M. et al. 2018, MOJAVE. XV. VLBA 15 GHz Total Intensity and Polarization Maps of 437 Parsec-scale AGN Jets from 1996 to 2017, *ApJS*, **234**, 12.

Little, A. and Bracewell, R. N. 1961, The Central Component of Centaurus A, *AJ*, **66**, 290.

Little, C. G. and Lovell, A. C. B. 1950, Origin of the Fluctuations in the Intensity of Radio Waves from Galactic Sources: Jodrell Bank Observations, *Nature*, **165**, 423.

Litvak, M. M. et al. 1966, Maser Model for Interstellar OH Microwave Emission, *Phys. Rev. Let.*, **17**, 821.

Livingston, M. S. 1954, *High-Energy Accelerators* (New York: Interscience Publishers).

Longair, M. 1966, On the Interpretation of Radio Source Counts, *MNRAS*, **133**, 421.

Longair, M. 2009, The Discovery of Pulsars and the Aftermath, *Proc. Am. Phil. Soc.*, **155**, 147.

Lopez-Corredoira, M. 2011, Pending Problems in QSOs, *Int. J. Astron. Astrophys.*, **1**, 73.

Lorimer, D. et al. 2007, Bright Millisecond Radio Burst of Extragalactic Origin, *Science*, **318**, 777.

Lovell, A. C. B. 1964, Joseph Lade Pawsey, *Mem. Roy. Soc.*, **228**, 229.

Lovell, A. C. B. 1969, Observation of Prolonged Radio Emission from a Red Dwarf Star, *Nature*, **222**, 1126.

Lovell, A. C. B. 1977, The Effects of Defense Science on the Advance of Astronomy, *J. Hist. Astron.*, **8**, 151.

Lovell, A. C. B. 1984, Impact of World War II on Radio Astronomy. In *Serendipitous Discoveries in Radio Astronomy*, ed. K.I. Kellermann and B. Sheets (Green Bank: NRAO/AUI), 89.

Lovell, A. C. B. 1987, *Voice of the Universe: Building the Jodrell Bank Telescope* (New York: Praeger).
Lovell, A. C. B. 1990, *Astronomer by Chance* (New York: Basic Books).
Low, F. 1964, Infrared Brightness Temperatures of Saturn, *AJ*, **69**, 550.
Low, F. 1966, The Infrared Brightness Temperature of Uranus, *ApJ*, **146**, 326.
Lynden-Bell, D., 1969, Galactic Nuclei as Collapsed Old Quasars, *Nature*, **223**, 690.
Lynden-Bell, D. and Rees, M. 1971, On Quasars, Dust and the Galactic Centre, *MNRAS*, **152**, 461.
Lynds, C. R. and Villere, G. 1965, On the Interpretation of the Integral Count–Apparent Magnitude Relation for the Haro-Luyten Objects, *ApJ*, **142**, 1296.
Lyne, A. and Bailes, M. 1992, No Planet Orbiting 1829-10, *Nature*, **355**, 325.
Lyne, A. et al. 2004, A Double Pulsar System: A Rare Laboratory for Relativistic Gravity and Plasma Physics, *Science*, **303**, 1153.
Maltby, P. and Moffet, A. T. 1965, Time Dependence of the Radio Emission from CTA 21 and CTA 102, *ApJ*, **142**, 1699.
Maltby, P., Matthews, T. A., and Moffet, A. T. 1963, Brightness Distribution in Discrete Radio Sources. IV. A Discussion of 24 Identified Sources, *ApJ*, **137**, 153.
Marcy, G. W. and Butler, R. P. 1998, Detection of Extrasolar Planets, *ARAA*, **36**, 57.
Margon, B. et al. 1979, Enormous Periodic Doppler Shifts in SS433, *ApJ*, **233**, 63.
Martyn, D.F. 1946, Temperature Radiation from the Quiet Sun in the Radio Spectrum, *Nature*, **158**, 632.
Masui, K. 2015, Dense Magnetic Plasma Associated with Fast Radio Burst, *Nature*, **528**, 523.
Mather, J.C. 2006, *Les Prix Nobel. The Nobel Prizes 2006*, ed. Karl Grandin (Stockholm: Nobel Foundation).
Matthews, T. A. and Sandage, A. 1963, Optical Identification of 3C 48, 3C 196, and 3C 286 with Stellar Objects, *ApJ*, **138**, 30.
Matthews, T. A., Bolton, J. G., Greenstein, J. L., Münch, G., and Sandage, A. R. 1960, Unpublished paper presented at the 107th meeting of the AAS, December 1960.
Matthews, T. A., Morgan, W. W., and Schmidt, M. 1964, A Discussion of Galaxies Identified with Radio Sources, *ApJ*, **140**, 35.
Matveyenko, L. I. 2013, Early VLBI in the USSR. In *Resolving the Sky: Radio Interferometry: Past, Present and Future*, ed. M. A. Garrett and J. C. Greenwood (Manchester: SKA Observatory), 43.
Matveyenko, L. I., Kardashev, N. S., and Sholomitsky, G. V. 1965, Large Baseline Radio Interferometers, *Radiophysica*, **8**, 651; English translation, 1966, *Sov. Radiophys.*, **8**, 46.
Maxwell, A., Swarup, G., and Thompson, A. R. 1958, The Radio Spectrum of Solar Activity, *Proc. IRE*, **46**, 142.
Mayer, C. H. 1959, Planetary Radiation at Centimeter Wavelengths, *AJ*, **64**, 43.
Mayer, C. H. 1984, Early Observations of Planetary Radio Emission. In *Serendipitous Discoveries in Radio Astronomy*, ed. K. I. Kellermann and B. Sheets (Green Bank: NRAO/AUI), 266.
Mayer, C. H., McCullough, T. P., and Sloanaker, R. M. 1957, Evidence for Polarized Radio Emission from the Crab Nebula, *ApJ*, **126**, 468.
Mayer, C. H., McCullough, T. P., and Sloanaker, R. M. 1958a, Observations of Venus at 3.15-CM Wave Length, *ApJ*, **127**, 1.
Mayer, C. H., McCullough, T. P., and Sloanaker, R. M. 1958b, Observation of Mars and Jupiter at a Wave Length of 3.15 cm, *ApJ*, **127**, 11.
Mayer, C. H., McCullough, T. P., and Sloanaker, R. M. 1958c, Observation of Mars and Jupiter at a Wave Length of 3.15 cm, *Proc. IRE*, **46**, 260.

Mayer, C. H., McCullough, T. P., and Sloanaker, R. M. 1962a, Evidence for Polarized 3.5-CM Radiation from the Radio Galaxy Cygnus A, *ApJ*, **135**, 656.

Mayer, C. H., McCullough, T. P., and Sloanaker, R. M. 1962b, Polarization of the Radio Emission of Taurus A, Cygnus A, and Centaurus A, *AJ*, **67**, 581.

Mayer, C. H., McCullough, T. P., and Sloanaker, R. M. 1964, Linear Polarization of the Centimeter Radiation of Discrete Sources, *ApJ*, **139**, 248.

Mayor, M. and Queloz, D. 1995, A Jupiter-Mass Companion to a Solar-Type Star, *Nature*, **378**, 355.

McClain, E. 1960, The 600-Foot Radio Telescope, *SciAm*, **202**, 45.

McClain, E. and Sloanaker, R. 1959, Preliminary Observations at 10-cm Wavelength Using the NRL 84-ft Radio Telescope. In IAU Symposium no. **9**, *Paris Symposium on Radio Astronomy*, ed. R. N. Bracewell (Stanford: Stanford University Press), 61.

McCrea, W. 1984, The Influence of Radio Astronomy on Cosmology. In *The Early Years of Radio Astronomy*, ed. W. T. Sullivan III (Cambridge: Cambridge University Press), 272.

McCready, L. L., Pawsey, J. L., and Payne-Scott, R. 1947, Solar Radiation at Radio Frequencies and Its Relation to Sunspots, *Proc. Roy. Soc. London A*, **190**, 357.

McGee, R. X. and Bolton, J. G. 1954, Probable Observation of the Galactic Nucleus at 400 Mc/s, *Nature*, **173**, 985.

McGee, R. X. et al. 1965, Anomalous Intensity Ratios of the Interstellar Lines of OH in Absorption and Emission, *Nature*, **208**, 1193.

McGovern, W. E., Gross, S. H., and Rasool, S. I. 1965, Rotation Period of the Planet Mercury, *Nature*, **208**, 375.

McGuire, B. et al. 2018, First Results of an ALMA Band 10 Spectral Line Survey of NGC 6334I: Detections of Glycolaldehyde (HC(O)CH2OH) and a New Compact Bipolar Outflow in HDO and CS, *ApJL*, **863**, L35.

McKellar, A. 1940, Evidence for the Molecular Origin of Some Hitherto Unidentified Interstellar Lines, *PASP*, **52**, 187.

McKellar, A. 1941, Molecular Lines from the Lowest States of Diatomic Molecules Composed of Atoms Probably Present in Interstellar Space, *Publ. Dominion Astrophysical Observatory*, **7**, 251.

McLaughlin, M. A. et al. 2006, Transient Radio Bursts from Rotating Neutron Stars, *Nature*, **439**, 817.

Menzel, D. 1937, Physical Processes in Gaseous Nebulae. I., *ApJ*, **85**, 330.

Merton, R. K. and Barber, E. 2004, *The Travels and Adventures of Serendipity* (Princeton: Princeton University Press).

Mezger, P. G. and Palmer, P. 1968, Radio Recombination Lines: A New Observational Tool in Astrophysics, *Science*, **160**, 29.

Michelson, A. A. 1890, On the Application of Interference Methods to Astronomical Measurements, *Phil. Mag. Ser.* **5**, 30, 1.

Michelson, A. A. 1920, On the Application of Interference Methods to Astronomical Measurements, *ApJ*, **51**, 257.

Michelson, A. A. and Pease, F. G. 1921, Measurement of the Diameter of α Orionis with the Interferometer, *ApJ*, **53**, 249.

Mickaliger, M. B. et al. 2012, A Giant Sample of Giant Pulses from the Crab Pulsar, *ApJ*, **760**, 64.

Miley, G. 1971, The Radio Structure of Quasars: A Statistical Investigation, *MNRAS*, **152**, 477.

Milgrom, M. 1983, A Modification of the Newtonian Dynamics as a Possible Alternative to the Hidden Mass Hypothesis, *ApJ*, **270**, 365.

Milgrom, M. 2002, Does Dark Matter Really Exist?, *SciAm*, **287**, 42.

Mills, B. Y. 1952a, The Positions of Six Discrete Sources of Cosmic Radio Radiation, *Austrl. J. Sci. Res. A*, **5**, 456.
Mills, B. Y. 1952b, The Distribution of the Discrete Sources of Cosmic Radio Radiation, *Austrl. J. Sci. Res. A*, **5**, 266.
Mills, B. Y. 1952c, Apparent Angular Sizes of Discrete Radio Sources: Observations at Sydney, *Nature*, **170**, 1063.
Mills, B. Y. 1956, Letter to the Editor, *SciAm*, **195**, 8.
Mills, B. Y. 1959, A Survey of Radio Sources at 3.5-m Wavelength. In IAU Symposium no. **9**, *Paris Symposium on Radio Astronomy*, ed. R. N. Bracewell (Stanford: Stanford University Press), 498.
Mills, B. Y. 1984, Radio Source Counts and the Log N–Log S Controversy. In *The Early Years of Radio Astronomy*, ed. W. T. Sullivan III (Cambridge: Cambridge University Press), 147.
Mills, B. Y. and Slee, O. B. 1957, A Preliminary Survey of Radio Sources in a Limited Region of the Sky at a Wavelength of 3.5m, *Austrl. J. Phys.*, **10**, 162.
Mills, B. Y. and Thomas, A. B. 1951, Observations of the Source of Radio-Frequency Radiation in the Constellation of Cygnus, *Austrl. J. Sci. Res. A*, **4**, 158.
Mills, B. Y., Slee, O. B., and Hill, E. R. 1958, A Catalogue of Radio Sources between Declinations +10° and –20°, *Austrl. J. Phys.*, **11**, 360.
Mills, B. Y., Slee, O. B., and Hill, E. R. 1960, A Catalogue of Radio Sources between Declinations –20° and –50°, *Austrl. J. Phys.*, **13**, 676.
Mills, B. Y., Slee, O. B., and Hill, E. R. 1961, A Catalogue of Radio Sources between Declinations –50° and –80°, *Austrl. J. Phys.*, **14**, 497.
Minkowski, R. 1942, The Crab Nebula, *ApJ*, **96**, 199.
Minkowski, R. 1959, Optical Observations of Nonthermal Galactic Radio Sources. In IAU Symposium No. **9**, *Paris Symposium on Radio Astronomy*, ed. R. N. Bracewell (Stanford: Stanford University Press), 315.
Minkowski, R. 1960a, A New Distant Cluster of Galaxies, *PASP*, **72**, 354.
Minkowski, R. 1960b, A New Distant Cluster of Galaxies, *ApJ*, **132**, 908.
Mirabel, I. F. and Rodriguez, L. F. 1994, A Superluminal Source in the Galaxy, *Nature*, **371**, 46.
Mitchell, D. L. and de Pater, I. 1994, Microwave Imaging of Mercury's Thermal Emission at Wavelengths from 0.3 to 20.5 cm, *Icarus*, **110**, 2.
Mitton, S. 2005, *Conflict in the Cosmos: Fred Hoyle's Life in Science* (Washington: Joseph Henry Press).
Miyoshi, M. et al. 1995, Evidence for a Black Hole from High Rotation Velocities in a Sub-Parsec Region of NGC 4258, *Nature*, **373**, 127.
Moffet, A. T. 1966, The Structure of Radio Galaxies, *ARAA*, **4**, 145.
Moran, J. M. 1967a, Observations of OH Emissions in the HII Region W3 with a 74,400λ Interferometer, *ApJ*, **148**, L69.
Moran, J. M. 1967b, Spectral Line Interferometry with Independent Time Standards at Stations Separated by 845 Kilometers, *Science*, **157**, 676.
Moran, J. M. et al. 1968, The Structure of the OH Source in W3, *ApJ*, **152**, L97.
Moran, J. M. et al. 1973, Very Long Baseline Interferometric Observations of the H_2O Sources in W49N, W3(OH), Orion A, and VY Canis Majoris, *ApJ*, **185**, 535.
Moran, J. M. et al. 1978, Evidence for the Zeeman Effect in the OH Maser Emission from W3(OH), *ApJ*, **224**, 67.
Morris, D. and Berge, G. L. 1962, Measurements of the Polarization and Angular Extent of the Decimetric Radiation of Jupiter, *ApJ*, **136**, 276.
Morris, D., Palmer, H. P., and Thompson, A. R. 1957, Five Radio Sources of Small Angular Diameter, *Obs.*, **77**, 103.

Morrison, D. and Klein, M. J. 1970, The Microwave Spectrum of Mercury, *ApJ*, **160**, 325.
Morrison, D. and Sagan, C. 1967, The Microwave Phase Effect of Mercury, *ApJ*, **150**, 1105.
Morrison, D., Sagan, C., and Pollack, J. 1969, Martian Temperatures and Thermal Properties, *Icarus*, **11**, 36.
Muhleman, D. O., Ekers, R. D., and Fomalont, E. B. 1970, Radio Interferometric Test of the General Relativistic Light Bending Near the Sun, *Phys. Rev. Let.*, **24**, 1377.
Muller, C. A. and Oort, J. H. 1951, The Interstellar Hydrogen Line at 1,420 Mc./sec and an Estimate of Galactic Rotation, *Nature*, **168**, 356.
Murray, B. C. and Wildey, R. L. 1963, Stellar and Planetary Observations at 10 Microns, *ApJ*, **137**, 692.
Nakagami, M. and Miya, K. 1939, On the Incident Angle of Short Radio Waves during the "Dellinger Effect," *J. Inst. Elec. Eng. Japan*, **59**, 176 (in Japanese).
Nakai, N., Inoue, M., and Miyoshi, M. 1993, Extremely-High-Velocity H_2O Maser Emission in the Galaxy, NGC 4258, *Nature*, **361**, 45.
Napier, P. J., Thompson, A. R., and Ekers, R. D. 1983, The Very Large Array: Design and Performance of a Modern Synthesis Radio Telescope, *Proc. IEEE*, **71**, 1295.
Noordam, J. 2013, The Dawn of the SKAI: What Really Happened. In *Resolving the Sky: Radio Interferometry: Past, Present, and Future*, ed. M. A. Garrett and J. C. Greenwood (Manchester: SKA Organization), 68.
Nordmann, C. 1902, A Search for Hertzian Waves Emanating from the Sun, *Comptes Rendus*, **134**, 273 (in French). English translation 1982 in *Classics in Radio Astronomy*, ed. W. T. Sullivan (Dordrecht: Reidel), 158.
Novikov, I. 2009, Cosmology in the Soviet Union in the 1960s. In *Finding the Big Bang*, ed. P. J. E. Peebles, L. A. Page, and R. B. Partridge (Cambridge: Cambridge University Press), 99.
Ohm, E. A. 1961, Receiving System, *Bell Syst. Tech. J.*, **40**, 1065.
Oke, J. B. 1963, Absolute Energy Distribution in the Optical Spectrum of 3C 273, *Nature*, **197**, 1040.
Oort, J. H. and Walraven, Th. 1956, Polarization and Composition of the Crab Nebula, *BAN*, **12**, 285.
Oort, J. H., Kerr, F. J., and Westerhout, G. 1958, The Galactic System as a Spiral Nebula, *MNRAS*, **118**, 379.
Öpik, E. J. 1962, Jupiter: Chemical Composition, Structure, and Origin of a Giant Planet, *Icarus*, **1**, 200.
Orchiston, W. 2004, From the Solar Corona to Clusters of Galaxies: The Radio Astronomy of Bruce Slee, *PASA*, **21**, 23.
Orchiston, W. 2005a, Dr. Elizabeth Alexander: First Female Radio Astronomer. In *The New Astronomy: Opening the Electromagnetic Window and Expanding Our View of Planet Earth*, ed. W. Orchiston (Dordrecht: Springer), 71.
Orchiston, W. 2005b, Sixty Years in Radio Astronomy: A Tribute to Bruce Slee, *J. Astron. Hist. Heritage*, **8**, 30.
Orchiston, W. and Slee, O. B. 2002, The Australasian Discovery of Solar Radio Emission, *AAO Newsletter*, November 2002, 25.
Orchiston, W. and Steinberg, J.-L. 2007, Highlighting the History of French Radio Astronomy. 2: The Solar Eclipse Observations of 1949–1954, *J. Astron. Hist. Heritage*, **10**, 11.
Orchiston, W., Nakamura, T., and Ishiguro, M. 2016, Highlighting the History of Japanese Radio Astronomy. 4: Early Solar Research at Osaka, *J. Astron. Hist. Heritage*, **19**, 240.
Orchiston, W. et al. 2009, Highlighting the History of French Radio Astronomy. 4: Early Solar Research at the École Normale Supérieure, Narcoussis and Nançay, *J. Astron. Hist. Heritage*, **12**, 175.

Ostriker, J. P., Peebles, P. J. E., and Yahil, A. 1974, The Size and Mass of Galaxies, and the Mass of the Universe, *ApJ*, **193**, L1.

Ozernoy, L. M. and Sazonov, V. N. 1969, The Spectrum and Polarization of a Source of Synchrotron Emission with Components Flying Apart at Relativistic Velocities, *Astrophys. Space Sci.*, **3**, 395.

Pacini, F. 1967, Energy Emission from a Neutron Star, *Nature*, **216**, 567.

Palmer, H. et al. 1967, Radio Diameter Measurements with Interferometer Baselines of One Million and Two Million Wavelengths, *Nature*, **213**, 789.

Parijisky, Yu. N. 1992, Radio Astronomy of the Next Century, *Astron. Astrophys. Trans.*, **1**, 85.

Partridge, R. B. 1980, Fluctuations in the Cosmic Microwave Background at Small Angular Scales. In *The Universė at Large Redshifts*, ed. J. Kalckar, O. Ulfbeck, and N. R. Nelson, *Physica Scripta*, 21, 624.

Pauliny-Toth, I. I. K. and Kellermann, K. I. 1966, Variations in the Radio-Frequency Spectra of 3C 84, 3C 273, 3C 279, and Other Radio Sources, *ApJ*, **146**, 634.

Pauliny-Toth, I. I. K. et al. 1972, The NRAO 5-GHz Radio Source Survey. II. The 140-Ft "Strong," "Intermediate," and "Deep" Source Surveys, *AJ*, **77**, 265.

Pauliny-Toth, I. I. K. et al. 1978, The 5 GHz Strong Source Surveys. IV. Survey of the Area between Declination 35 and 70 Degrees and Summary of Source Counts, Spectra and Optical Identifications, *AJ*, **83**, 451.

Pawsey, J. L. 1946, Observations of Million Degree Thermal Radiation from the Sun at a Wavelength of 1.5 Metres, *Nature*, **158**, 633.

Pawsey, J. L. 1957, Preliminary Statistics of Discrete Sources Obtained with the "Mills Cross," In IAU Symposium no. **4**, *Radio Astronomy*, ed. H. C. Van de Hulst (Cambridge: Cambridge University Press), 228.

Pawsey, J. L. and Bracewell, R. N. 1955, *Radio Astronomy* (Oxford: Clarendon Press).

Pawsey, J. L., Payne-Scott, R., and McCready, L. L. 1946, Radio Frequency Energy from the Sun, *Nature*, **157**, 158.

Payne-Gaposchkin, C. 1954, *Introduction to Astronomy* (New York: Prentice-Hall).

Payne-Scott, R., Yabsley, D. E., and Bolton, J. G. 1947, Relative Times of Arrival of Bursts of Solar Noise on Different Radio Frequencies, *Nature*, **160**, 256.

Peale, S. J. and Gold, T. 1965, Rotation of the Planet Mercury, *Nature*, **206**, 1240.

Pearson, T. J. 1974, Variation of Radio Source Counts With Direction, for the 3CR and 4C Surveys, *MNRAS*, **166**, 249.

Pearson, T. J. and Kus, A. J. 1978, The 5C 6 and 5C 7 Surveys of Radio Sources, *MNRAS*, **182**, 273.

Peebles, P. J. E. 2014, Discovery of the Hot Big Bang: What Happened in 1948, *Eur. Phys. J. H*, **39**, 205.

Peebles, P. J. E., Page, L. A. and Partridge, R. B. (eds.) 2009, *Finding the Big Bang* (Cambridge: Cambridge University Press).

Penzias, A. A. 1965, Helium-Cooled Reference Noise Source in a 4-kMc Waveguide, *Rev. Sci. Instr.*, **36**, 68.

Penzias, A. A. 1992, The Origin of the Elements. In *Nobel Lectures in Physics 1971–1980*, ed. S. Lundqvist (Singapore: World Scientific), 444.

Penzias, A. A. 2009, Encountering Cosmology. In *Finding the Big Bang*, ed. J. E. Peebles, L. A. Page, and R. B. Partridge (Cambridge: Cambridge University Press), 144.

Penzias, A. A. and Wilson, R. W. 1965, A Measurement of Excess Antenna Temperature at 4080 Mc/s, *ApJ*, **142**, 419.

Penzias, A. A. and Wilson, R. W. 1967, A Measurement of the Background Temperature at 1415 MHz, *AJ*, **72**, 315.

Penzias, A. A., Schraml, J., and Wilson, R. 1969, Observational Constraints on a Discrete-Source Model to Explain the Micro-Wave Background, *ApJ*, **157**, L49.
Perkins, F., Gold, T., and Salpeter, E. E. 1966, Maser Action in Interstellar OH, *ApJ*, **145**, 361.
Perley, R. A. and Meisenheimer, K. 2017, High-Fidelity VLA Imaging of the Radio Structure of 3C 273, *A&A*, **601**, 35.
Perley, R. A., Dreher, J. W., and Cowan, J. J. 1984, The Jet and Filaments in Cygnus A, *ApJ*, **285**, 35.
Perlmutter, A. et al. 1999, Measurements of Omega and Lambda from 42 High Redshift Supernovae, *ApJ*, **517**, 565.
Pesce, D. W. et al. 2020, The Megamaser Cosmology Project. XIII. Combined Hubble Constant Constraints, *ApJ*, **891**, L1.
Petroff, E. et al. 2015, Identifying the Source of Perytons at the Parkes Radio Telescope, *MNRAS*, **451**, 393.
Pettengill, G. H. 1984, Discovery of Mercury's Rotation. In *Serendipitous Discoveries in Radio Astronomy*, ed. K. I. Kellermann and B. Sheets (Green Bank: NRAO/AUI), 275.
Pettengill, G. H. and Dyce, R. B. 1965, A Radar Determination of the Rotation of the Planet Mercury, *Nature*, **206**, 1240.
Pettengill, G. H. and Price, R. 1961, Radar Echoes from Venus and a New Determination of the Solar Parallax, *Planet. Space Sci.*, **5**, 71.
Pettengill, G. H. et al. 1962, A Radar Investigation of Venus, *AJ*, **67**, 181.
Pettit, E. and Nicholson, S. B. 1936, Radiation from the Planet Mercury, *ApJ*, **83**, 84.
Pfeiffer, J. 1956, *The Changing Universe: The Story of the New Astronomy* (New York: Random House).
Pick, M. et al. 2011, Highlighting the History of French Radio Astronomy. 6: The Multi-Element Grating Arrays at Nançay, *J. Astron. Hist. Heritage*, **14**, 57.
Piddington, J. H. and Minnett, H. C. 1949, Microwave Thermal Radiation from the Moon, *Austrl. J. Sci. Res. A*, **2**, 63.
Piddington, J. H. and Minnett, H. C. 1951, Observations of Galactic Radiation at Frequencies of 1200 and 3000 Mc/s, *Austrl. J. Sci. Res. A*, **4**, 495.
Pilkington, J. D. H. and Scott, J. F. 1965, A Survey of Radio Sources between Declinations 20° and 40°, *MNRAS*, **60**, 183.
Pilkington, J. D. H. et al. 1968, Observations of Some Further Pulsed Radio Sources, *Nature*, **218**, 126.
Pollack, J. B. and Sagan, C. 1965, The Microwave Phase Effect of Venus, *Icarus*, **4**, 62.
Pooley, G. G. 1968, Counts of Radio Sources at 2,700 MHz, *Nature*, **218**, 152.
Porcas, R. W. et al. 1980, Radio Positions and Optical Identifications for Radio Sources Selected at 966 MHz – II, *MNRAS*, **191**, 607.
Porcas, R. W. et al. 1981, VLBI Structures of the Images of the Double QSO 0957+561, *Nature*, **289**, 758.
Pound, R. V. and Rebka, G. A. 1960, Apparent Weight of Photons, *Phys. Rev. Let.*, **4**, 337.
Price, R. M. 1969, A Measurement of the Sky Brightness Temperature at 408 MHz, *Austrl. J. Phys.*, **22**, 641.
Price, R. M. 1984, The First Years at Parkes. In *Serendipitous Discoveries in Radio Astronomy*, ed. K. I. Kellermann and B. Sheets (Green Bank: NRAO/AUI), 300.
Price, R. et al. 1959, Radar Echoes from Venus, *Science*, **129**, 751.
Radhakrishnan, V. and Roberts, J. A. 1960, Polarization and Angular Extent of the 960-Mc/sec Radiation from Jupiter, *Phys. Rev. Let.*, **4**, 493.
Read, R. B., 1962, Accurate Measurement of the Declinations of Radio Sources, PhD Dissertation, California Institute of Technology.

Read, R. B. 1963, Accurate Measurement of the Declinations of Radio Sources, *ApJ*, **138**, 1.
Readhead, A. C. S. and Wilkinson, P. N. 1978, The Mapping of Compact Radio Sources from VLBI Data, *ApJ*, **223**, 25.
Reasenberg, R. D. et al. 1979, Viking Relativity Experiment: Verification of Signal Retardation by Solar Gravity, *ApJ*, **234**, L219.
Reber, G. 1940a, Cosmic Static, *ApJ*, **91**, 621.
Reber, G. 1940b, Cosmic Static, *Proc. IRE*, **28**, 68.
Reber, G. 1942, Cosmic Static, *Proc. IRE*, **30**, 367.
Reber, G. 1944, Cosmic Static, *ApJ*, **100**, 279.
Reber, G. 1946, Solar Radiation at 480 Mc, *Nature*, **158**, 945.
Reber, G. 1948a, Cosmic Static, *Proc. IRE*, **36**, 1215.
Reber, G. 1948b, Cosmic Radio Noise, *Radio-Electronic Eng.*, **11**, 3.
Reber, G. 1948c, Solar Intensity at 408 Mc, *Proc. IRE*, **36**, 88.
Reber, G. 1949a, Galactic Radio Waves, *Sky Tel.*, **8**, 139.
Reber, G. 1949b, Radio Astronomy, *SciAm*, **181**, 34.
Reber, G. 1950, Galactic Radio Waves, *Astron. Soc. Pacific Leaflet*, No. 259.
Reber, G. 1958, Early Radio Astronomy in Wheaton, Illinois, *Proc. IRE*, **46**, 15.
Reber, G. 1964, Reversed Bean Vines, *J. Genet.*, **59**, 37.
Reber, G. 1966, Ground-Based Astronomy: The NAS 10-Year Program, *Science*, **152**, 150.
Reber, G. 1982, A Timeless, Boundless, Equilibrium Universe, *Publ. Astron. Soc. Austrl.*, **4**, 482.
Reber, G. 1984, Radio Astronomy between Jansky and Reber. In *Serendipitous Discoveries in Radio Astronomy*, ed. K. I. Kellermann and B. Sheets (Green Bank: NRAO/AUI), 71.
Reber, G. and Beck, E. 1951, The Measurement of 65 Centimeter Radiation during the Total Solar Eclipse of September 12, 1960, *AJ*, **56**, 47.
Reber, G. and Ellis, W. 1956, Cosmic Radio-Frequency Radiation Near One Megacycle, *J. Geophys. Res.*, **61**, 1.
Reber, G. and Greenstein, J. L. 1947, Radio Frequency Investigations of Astronomical Interest, *Obs.*, **67**, 15.
Rees, M. J. 1966, Appearance of Relativistically Expanding Radio Sources, *Nature*, **211**, 468.
Rees, M. J. 1967, Studies in Radio Source Structure: I. A Relativistically Expanding Model for Variable Quasi-Stellar Radio Sources, *MNRAS*, **135**, 345.
Rees, M. J. 1977, A Better Way of Searching for Black-Hole Explosions, *Nature*, **266**, 333.
Rees, M. J. and Sciama, D. W., 1968, Large-Scale Density Inhomogeneities in the Universe, *Nature*, **217**, 511.
Rees, M. J. and Setti, G. 1968, Model for the Evolution of Extended Radio Sources, *Nature*, **219**, 127.
Refsdal, S. 1964, On the Possibility of Determining Hubble's Parameter and the Masses of Galaxies from the Gravitational Lens Effect, *MNRAS*, **128**, 307.
Refsdal, S. and Surdej, J. 1994, Gravitational Lenses, *Rpts. Prog. Phys.*, **57**, 117.
Reid, M. J. and Honma, M. 2014, Microarcsecond Radio Astrometry, *ARAA*, **52**, 339.
Reid, M. J., Pesce, D. W., and Riess, A. G. 2019, An Improved Distance to NGC 4258 and Its Implications for the Hubble Constant, *ApJ*, **886**, 27.
Reifenstein, E. C. et al. 1970, A Survey of H109α Recombination Line Emission in Galactic HII Regions of the Northern Sky, *A&A*, **4**, 357.
Reiss, A. G. et al. 1998, Observational Evidence from Supernovae for an Accelerating Universe and a Cosmological Constant, *AJ*, **116**, 1009.
Rickard, J. R. and Cronyn, W. M. 1979, Interstellar Scattering, the North Polar Spur, and a Possible New Class of Compact Galactic Radio Sources, *ApJ*, **228**, 755.

Rieke, G. and Low, F. 1973, Infrared Maps of the Galactic Nucleus, *ApJ*, **184**, 415.
Roberts, J. A. and Stanley, G. J. 1959, Radio Emission from Jupiter at a Wavelength of 31 Centimeters, *PASP*, **71**, 485.
Roberts, M. S. 1966, A High-Resolution 21-cm Hydrogen-Line Survey of the Andromeda Nebula, *ApJ*, **144**, 639.
Roberts, M. S. 1975a, Radio Observations of Neutral Hydrogen in Galaxies. In *Galaxies and the Universe*, ed. A. Sandage, M. Sandage, and J. Kristian (Chicago: University of Chicago Press), 321.
Roberts, M. S. 1975b, The Rotation Curve of Galaxies. In IAU Symposium no. **69**, *Dynamics of Stellar Systems*, ed. A. Hayli (Dordrecht: Reidel), 331.
Roberts, M. S. 2008, M31 and a Brief History of Dark Matter. In ASPC **395**, *A Celebration of NRAO's 50th Anniversary, 2008*, ed. A. H. Bridle, J. J. Condon, and G. C. Hunt (San Francisco: Astronomical Society of the Pacific), 283.
Roberts, M. S. and Rots, A. H., 1973, Comparison of Rotation Curves of Different Galaxy Types, *A&A*, **26**, 483.
Roberts, M. S. and Whitehurst, R. N. 1975, The Rotation Curve and Geometry of M31 at Large Galactocentric Distances, *ApJ*, **201**, 327.
Robertson, D. S., Carter, W. E., and Dillinger, W. H. 1991, New Measurement of Solar Gravitational Deflection of Radio Signals Using VLBI, *Nature*, **349**, 768.
Robertson, P. 1992, *Beyond Southern Skies: Radio Astronomy and the Parkes Telescope* (Cambridge: Cambridge University Press).
Robertson, P. 2017, *Radio Astronomer: John Bolton and a New Window on the Universe* (Sydney: New South).
Robertson, P., Orchiston, W., and Slee, B. 2014, John Bolton and the Discovery of Discrete Radio Sources, *J. Astron. Hist. Heritage*, **17**, 283.
Robinson, B. J. et al. 1964, An Intense Concentration of OH Near the Galactic Centre, *Nature*, **202**, 989.
Rogers, A. A. E. and Morrison, P. 1972, Long-Baseline Interferometry, *Science*, **175**, 218.
Rogers, A. E. E. et al. 1966, Interferometric Study of Cosmic Line Emission at OH Frequencies, *Phys. Rev. Let.*, **17**, 50.
Roll, P. G. and Wilkinson, D. T. 1966, Cosmic Background Radiation at 3.2 cm: Support for Cosmic Black-Body Radiation, *Phys. Rev. Let.*, **16**, 405.
Rowson, B. 1962, High Resolution Observations with a Tracking Radio Interferometer, *MNRAS*, **125**, 177.
Rubin, V. C. 2000, One Hundred Years of Rotating Galaxies, *PASP*, **112**, 747.
Rubin, V. C. and Ford, W. K. 1970, Rotation of the Andromeda Nebula from a Spectroscopic Survey of Emission Regions, *ApJ*, **159**, 379.
Rudy, D. J. et al. 1987, Mars: VLA Observations of the Northern Hemisphere and the North Polar Region at Wavelengths of 2 and 6 cm, *Icarus*, **71**, 159.
Ryle, M. 1949, Evidence for the Stellar Origin of Cosmic Rays, *Proc. Phys. Soc. A*, **62**, 491.
Ryle, M. 1950, Radio Astronomy, *Rpts. Prog. Phys.*, **13**, 184.
Ryle, M. 1955, Halley Lecture: Radio Stars and Their Cosmological Significance, *Obs.*, **74**, 137.
Ryle, M. 1956a, Radio Galaxies, *SciAm*, **195**, 205.
Ryle, M. 1956b, Letter to the Editor, *SciAm*, **195**, 10.
Ryle, M. 1957, The Spatial Distribution of Radio Stars. In IAU Symposium no. **4**, *Radio Astronomy*, ed. H. C. Van de Hulst (Cambridge: Cambridge University Press), 110.
Ryle, M. 1958, Bakerian Lecture: The Nature of Cosmic Radio Sources, *Proc. R. Soc. London*, **248**, 289.
Ryle, M. 1959, The Nature of Radio Sources. In IAU Symposium no. **9**, *Paris Symposium on Radio Astronomy*, ed. R. N. Bracewell (Stanford: Stanford University Press), 523.

Ryle, M. 1961, Radio Astronomy and Cosmology, *Nature*, **190**, 852.
Ryle, M. 1962, The Radio Luminosity Function and the Number Flux-Density Relationship for the Discrete Sources. In IAU Symposium no. **15**, *Problems of Extra-Galactic Research*, ed. G. C. McVittie (New York: Macmillan), 326.
Ryle, M. 1963, Radio Astronomical Tests of Cosmological Models. In *Radio Astronomy Today*, ed. H. P. Palmer and M. I. Large (Manchester: Manchester University Press), 228.
Ryle, M. 1968, Counts of Radio Sources, *ARAA*, **6**, 249.
Ryle, M. and Clark, R. W. 1961, An Examination of the Steady-State Model in the Light of Some Recent Observations of Radio Sources, *MNRAS*, **122**, 349.
Ryle, M. and Neville, A. C. 1962, A Radio Survey of the North Polar Region with a 4.5 Minute of Arc Pencil-Beam System, *MNRAS*, **125**, 39.
Ryle, M. and Scheuer, P. A. G. 1955, The Spatial Distribution and the Nature of Radio Stars, *Proc. R. Soc. London*, **230**, 448.
Ryle, M. and Smith, F. G. 1948, A New Intense Source of Radio-Frequency Radiation in the Constellation of Cassiopeia, *Nature*, **162**, 462.
Ryle, M. and Vonberg, D. D. 1946, Solar Radiation on 175 Mc./s, *Nature*, **158**, 339.
Ryle, M., Smith, F. G., and Elsmore, B. 1950, A Preliminary Survey of the Radio Stars in the Northern Hemisphere, *MNRAS*, **110**, 50.
Ryle, M. et al. 1965, High-Resolution Observations of the Radio Sources in Cygnus and Cassiopeia, *Nature*, **205**, 1259.
Sachs, R. K. and Wolfe, A. M. 1967, Perturbations of a Cosmological Model and Angular Variations of the Microwave Background, *ApJ*, **147**, 73.
Sagan, C. 1960, The Surface Temperature of Venus, *AJ*, **65**, 352.
Saha, M. N. 1946, Origin of Radio Waves from the Sun and the Stars, *Nature*, **158**, 717.
Sandage, A. 1961, The Ability of the 200-Inch Telescope to Discriminate between Selected World Models, *ApJ*, **133**, 355.
Sandage, A. 1965, The Existence of a Major New Constituent of the Universe: The Quasi-Stellar Galaxies, *ApJ*, **141**, 1560.
Sandage, A. R. 1970, Cosmology: A Search for Two Numbers, *Phys. Today*, **23**, 34.
Sandage, A. R. 1999, The First 50 Years at Palomar, *ARAA*, **37**, 445.
Sander, K. F. 1946, Measurement of Galactic Noise at 60 Mc/s, *J. Inst. Elec. Eng.*, **93**, 1487.
Sander, W. 1963, *The Planet Mercury* (London: Faber and Faber).
Scheuer, P. A. G. 1957, A Statistical Method for Analyzing Observations of Faint Radio Stars, *Math. Proc. Cambridge Phil. Soc.*, **53**, 764.
Schiaperelli, G. 1890, Sulla Rotazione di Mercurio, *Astron. Nach.*, **123**, 241.
Schilizzi, R. D. et al. 2023, *The Square Kilometre Array: A Mega-Science Project in the Making 1993–2012* (Cham: Springer).
Schisler, C. 2008, An Independent 1967 Discovery of Pulsars. In *40 Years of Pulsars: Millisecond Pulsars, Magnetars, and More*, ed. C. G. Bassa et al. (New York: American Institute of Physics), 462.
Schmidt, M. 1963, 3C 273: A Star-Like Object with Large Redshift, *Nature*, **197**, 1040.
Schmidt, M. 1968, Space Distribution and Luminosity Functions of Quasi-Stellar Radio Sources, *ApJ*, **151**, 393.
Schmidt, M. 1970, Space Distribution and Luminosity Functions of Quasars, *ApJ*, **162**, 371.
Schmidt, M. 1984, Discovery of Quasars. In *Serendipitous Discoveries in Radio Astronomy*, ed. K. I. Kellermann, and B. Sheets (Green Bank: NRAO/AUI), 171.
Schmidt, M. 2011, The Discovery of Quasars, *Proc. Am. Phil. Soc.*, **155**, 142.

Schmidt, M. and Green, R. 1983, Quasar Evolution Derived from the Palomar Bright Quasar Survey and Other Complete Quasar Surveys, *ApJ*, **269**, 352.

Schmidt, M., Schneider, D. P., and Gunn, J. E. 1995, Spectrscopic CCD Surveys for Quasars at Large Redshift. IV. Evolution of the Luminosity Function from Quasars Detected by Their Lyman-Alpha Emission, *AJ*, **110**, 68.

Schott, E. 1947, Radiation from the Sun, *Physikalische Blätter*, **3**, 159 (in German).

Schott, G. A. 1912, *Electromagnetic Radiation and Mechanical Reactions Arising from It* (Cambridge: Cambridge University Press).

Schwinger, J. 1949, On the Classical Radiation of Accelerated Electrons, *Phys. Rev.*, **75**, 1912.

Scott, P. F. and Ryle, M. 1961, The Number-Flux Density Relation for Radio Sources Away from the Galactic Plane, *MNRAS*, **122**, 389.

Scott, P. F., Ryle, M., and Hewish, A. 1961, First Results of Radio Star Observations Using the Method of Aperture Synthesis, *MNRAS*, **122**, 95.

Seaquist, E. R. 1967, Radio Emission from Stellar Coronas, *ApJ*, **148**, 23.

Seielstad, G. A., Sramek, R. A., and Weiler, K. W. 1970, Measurement of the Deflection of 9.602-GHz Radiation from 3C279 in the Solar Gravitational Field, *Phys. Rev. Let.*, **24**, 1373.

Seling, T. V. 1970, Observations of Saturn at λ3.75 cm, *AJ*, **75**, 67.

Seyfert, C. 1943, Nuclear Emission in Spiral Nebulae, *ApJ*, **97**, 28.

Shain, C. A. 1955, Location on Jupiter of a Source of Radio Noise, *Nature*, **176**, 836.

Shain, C. A. 1956, 18.3 Mc/s Radiation from Jupiter, *Austrl. J. Phys.* **9**, 61.

Shakeshaft, J. et al. 1955, A Survey of Radio Sources between Declinations –38° and +83°, *Mem. RAS*, **67**, 106.

Shapiro, I. I. 1964, Fourth Test of General Relativity, *Phys. Rev. Let.*, **13**, 789.

Shapiro, I. I. 1968, Fourth Test of General Relativity: Preliminary Results, *Phys. Rev. Let.*, **20**, 1265.

Shapiro, I. I. 1971, Fourth Test of General Relativity: New Radar Result, *Phys. Rev. Let.*, **26**, 1132.

Shapiro, I. I. et al. 2004, Measurement of the Solar Gravitational Deflection of Radio Waves Using Geodetic Very-Long-Baseline Interferometry Data, 1979–1999, *Phys. Rev. Let.*, **92**, 11101.

Shaver, P. A. and Pierre, M. 1989, Large-Scale Anisotropy in the Sky Distribution of Extragalactic Radio Sources, *A&A*, **220**, 35.

Shaver, P. A. et al. 1996, Decrease in the Space Density of Quasars at High Redshift, *Nature*, **384**, 439.

Shimmins, A. J., Bolton, J. G., and Wall, J. V. 1968, Counts of Radio Sources at 2,700 MHz, *Nature*, **217**, 818.

Shimmins, A. J., Clarke, M. E., and Ekers, R. D. 1966, Accurate Positions of 644 Radio Sources, *Austrl. J. Phys.*, **19**, 649.

Shklovsky, I. S. 1946, On the Radiation of Radio Waves by the Galaxy and the Upper Layers of the Solar Atmosphere, *Astron. Zh.*, **23**, 333.

Shklovsky, I. S. 1949, Monochromatic Radio Emission from the Galaxy and the Possibility of Its Observation, *Astron. Zh.*, **26**, 10. English translation in W. T. Sullivan, III 1982, *Classics in Radio Astronomy* (Cambridge: Cambridge University Press), 318.

Shklovsky, I. S. 1952, On the Nature of Radio Emission from the Galaxy, *Astron. Zh.*, **29**, 418.

Shklovsky, I. S. 1953a, On the Nature of the Radiation from the Crab Nebula, *Akad. Nauk Doklady*, **90**, 983 (in Russian). English translation 1979 in *A Source Book in Astronomy and Astrophysics, 1900–1975*, ed. K. R. Lang and O. Gingerich (Cambridge, MA: Harvard University Press), 490.

Shklovsky, I. S. 1953b, Possibility of Observing Monochromatic Radio Emission from Interstellar Molecules, *Dokl. Akad. Nauk SSSR*, **92**, 25 (in Russian).
Shklovsky, I. S. 1960a, Secular Variation of the Flux and Intensity of Radio Emission from Discrete Sources, *Sov. Astron.*, **4**, 243; English translation of *Astron. Zh.*, **37**, 256.
Shklovsky, I. S. 1960b, *Cosmic Radio Waves*, English ed., translated by R. B. Rodman and C. M. Varsavsky (Cambridge: Harvard University Press).
Shklovsky, I. S. 1962, Radio Galaxies. *Sov. Phys. Usp.*, **5**, 365; Russian original: *Uspekhi Fizicheskikh Nauk*, **77**, 60.
Shklovsky, I. S. 1965a, Possible Secular Variation of the Flux and Spectrum of Radio Emissions of Source 1934–63, *Nature*, **206**, 176.
Shklovsky, I. S. 1965b, A Possible Secular Variation of the Flux and Spectrum of the Radio Source 1934–63, *Sov. Astron.*, **9**, 22; Russian original: *Astron. Zh.*, **42**, 30.
Shklovsky, I. S. 1966, *Astron. Circ.* No. **364**, USSR Acad. Sci.
Shklovsky, I. S. 1982, On the History of the Development of Radio Astronomy in the USSR. In *News on Life, Science, and Technology*, No. **11** (Moscow: Izd. Znanie), 82 (in Russian).
Shmaonov, T. 1957, *Prbori I Tekhnika Experimenta*, **1**, 84 (in Russian).
Shobbrook, R. R. and Robinson, B. J. 1967, 21 cm Observations of NGC 300, *Austrl. J. Phys.*, **20**, 131.
Sholomitsky, G. B. 1965a, Variability of Radio Source CTA-102, *IAU Info. Bull. Var. Stars*, **No. 83**.
Sholomitsky, G. B. 1965b, Fluctuations in the 32.5-cm Flux of CTA 102, *Sov. Astron.*, **9**, 516; Russian original: *Astron. Zh.*, **42**, 673.
Sholomitsky, G. B. 1966, Flux Density Variations of CTA 102 at the Frequency 920 Mc/s, *Astron. Circ.* (Astronomical Council, USSR Academy of Sciences), **No. 359** (in Russian).
Silk, J. 1967, Fluctuations in the Primordial Fireball, *Nature*, **215**, 1155.
Singal, J. et al. 2018, The Radio Synchrotron Background: Conference Summary and Report, *PASP*, **130**, 036001.
Slee, B. 1994, Memories of the Dover Heights Field Station, *Austrl. J. Phys.*, **47**, 517.
Slipher, V. M. 1914, The Detection of Nebular Rotation, *Lowell Obs. Bull.*, **2**, 66.
Slysh, V. I. 1963, Angular Size of Radio Stars, *Nature*, **199**, 682.
Smith, H. and Hoffleit, D. 1963, Light Variations in the Superluminous Radio Galaxy 3C273, *Nature*, **198**, 630.
Smith, W. B. 1963, Radar Observations of Venus, 1961 and 1959, *AJ*, **68**, 158.
Smolders, A. B. and van Haarlem, M. P. 1999, *Perspectives on Radio Astronomy: Technologies for Large Antenna Arrays* (Dwingeloo: ASTRON).
Smoot, G. F. 2006, Nobel Lecture: Cosmic Microwave Background Radiation Anisotropies: Their Discovery and Utilization, *The Nobel Prize in Physics. NobelPrize.org*, available at: www.nobelprize.org/prizes/physics/2006/summary/ (last accessed 1 December 2022).
Smoot, G. F. and Davidson, K. 1993, *Wrinkles in Time* (New York: Avon Books).
Smoot, G. F., Gorenstein, M. V., and Muller, R. A. 1977, Detection of Anisotropy in the Cosmic Blackbody Radiation, *Phys. Rev. Let.*, **39**, 898.
Snyder, L. E. and Buhl, D. 1969a, On the Possibility of Detecting Interstellar Water Vapor Using a Radio Telescope, *BAAS*, **1S**, 204.
Snyder, L. E. and Buhl, D. 1969b, Water-Vapor Clouds in the Interstellar Medium, *ApJ*, **155**, L65.
Snyder, L. E. and Buhl, D. 1974, Detection of Possible Maser Emission Near 3.48 Millimeters from an Unidentified Molecular Species in Orion, *ApJ*, **189**, L31.

Snyder, L. E. et al. 1969, Microwave Detection of Interstellar Formaldehyde, *Phys. Rev. Let.*, **22**, 679.

Sofue, Y. and Rubin, V. 2001, Rotation Curves of Spiral Galaxies, *ARAA*, **39**, 137.

Sorochenko, R. L. and Borodzich, E. V. 1966, Detection of Radio Emission from the Excited Hydrogen Line in in NGC 6618 (Omega Nebula), *Sov. Phys. Dokl.*, **10**, 588; Russian original: *Dokl. Akad. Nauk SSSR*, **163**, 603, 1965.

Southworth, G. C. 1945, Microwave Radiation from the Sun, *J. Franklin Institute*, **239**, 285.

Southworth, G. C. 1956, Early History of Radio Astronomy, *Scientific Monthly*, **82**, 55.

Southworth, G. C. 1962, *Forty Years of Radio Research* (New York: Gordon and Breach).

Spitler, L. G. et al. 2014, Fast Radio Burst Discovered in the Arecibo Pulsar ALFA Survey, *ApJ*, **101**, 1.

Spitler, L. G. et al. 2016, A Repeating Fast Radio Burst, *Nature*, **531**, 202.

Spitzer, L. and Baade, W. 1951, Stellar Populations and Collisions of Galaxies, *ApJ*, **113**, 413.

Sramek, R. A. 1970, A Measurement of the Gravitational Deflection of Microwave Radiation Near the Sun, 1970 October, *ApJ*, **167**, L5.

Staelin, D. H. and Reifenstein, III, E. C. 1968, Pulsating Radio Sources Near the Crab Nebula, *Science*, **162**, 1481.

Staff of the Millstone Radar Observatory 1961, The Scale of the Solar System, *Nature*, **190**, 592.

Stanley, G. J. 1994, Recollections of John G. Bolton at Dover Heights and Caltech, *Austrl. J. Phys.*, **47**, 507.

Stanley, G. J. and Slee, O. B. 1950, Galactic Radiation at Radio Frequencies II. The Discrete Sources, *Austrl. J. Sci. Res. A*, **3**, 234.

Stebbins, J. and Whitford, A. 1947, Six-Color Photometery of Stars. V. Infrared Radiation from the Region of the Galactic Center, *ApJ*, **106**, 235.

Stephan, K. D. 1990, How Ewen and Purcell Discovered the 21-cm Interstellar Hydrogen Line, *IEEE Trans. Ant. Prop.*, **41**, 7.

Stephenson, C. G. and Sanduleak, N. 1977, New H-Alpha Emission Stars in the Milky Way, *ApJS*, **33**, 459.

Stratton, F. J. M. 1946, Untitled Letter, *Nature*, **157**, 48.

Struve, O. 1949, Progress in Radio Astronomy, *Sky. Tel.*, **9**, 55.

Sullivan, W. 1964, Mars Gives a Hint of Radiation Belt, *New York Times*, 29 August 1964, p. 46.

Sullivan, W. T. III. 1978, A New Look at Karl Jansky's Original Data, *Sky Tel.*, **56**, 101.

Sullivan, W. T. III. 1982, *Classics in Radio Astronomy* (Dordrecht: Reidel).

Sullivan, W. T. III. (ed.) 1984a, *The Early Years of Radio Astronomy* (Cambridge: Cambridge University Press).

Sullivan, W. T. III. 1984b, Karl Jansky and the Beginnings of Radio Astronomy. In *Serendipitous Discoveries in Radio Astronomy*, ed. K. I. Kellermann and B. Sheets (Green Bank: NRAO/AUI), 39.

Sullivan, W. T. III. 2009, *Cosmic Noise: A History of Early Radio Astronomy* (Cambridge: Cambridge University Press).

Swarup, G. 1981, Proposal for an International Institute for Space Science and Electronics and for a Giant Equatorial Radio Telescope, *Bull. Astron. Soc. India*, **9**, 269.

Tanaka, H. 1984, Development of Solar Radio Astronomy in Japan Up Until 1960. In *The Early Years of Radio Astronomy*, ed. W. T. Sullivan III (Cambridge: Cambridge University Press), 335.

Tarter, J. 2001, The Search for Extraterrestrial Intelligence, *ARAA*, **39**, 511.

Taylor, A. R. and Braun, R. (eds.) 1999, *Science with the Square Kilometre Array: A Next Generation World Radio Observatory* (Dwingeloo: ASTRON).

Taylor, J. H. 1974, A Sensitive Method for Detecting Dispersed Radio Emission, *A&A Suppl.*, **15**, 367.

Taylor, J. H. 1994a, Binary Pulsars and Relativistic Gravity, *Rev. Mod. Phys.*, **66**, 711.

Taylor, J. H. 1994b, Binary Pulsars and Relativistic Gravity, *BAAS*, **26**, 971.

Taylor, J. H. and Hulse, R. A. 1974, Binary Pulsar, *IAU Circular*, No. 2704.

Taylor, J. H. and McCulloch, P. M. 1980, Evidence for the Existence of Gravitational Radiation from Measurements of the Binary Pulsar PR 1913+16, *Ann. NY Acad. Sci.*, **336**, 442.

Taylor, J. H. and Weisberg, J.M. 1982, A New Test of General Relativity; Gravitational Radiation and the Binary Pulsar PSR 1913+16, *ApJ*, **253**, 908.

Taylor, J. H., Fowler, L. S., and McCulloch, P. M. 1979, Measurements of the General Relativistic Effects in the Binary Pulsar PSR1913+16, *Nature*, **277**, 437.

Terrell, J. 1964, Quasi-Stellar Diameters and Intensity Fluctuations, *Science*, **145**, 918.

Thaddeus, P. 1972, The Short-Wavelength Spectrum of the Microwave Background, *ARAA*, **10**, 305.

Thaddeus, P. and Clauser, J. P. 1966, Cosmic Microwave Radiation at 2.63 mm from Observations of Interstellar CN, *Phys. Rev. Let.*, **16**, 819.

Thompson, A. R. and Frater, R. H. 2010, Ronald Bracewell: An Appreciation, *J. Astron. Hist. Heritage*, **13**, 172.

Thompson, A. R., Moran, J. M., and Swenson, G. W. 2017, *Interferometry and Synthesis in Radio Astronomy* (Cham: Springer).

Thornton, D. et al. 2013, A Population of Fast Radio Bursts at Cosmological Distances, *Science*, **341**, 53.

Townes, C. H. 1946, Interpretation of Cosmic Noise: Radio Waves from Extraterrestrial Sources, *Phys. Rev.*, **105**, 235.

Townes, C. H. 1957, Microwave and Radio-Frequency Resonance Lines of Interest to Radio Astronomy. In IAU Symposium No. **4**, *Radio Astronomy*, ed. H. C. van de Hulst (Cambridge: Cambridge University Press), 92.

Townes, C. H. 1983, At What Wavelength Should We Search for Signals from Extraterrestrial Intelligence?, *PNAS*, **80**, 1147.

Townes, C. H. 1999, *How the Laser Happened* (New York: Oxford University Press).

Trafton, L. M. 1964, The Thermal Opacity in the Major Planets, *ApJ*, **140**, 1340.

Trimble, V. and Zaich, P. 2006, Productivity and Impact of Radio Telescopes, *PASP*, **118**, 933.

Tyson, J. A. and Giffard, R. P. 1978, Gravitational-Wave Astronomy, *ARAA*, **16**, 521.

Van Allen, J. A. 1966, Spatial Distribution and Time Decay of the Intensities of Geomagnetically Trapped Electrons from the High Altitude Nuclear Burst of July 1962. In *Radiation Trapped in the Earth's Magnetic Field*, ed. B. M. McCormac (New York: Gordon and Breach), 575.

Van Allen, J. A. et al. 1958, Observation of High Intensity Radiation by Satellites 1958 Alpha and Gamma, *Jet Propulsion*, **28**, 588.

van de Hulst, H. C. 1945, The Origin of Radio Waves from Space, *Nederlandsch Tijdscrift voor Natuurkunde*, **11**, 210 (in Dutch). English translation 1982 in *Classics in Radio Astronomy*, ed. W. T. Sullivan III (Dordrecht: Reidel).

van de Hulst, H. C. 1957, Studies of the 21-cm Line and their Interpretation. In IAU Symposium No. **4**, *Radio Astronomy*, ed. H. C. van de Hulst (Cambridge: Cambridge University Press), 3.

van de Hulst, H. C., Muller, C. A., and Oort, J. H. 1954, The Spiral Structure of the Outer Part of the Galactic System Derived from the Hydrogen Emission at 21 cm Wavelength, *BAN*, **12**, 117.

van Woerden, H. and Strom, R. 2006, The Beginnings of Radio Astronomy in the Netherlands, *J. Astron. Hist. Heritage*, **9**, 3.

Verde, L., Treu, T., and Riess, A. G. 2019, Tensions between the Early and Late Universe, *Nat. Astron.*, **3**, 891.

Verschuur, G. L. 1969, Measurements of Magnetic Fields in Interstellar Clouds of Neutral Hydrogen, *ApJ*, **156**, 861.

Vessot, R. F. C. et al. 1980, Test of Relativistic Gravitation with a Space-Borne Hydrogen Maser, *Phys. Rev. Let.*, **45**, 2081.

Victor, W. K. and Stevens, R. 1961, Exploration of Venus by Radar, *Science*, **134**, 46.

Vladimirsky, V. V. 1948, Influence of the Terrestrial Magnetic Field on Large Auger Showers, *J. Exp. Theor. Phys.*, **18**, 393 (in Russian).

von Hoerner, S. 1964, Requirements for Cosmological Studies in Radio Astronomy, *IEEE Trans. Mil. Elec.*, **8**, 282.

Von Klüber, H. 1960, The Determination of Einstein's Light-Deflection in the Gravitational Field of the Sun, *Vistas Astron.*, **3**, 47.

Wade, C. M. 1984, The Discovery of Radio Novae. In *Serendipitous Discoveries in Radio Astronomy*, ed. K. I. Kellermann and B. Sheets (Green Bank: NRAO/AUI), 291.

Wade, C. M. and Hjellming, R. 1970, Detection of Radio Emission from Antares, *ApJ*, **163**, 105.

Walker, R. G. and Sagan, C. 1960, The Ionospheric Model of the Venus Microwave Emission: An Obituary, *Icarus*, **5**, 105.

Walsh, D. 1989, 0957+561: The Unpublished Story. In Lecture Notes in Physics **330**, *Gravitational Lenses*, ed. J. M. Moran, J. M. Hewitt, and K. Y. Lo (Berlin: Springer), 11.

Walsh, D., Carswell, R. F., and Weymann, R. J. 1979, 0957+561 A, B: Twin Quasistellar Objects or Gravitational Lens, *Nature*, **279**, 381.

Waluska, E. R. 2007, Quasars and the Caltech-Carnegie Connection, *J. Astron. Hist. Heritage*, **10**, 79.

Weaver, H. et al. 1965, Observations of a Strong Unidentified Microwave Line and of Emission from the OH Molecule, *Nature*, **208**, 29.

Webb, H. D. 1946, Project Diana, *Sky Tel.*, **5**, 3.

Webb, S. 2015, *If the Universe Is Teeming with Aliens ... Where Is Everybody?* (Cham: Springer).

Weber, J. 1960, Detection and Generation of Gravitational Waves, *Phys. Rev.*, **117**, 306.

Weber, J. 1968, Gravitational Waves, *Phys. Today*, **21**, 34.

Weber, J. 1969, Evidence for Discovery of Gravitational Radiation, *Phys. Rev. Let.*, **22**, 1320.

Weber, J. 1970, Gravitational Radiation Experiments, *Phys. Rev. Let.*, **24**, 276.

Weiler, K. W. et al. 1974, A Measurement of Solar Gravitational Microwave Deflection with the Westerbork Synthesis Telescope, *A&A*, **30**, 241.

Weiler, K. W. et al. 1975, Dual-Frequency Measurement of the Solar Gravitational Microwave Deflection, *Phys. Rev.*, **35**, 134.

Weinreb, S. 1961, Digital Radiometer, *Proc. IRE*, **49**, 1099.

Weinreb, S. et al. 1963, Radio Observations of OH in the Interstellar Medium, *Nature*, **200**, 829.

Weinreb, S. et al. 1965, Observations of Polarized OH Emission, *Nature*, **208**, 440.

Weisberg, J. M. and Huang, Y. 2016, Relativistic Measurements from Timing the Binary Pulsar PSR 1913+16, *ApJ*, **829**, 55.

Weiss, H. G. 1965, The Haystack Microwave Research Facility, *IEEE Spectrum*, **2**, 50.

Westerhout, G. and Oort, J. H. 1951, A Comparison of the Intensity Distribution of the Radio Frequency Radiation with a Model of the Galactic System, *BAN*, **11**, 323.

Whipple, F. L. and Greenstein, J. L. 1937, On the Origin of Interstellar Radio Disturbances, *PNAS*, **23**, 177.

Whiteoak, J. 1994, Early Polarization Research at Parkes. In *Parkes: Thirty Years of Radio Astronomy*, ed. D. E. Goddard and D. K. Milne (Melbourne: CSIRO Publishing), 75.

Whitney, A. R. et al. 1971, Quasars Revisited: Rapid Time Variations Observed via Very-Long-Baseline Interferometry, *Science*, **173**, 225.

Wielebinski, R. 2012, A History of Radio Astronomy Polarisation Measurements, *J. Astron. Hist. Heritage*, **15**, 76.

Wild, J. P. 1950, Observations of the Spectrum of High-Intensity Solar Radiation at Metre-Wavelengths. II. Outbursts, *Austrl. J. Sci. Res. A*, **3**, 399.

Wild, J. P. 1952, The Radio-Frequency Line Spectrum of Atomic Hydrogen and Its Applications in Astronomy, *ApJ*, **115**, 206.

Wild, J. P. 1967, The Radioheliograph and the Radio Astronomy Programme of the Culgoora Observatory, *PASA*, **1**, 38.

Wild, J. P. and McCready, L. L. 1950, Observations of the Spectrum of High-Intensity Solar Radiation at Metre Wavelengths. I. The Apparatus and Spectral Types of Solar Burst Observed, *Austrl. J. Sci. Res. A*, **3**, 387.

Wild, J. P., Smerd, S. F. and Weiss, A. A. 1963, Solar Bursts, *ARAA*, **1**, 231.

Wildt, R. 1940, Note on the Surface Temperature of Venus, *AJ*, **91**, 266.

Wilkinson, D. T. 2009, Measuring the Cosmic Microwave Background Radiation. In *Finding the Big Bang*, ed. P. J. E. Peebles, L. A. Page, and R. B. Partridge (Cambridge: Cambridge University Press), 200.

Wilkinson, D. T. and Peebles, P. J. E. 1984, Discovery of the 3 K Radiation. In *Serendipitous Discoveries in Radio Astronomy*, ed. K. I. Kellermann and B. Sheets (Green Bank: NRAO/AUI), 175.

Wilkinson, P. N. et al. 2004, The Exploration of the Unknown, *New Astron. Rev.*, **48**, 1551.

Will, C. M. 1983, Testing General Relativity: 20 Years of Progress, *Sky Tel.*, **66**, 294.

Will, C. M. 2015, The 1919 Measurement of the Deflection of Light, *Quantum Gravity*, **32**, 124001.

Wilsing, J. and Scheiner, J. 1896, On an Attempt to Detect Electrodynamic Solar Radiation and on the Change in Contact Resistance When Illuminating Two Conductors by Electric Radiation, *Annalen der Physik und Chemie*, **59**, 782 (in German). English translation 1982, in *Classics in Radio Astronomy*, ed. W. T. Sullivan III (Dordrecht: Reidel), 147.

Wilson, R. W. 1980, History of the Discovery of the Cosmic Microwave Background, *Physica Scripta*, **21**, 599.

Wilson, R. W. 1984, Discovery of the Cosmic Microwave Background. In *Serendipitous Discoveries in Radio Astronomy*, ed. K. I. Kellermann and B. Sheets (Green Bank: NRAO/AUI), 185.

Wilson, R. W. 1992, The Cosmic Microwave Background Radiation. In *Nobel Lectures in Physics 1971–1980*, ed. S. Lundqvist (Singapore: World Scientific), 463.

Wilson, R. W. 2009, Two Astronomical Discoveries. In *Finding the Big Bang*, ed. P. J. E. Peebles, L. A. Page, and R. B. Partridge (Cambridge: Cambridge University Press), 157.

Wilson, R. W., Jefferts, K. B., and Penzias, A. A. 1970, Carbon Monoxide in the Orion Nebula, *ApJ*, **161**, L43.

Winn, J. N. and Fabrycky, D. C. 2015. The Occurrence and Architecture of Exoplanetary Systems, *ARAA*, **53**, 409.

Wolfe, A. M. and Burbidge, G. R. 1969, Discrete Source Models to Explain the Microwave Background Radiation, *ApJ*, **156**, 345.
Wolszczan, A. 1994, Confirmation of Earth-Mass Planets Orbiting the Millisecond Pulsar PSR B1257+12, *Science*, **264**, 538.
Wolszczan, A. and Frail, D. A. 1992, A Planetary System around the Millisecond Pulsar PSR 1257+12, *Nature*, **355**, 145.
Woltjer, L. 1968, The Nature of Pulsating Radio Sources, *ApJ*, **152**, 179.
Woolf, N. J. 2009, Conversations with Dicke. In *Finding the Big Bang*, ed. P. J. E. Peebles, L. A. Page, and R. B. Partridge (Cambridge: Cambridge University Press), 74.
Yahil, A. 1972, Observed Anisotropy in the Distribution of Radio Sources, *ApJ*, **178**, 45.
Yodh, G. B. and Wallis, R. F. 2001, Joseph Weber, Obituary, *Phys. Today*, **54**, 74.
Zhu, W. and Dong, S. 2021, Exoplanet Statistics and Theoretical Implications, *ARAA*, **59**, 291.
Zuckerman, B., Lilley, A. E., and Penfield, H., 1965, OH Emission in the Direction of Radio Source W49, *Nature*, **208**, 441.
Zwicky, F. 1933, Die Rotverschiebung von Extragalaktischen Nebeln, *Helvetica Physica Acta*, **6**, 110.
Zwicky, F. 1937a, Nebulae as Gravitational Lenses, *Phys. Rev.*, **51**, 290.
Zwicky, F. 1937b, On the Probability of Detecting Nebulae Which Act as Gravitational Lenses, *Phys. Rev.*, **51**, 679.
Zwicky, F. 1937c, On the Masses of Nebulae and of Clusters of Nebulae, *ApJ*, **86**, 217.
Zwicky, F. 1939, On the Theory and Observation of Highly Collapsed Stars, *Phys. Rev.*, **55**, 726.
Zwicky, F. 1963, New Types of Objects, *AJ*, **68**, 301.
Zwicky, F. 1965, Blue Compact Galaxies, *ApJ*, **142**, 1293.
Zwicky, F. 1969, *Discovery, Invention, Research* (New York: Macmillan).
Zwicky, F. and Zwicky, M. 1971, *Catalogue of Selected Compact Galaxies and of Post-Eruptive Galaxies* (Guemligen: Zwicky), xix.

Suggested Reading

Burke, B. F., Graham-Smith, F., and Wilkinson, P. N. 2019, *An Introduction to Radio Astronomy*, 4th ed. (Cambridge: Cambridge University Press).
Condon, J. J. and Ransom, S. M. 2016, *Essential Radio Astronomy* (Princeton: Princeton University Press).
Graham-Smith, F. 2013, *Unseen Cosmos* (Oxford: Oxford University Press).
Harwit, M. 2013, *In Search of the True Universe* (Cambridge: Cambridge University Press).
Harwit, M. 2019, *Cosmic Discovery: The Search, Scope, and Heritage of Astronomy* (Cambridge: Cambridge University Press).
Harwit, M. 2021, *Cosmic Messengers: The Limits of Astronomy in an Unruly Universe* (Cambridge: Cambridge University Press).
Kellermann, K. I. (ed.) 2012, *A Brief History of Radio Astronomy in the USSR* (Dordrecht: Springer).
Kellermann, K. I. and Sheets, B. (eds.) 1984, *Serendipitous Discoveries in Radio Astronomy* (Green Bank: NRAO/AUI).
Kellermann, K. I., Bouton, E. N., and Brandt, S. S. 2020, *Open Skies: The National Radio Astronomy Observatory and Its Impact on US Radio Astronomy* (Cham: Springer).
Kurczy, S. 2021, *The Quiet Zone: Unraveling the Mystery of a Town Suspended in Silence* (New York: William Morrow).
Munns, D. P. 2013, *A Single Sky* (Cambridge: MIT Press).
Orchiston, W., Robertson, P., and Sullivan, W. 2021, *Golden Years of Australian Radio Astronomy: An Illustrated History* (Cham: Springer).
Pawsey, J. L. and Bracewell, R. N. 1955, *Radio Astronomy* (Oxford: Clarendon Press).
Pfeiffer, J. 1956, *The Changing Universe* (New York: Random House).
Roberts, R. M. 1989, *Serendipity: Accidental Discoveries in Science* (New York: Wiley).
Shklovsky, I. S. 1960, *Cosmic Radio Waves* (Cambridge: Harvard University Press).
Sullivan, W. T. III (ed.) 1982, *Classics in Radio Astronomy* (Dordrecht: Reidel).
Sullivan, W. T. III (ed.) 1984, *The Early Years of Radio Astronomy* (Cambridge: Cambridge University Press).
Sullivan, W. T. III. 2009, *Cosmic Noise: A History of Early Radio Astronomy* (Cambridge: Cambridge University Press).
Verschuur, G. 2007, *The Invisible Universe: The Story of Radio Astronomy* (New York: Springer).

Index

Printed in the United States
by Baker & Taylor Publisher Services